Crude Unit Corrosion Guide

A Manual for Plant Operators, Process Engineers, Metallurgists, Maintenance Engineers, Inspectors, and Equipment Specialists

3rd Edition

Joerg Gutzeit, M.S.Chem.Eng.

NACE INTERNATIONAL

NACE International
The Worldwide Corrosion Authority

Printed in the United States of America

ISBN: 978-1-57590-330-9

NACE International
The Worldwide Corrosion Authority
15835 Park Ten Place
Houston, TX 77084
nace.org

Dedication

The *Crude Unit Corrosion Guide* is dedicated to Norman S. Mack, former colleague and chief metallurgist at Amoco's Texas City refinery, in appreciation of past assistance.

Dedication

Preface to the Revised 3rd Edition

The *Crude Unit Corrosion Guide* was written for plant operators, process engineers, metallurgists, maintenance engineers, inspectors, and all others who deal with corrosion and fouling issues on a daily basis. In the past, most refineries also employed corrosion engineers. That position has all but disappeared as the experienced practitioners moved on, retired, or were replaced by engineering contractors, equipment builders, and process additives suppliers.

Like the earlier editions of the *Guide,* the Revised 3rd Edition is based primarily on the author's experience dealing with all aspects of corrosion and fouling in crude units. The *Guide* also reflects the Industry's consensus experiences reported at past meetings of the NACE[1] STG34 Committee on Petroleum Refining and Gas Processing (formerly T8 Committee on Refining Industry Corrosion) and the API[2] Subcommittee on Corrosion and Materials. Diluent recovery units (DRUs) and vacuum distillation units (VDUs) of bitumen upgraders are quite similar to crude units with similar corrosion and fouling issues and are now included in the discussions where appropriate. In addition to corrosion and fouling issues with the usual 3 or 4 crude types, the *Guide* now covers various bitumen-derived crudes, including synthetic crude, dilbit crude, synbit crude, and shale crude all of which have come to market in recent years.

In many ways, refinery corrosion and fouling control remained more "Art than Science." Despite college-level courses in corrosion and materials science and ongoing efforts by NACE, API, and ASM[3], few resources for practical training on plant corrosion and fouling control are available. In recent years, process simulation (PS) and risk-based inspection management (RBIM) software has been very helpful for identifying and ranking critical components on which to concentrate future inspection and maintenance efforts. However, a certain amount of expertise is still required to facilitate reliable data input and assessment. The *Guide* attempts to present in compact form what the author considers the most important aspects of corrosion and fouling control in the atmospheric and vacuum sections of crude units and the corresponding sections in bitumen upgraders. Various chapters deal with general corrosion characteristics of different feedstocks; materials of construction (metallurgy); corrosion and fouling monitoring; feedstock preparation by tank settling, desalting, and caustic injection; and the types of damage seen in major components. Other chapters cover specific topics, including preheat exchanger fouling; high-temperature corrosion and fouling by organic sulfur compounds and naphthenic acids in fired heaters, transfer lines, and distillation columns; low-temperature corrosion and fouling by inorganic chlorides in the top of columns and in overhead systems; organic chloride problems; and various issues with neutralizer, corrosion inhibitor, and water injection systems.

While the *Guide* is not a formal literature review per se, important reports, research papers, and monographs published during the past 60 years or so are cited in the text, or listed at the end of each chapter. Many are out of print, but

1. NACE International (formerly National Association of Corrosion Engineers), Houston, TX.
2. American Petroleum Institute, Washington, DC.
3. ASM International (formerly American Society for Materials), Metals Park, OH.

worth pursuing in university, public, or corporate libraries. All data and illustrations are by the author, unless otherwise credited. NACE members are reminded that they can obtain, at no charge, copies of conference proceedings, reports, and standards published by NACE since 1996. Also, the discussions on refinery corrosion and fouling issues during the biannual NACE STG34 (formerly T8) *Technical Information Exchange* were recorded since 1957 and are available up to 2008 on a searchable CD disk.[4] Annual updates can be downloaded from NACE.

May 2016
Joerg Gutzeit

4. *REFIN•COR Vers.10.0.*

Acknowledgments

The author had the great privilege and good fortune to have met, worked with, and learned from people who laid out the basic principles of refinery metallurgy and corrosion control. Many are no longer with us, but their contributions were invaluable and should not be forgotten. They are listed below – in no particular order – and include Henry McConomy (author of the McConomy Curves); Al Couper (sulfidic and H_2/H_2S corrosion and coauthor of the Couper-Gorman Curves); Carl Samans (polythionic acid cracking); Professor Norman Hackerman (organic corrosion inhibitors); Andy Foroulis (sulfidic corrosion); Bill Bonner (H_2 blistering); Bob Merrick (H_2 damage and stress cracking); Woody Backensto (H_2/H_2S corrosion); Fred Hamel (H_2/H_2S corrosion); Chuck Robertson (refinery metallurgy and inspections); Larry Scharfstein (refinery metallurgy); and Dick Schillmoller (application of nickel alloys).

Thanks are also due to present and former colleagues on various API and NACE committees – again listed in no particular order – for sharing their corrosion experiences and data over many years. They include Norm Mack, Dave Hansen, Tim Munsterman, Dick Horvath, Jorge Hau, Les Overstreet, Mike Humphries, Dick White, Ernie Ehmke, Bob Piehl, John Coombs, Emil Bereczky, Don Truax, Ara Bagdasarian, Dannie Clarida, Don Bird, Bryan Hopkinson, Brian Tkachyk, Gerry Sorell, Bill Neill, Lee Craig, Jay Cantwell, Kirby Miller, Nile Coble, Gerry Wilks, Keith Lewis, Walt Jenkins, Bob Smallwood, Stew McHie, George Moller, Ron Schutz, Conrad Polk, Paul Dillon, Alex Shparber, Leroy Fisher, Bill Fort III, Bob and Ed Hildebrand, Mike Winters, Greg Kobrin, Russ Kane, Ray Konet, Dick Setterlund, Phil Petersen, Jim Rue, Cathy Shargay, Gene Rak, Dick Mueller, Ted Nichol, Warren Pollock, and many others.

About the Author

Joerg Gutzeit is an independent consultant specializing in refinery process corrosion and fouling control since 1992. Previously, he held various technical and supervisory positions in refinery metallurgy and corrosion groups with two major international oil companies: Atlantic Refining /Arco in Philadelphia, Pennsylvania, and American Oil/Amoco in Whiting, Indiana, and Naperville, Illinois. Throughout his career, the author investigated many corrosion, metallurgy, and fouling issues both in the laboratory and in the field. The *Guide* includes his experiences as a consultant in dealing with crude units, DRUs, and VDUs at many different client companies and locations worldwide. The author received B.S. and M.S. degrees in Chemical Engineering from Drexel University in Philadelphia, Pennsylvania. He is an active member of NACE for many years and was Amoco's representative on the API Subcommittee on Corrosion and Materials. He is the author or coauthor of some 38 technical presentations/journal articles; 7 chapters in various monographs dealing with refinery corrosion and fouling issues; and the *Guideline Manual on Cleaning of Process Equipment and Piping*. Joerg received a NACE Technical Achievement Award and API Resolution of Appreciation for distinguished service to the petroleum industry. At the request of the U.S. National Academy of Sciences/National Research Council, he was one of two technical consultants for an Egyptian research project on crude unit corrosion between 1981 and1983.

List of Abbreviations and Acronyms

API	American Petroleum Institute
°API (API gravity)	Specific gravity of crude in API degrees
APS	Atmospheric pipe still (crude unit atmospheric section)
bar	Metric unit of pressure approximately equal to the atmospheric pressure at sea level
BBL	Petroleum barrel (42gal)
BP	Boiling point
BPD	Barrels per day
BS	Bottom sediment in crude
BS&W	Bottom sediment and water in crude
CA	Corrosion allowance
C (°C)	Centigrade (or Celsius) degrees
C–degrees (or F-degrees)	Units of temperature offset in degrees C or F, respectively
CE	Carbon equivalent in steels
CI	Cast iron
CPE	Cold preheat exchanger (upstream of the desalter)
CR	Corrosion rate
CS	Carbon steel
CSCC	Chloride stress corrosion cracking
CW	Cooling water
DRU	Diluent recovery unit of bitumen upgrader
F (°F)	Fahrenheit degrees
FPS	Foot per second
Gal (gal)	Gallons
GPM	Gallons per minute
GPM/BBL	Gallons per minute/barrel
GPD	Gallons per day
HGO	Heavy gas oil fraction
HPE	Hot preheat exchanger (downstream of desalter)
hr	Hour
HTSC	High-temperature sulfidic corrosion above 230 °C (450 °F)
HAGO	Heavy atmospheric gas oil fraction
HVGO	Heavy vacuum gas oil fraction
IC	Iron (ion) count

ID	Inside diameter
IPS	International pipe standard
Kg	Kilogram
KW	Kilowatt
LACT unit	Lease automatic custody transfer unit
LAGO	Light atmospheric gas oil fraction
Lb (Lb)	Pound
LMTD	Log mean temperature difference for heat exchanger
LVGO	Light vacuum gas oil fraction
MBPD	Thousand barrels per day
MMBPD	Million barrels per day
MTPD	Thousand metric tons per day
MOC	Material(s) of construction (metallurgy)
MP	Melting point
m/sec	Meters per second
Mg/L	Milligrams per liter (ppm)
mpy	Corrosion rate in mils (0.001in.) per year
MSCF	Thousand standard cubic feet
NAC	Naphthenic acid corrosion
NRS	Nonreactive sulfur (compounds)
OAG	Oil and grease content of API separator streams
OD	Outside diameter
OVHD	Overhead
PAR	Pump-around
PASCC	Polythionic acid stress corrosion cracking
PE	Preheat exchanger
PPM	Parts per million (mg/L)
PSI (psi)	Pounds per square inch
PSIA (psia)	Pounds per square inch, absolute
PTB	Pounds per thousand barrels
PWHT	Postweld heat treatment
RS	Reactive sulfur (compounds)
TCW	Treated cooling water
TPD (T/d)	Metric tons per day
UT	Ultrasonic thickness testing
VDU	Vacuum distillation unit of bitumen upgrader
VPS	Vacuum pipe still (vacuum section of crude unit)
WC	Water cooled
WSIM	Water separometer index, modified (in jet fuel)
WT	Wall thickness

Table of Contents

1 Feed Stocks

In this Chapter

1.1 First Glance

Crude oil (commonly called "crude" in singular or plural form) originated from organic matter in marine and other aquatic deposits over a period of many thousands to millions of years. The oil migrated from the deposit sites to geological traps or reservoirs. During this process, the oil composition changed: the oil matured (aged) and became heavier, while evolving gases, such as methane (CH_4), carbon dioxide (CO_2), hydrogen sulfide (H_2S), and water (H_2O) vapor.[1] Crudes range in consistency from waterlike liquids to tarlike solids, and in color from clear to black. Since the nature of the original deposits and the subsequent aging processes varied from reservoir to reservoir, crude compositions varied accordingly. The principal components of crude are aromatic, naphthenic, paraffinic, and asphaltic hydrocarbon species; aromatic or naphthenic ring structures make up the bulk of crude.

1. To avoid confusion with common chemical names, chemical formulas are provided in parenthesis throughout the text whenever practical.

1.2 Crude Sources

Crudes are usually named after the reservoir, geographic area, or country where they are produced. In practice, a crude source may encompass several different fields: Arabian Light, for example, comes from numerous wells in an area covering hundreds of square kilometers (square miles). However, the name by itself does not guarantee a product of constant composition or properties; crude from any given reservoir is bound to change over time. In addition, there can be major differences between individual crude cargoes to a refinery, depending on how the crude was produced, recovered, treated, and shipped. When purchasing crude, many plants still rely on crude assays obtained years ago instead of insisting on up-to-date crude analyses. As a result, crude unit operators are often surprised to find that processing, corrosion, or fouling characteristics have changed even though a given crude slate (at least in name) has remained the same. Overall, crudes have not only become heavier, but also more corrosive. Over 70% of the world's estimated crude reserves now have API gravities below 20, including heavy crude, extra heavy crude, and bitumen.

1.3 Processing and Corrosion Characteristics of Crude

Other than source, important crude characteristics affecting corrosion and fouling properties and therefore purchasing decisions and price include API gravity, UOP characterization factor K, sulfur content, total acid number (TAN), and salt content.

1.3.1 API Gravity. °API (commonly called API gravity) defines crude in terms of weight per unit volume and is useful, because crudes are sold on the basis of volume delivered, while pumping cost, freight rates, and tanker cargoes are determined on the basis of weight. More important, however, API gravity also defines the processing, corrosion, and fouling characteristics of crude. As a general rule, the lower the API gravity, the less desirable and less expensive crude is for several reasons: Pumping costs increase inversely with API gravity and heat and various chemicals have to be applied to aid pumping and reduce the amount of water accompanying all crudes. Water not only represents useless cargo, but also increases corrosion, fouling, and maintenance costs of pipelines, tank farms, and tanker cargo holds. At the refinery, low API gravity crudes are more expensive to process, because of increased tank settling and desalting costs and increased corrosion, fouling, and maintenance costs of preheat exchangers (PHEs), fired heaters, columns, and overhead systems.

API gravity is a dimensionless[2] number and is usually measured with a hydrometer as detailed in *ASTM[3] D-287 Standard Test Method for API Gravity of Crude Petroleum and Petroleum Products (Hydrometer Method).* The hydrometer has a modified Baumé scale designed so that most crudes fall between 10 and 70 degrees. API gravity of crudes ranges from less than 12 (specific gravity over 0.986) to more than 30 (specific gravity below 0.875) and higher, as shown in Table 1-1. By definition, API gravity is a measure of how light or heavy crude is compared to water. In comparison to crude, water has an API gravity of 10 (specific gravity of 1.000). If the API gravity is greater than 10, the crude is lighter than water and floats on top of water. If the API gravity is less than 10, the crude is heavier than water and settles out at the bottom of separators, tanks, and piping. Thus API gravity determines how difficult it is to separate water from a crude in order to meet pipeline BS&W[4] specifications. With high API gravity (light) crude, water can be readily removed in upstream and midstream operations in simple water-settling drums or tanks. With low API gravity (heavy) crude, heater-treaters and desalters have to be deployed along with water-settling drums or tanks.

In downstream operations, API gravity determines primarily the ease of water and salt removal in the crude receiving tanks. Most, if not all, crudes require additional desalting at the plant, and API gravity determines the optimum desalting, dewatering, and solids removal efficiencies (in %) obtainable with single or double desalting. Since API gravity determines the amount of entrained salt left in crude, API gravity also affects the extent and severity of low-temperature corrosion and fouling in preheat exchangers, flash drums, and flash columns upstream of the fired heater and, by extension, the severity of low-temperature corrosion and fouling in the top and overhead system of the atmospheric and vacuum columns downstream of the fired charge heater.

2. Degrees API gravity = 141.5 / (specific gravity 60°/60°F) - 131.5.
3. ASTM International (ASTM), formerly known as the American Society for Testing and Materials.
4. Bottom sediment and water.

1.3.2 UOP Characterization Factor K. The UOP (or Watson) characterization factor K defines crudes in terms of the predominant hydrocarbon species. While primarily of interest from a processing point of view[5], it can also affect corrosion-related decisions. UOP K factors for various types of crude normally range from approximately 10 to 13. K factors at the high end of the range (12.5 to 13) characterize crudes containing primarily paraffinic species, K factors in the mid range (11.5 to 12.5) characterize crudes containing primarily aromatic species, K factors at the low end of the range (11 to 11.5) characterize crudes containing naphthenic species, while K factors below 11 characterize crudes containing asphaltic species.

TABLE 1-1 API GRAVITY OF VARIOUS CRUDES

API Gravity	Specific Gravity	Weight/BBL, kg (lb)	Type	Examples
>30	<0.875	<139 (305)	Light and ultralight crude	Alaska North Slope, Edmonton Lt. Sour, ND Bakken Shale Crude, TEF Shale Crude
20 - 30	0.875 - 0.935	139 - 148 (305 - 325)	Intermediate crude	East Texas Medium, Lloydminster Hvy Sour, Athabasca Oil Sands Synthetic Crude, Synbit
12 - 20	0.935 - 0.986	148 - 156 (325 - 343)	Heavy, low resid crude	Venezuela Mesa, Athabasca Oil Sands Dilbit, Orinoco Tar Sands Merey
<12	>0.986	>156 (343)	Extra heavy crude	Athabasca Oil Sands Bitumen, Orinoco Tar Sands Bitumen

1.3.3 Sulfur Content. Sulfur content expressed as % S (or % total S) characterizes crudes in terms of potential high-temperature sulfidic corrosion (HTSC) by various organic sulfur compounds above approximately 230 °C (450 °F)[6]. As a general rule, the higher the sulfur content, the less desirable and less expensive crude is. High-sulfur crudes require alloy upgrading of many high-temperature circuits in refineries and bitumen upgraders. Aside from corrosion considerations, high-sulfur crudes tend to be heavy crudes and can be expected to be more difficult to transport, store, and desalt compared to low-sulfur crudes. They also require additional and often expensive downstream processing in order to yield products that meet sulfur and emission specifications.

1.3.4 Total Acid Number. Total acid number or TAN expressed as mg KOH/g crude characterizes crudes in terms of potential naphthenic acid corrosion (NAC) above approximately 230 °C (450 °F)[7]. As a general rule, the higher the TAN, the less desirable and less expensive crudes are. High-TAN crudes usually require alloy upgrading of many high-temperature circuits in refineries and bitumen upgraders beyond that required for high-sulfur crudes. High-TAN crudes may also require alloy upgrading of column tops and overhead systems to control corrosion by light organic acids, such as acetic acid (CH_3-COOH), formic acid (H-COOH), or proprionic acid (CH_3CH_2-COOH). These acids can form when naphthenic acids [R-$(CH_2)_n$–COOH] decompose in the fired heaters of the atmospheric and vacuum sections. Aside from corrosion considerations, high-TAN crudes are often more difficult to desalt and dewater than low-TAN crudes because sodium coordination complexes ("naphthenate soaps"), such as [R-$(CH_2)_n$-COO$]_2$-Na or [R-$(CH_2)_n$-COO$]_6$-O-Na_3 tend to form and cause foaming in the desalter.

1.3.5 Salt Content. Inorganic salt content (expressed as mg/L or PTB[8] chlorides) characterizes crudes in terms of potential corrosion and fouling below approximately 230 °C (450 °F). Inorganic chloride salts in crude include sodium chloride (NaCl), magnesium chloride ($MgCl_2$), and calcium chloride ($CaCl_2$) that originate with formation water[9], production brines, water floods, surface waters, and tanker ballast from upstream production and midstream treating and transportation operations. Upon passing through the hot preheat exchangers (HPE)[10] and fired crude heater, magnesium chloride ($MgCl_2$) and calcium chloride ($CaCl_2$) partially hydrolyze to hydrogen chloride (HCl). Subsequent

5. K = T.B.P.$^{1/3}$/ (specific gravity 60/60F) where T.B.P. is the average molal boiling point, °R.
6. Historically, the industry has used 260 °C (500 °F) as the temperature where corrosion by organic sulfur compounds becomes a major concern. The lower limit of 230 °C (450 °F) is a more realistic approach and agrees with the lower limit established for naphthenic acid corrosion.
7. Neutralization number is another term used to define naphthenic acid content of crude.
8. Pounds/1,000 barrels of crude (1PTB = 2.85 ppm).
9. Water produced in combination with crude.
10. Downstream of the desalter, where used.

reaction of HCl with ammonia (NH_3) or liquid organic neutralizers then yields ammonium hydrochloride ($NH_3 \bullet HCl$) or neutralizer hydrochlorides ($R\text{-}NH_2 \bullet HCl$), respectively, and leads to a variety of corrosion and fouling issues in column tops and overhead systems. As a general rule, the higher the salt content, the less desirable and less expensive crudes are, because crude desalting by water extraction becomes more critical. Often, extended tank settling, double desalting, and caustic addition are required to sufficiently lower chloride levels. Alloy upgrading of many low-temperature circuits, including the upper distillation columns and the overhead systems in refineries and bitumen upgraders, often becomes necessary and injection of various process chemical additives is often necessary.

1.4 Hydrocarbon Components of Crude

Crude consists of relatively few homologous series of hydrocarbon molecules containing 84-87% carbon (C), 11-14% hydrogen (H), 0-6% sulfur (S), and up to 2% each of nitrogen (N) and oxygen (O). The blend of hydrocarbon molecules may include from 1 to 70 or more carbon atoms and crude properties depend primarily on the number and arrangement of the carbon and hydrogen atoms in the molecules. For example, hydrocarbons containing less than 4 carbon atoms are usually in the form of gases, those with 5 to 19 carbon atoms are usually liquids, and those with over 20 carbon atoms are solids. The relative amounts and types of the major constituents of crude are listed in Table 1-2; unsaturated olefins (R–C=C-R) are rarely found in crude.

TABLE 1-2 HYDROCARBON COMPOSITION OF CRUDE

Category	Type	Chemistry	Amount, wt%
Alkanes, paraffins, asphalts	Saturated n-alkanes and iso-alkanes	To C_{78}	15 - 60
Aromatics	Unsaturated structures with one or more aromatic nuclei	C_nH_m	3 - 30
Waxes	Unbranched n-alkanes	To C_{30}	?
Naphthenes	Cycloparaffins or cycloalkanes	C_nH_{2n}	30 - 60
Hydrocarbon gases	Methane butane, propane	CH_3 to C_3H_8	?
Unsaturated olefins	?	R-C=C-R	Rarely
Heteroatomic compounds	Organic sulfur, nitrogen, and oxygen compounds	Many types	Balance

1.5 Principal Crude Types

As far as corrosion and fouling in crude refineries are concerned, the six principal types of crude charged are sweet crude, sour crude, acidic crude, bitumen-derived crude, shale crude, and opportunity crude. Bitumen upgraders receive primarily diluted bitumen (dilbit) for conversion into crude.

1.5.1 Sweet Crude. The term "sweet crude" originated from the fact that a low level of organic sulfur provides crude with a mildly sweet taste and pleasant smell. Crude traders designate crude with less than 0.4% S as sweet. From a corrosion point of view, the term usually describes any low-sulfur crude with less than 0.6% S as opposed to high-sulfur or "sour crude" with more than 0.6% S. Sweet crude may also contain small amounts of free[11] hydrogen sulfide (H_2S) and dissolved carbon dioxide (CO_2). High-quality, light sweet crude is the most sought-after crude as it contains a large hydrocarbon fraction that can be directly distilled into gasoline (light naphtha), kerosene, diesel, and various gas oil fractions without additional processing. *West Texas Intermediate (WTI)* crude contains approximately 0.25% S and is traded on the New York Mercantile Exchange (NYMEX) as a benchmark for crude pricing, while *North Sea Brent* crude with approximately 0.35% S is traded as a benchmark on the London ICE Futures Europe Exchange for refining in Northern Europe.

11. Dissolved as molecular H_2S instead of dissociated into HS^-.

1.5.2 Sour Crude. The term sour crude refers to crudes that contain significant amounts of free hydrogen sulfide (H_2S) regardless of organic sulfur content. As originally defined, sour crudes contain and evolve more than $4{\cdot}10^{-4}$ m^3/100L (0.05ft^3/100gal) free H_2S. The industry currently uses the term sour crude simply to designate any crude with more than 0.6% S. While sour crude has an obnoxious odor, free H_2S presents no high-temperature corrosion issues until the crude contains and evolves at least 10 times that amount, or $4{\cdot}10^{-3}$ m^3/100L (0.5ft^3/100gal). In contrast to H_2S, organic sulfur compounds in crudes can attack components fabricated from carbon and low-alloy Cr-Mo steels at temperatures above approximately 230 °C (450 °F). The resultant complex iron sulfide scales can become pyrophoric when exposed to the atmosphere during shutdown.

Crudes with high organic sulfur content but an insignificant amount of free H_2S include certain Californian and Venezuelan crudes. Crudes with both high organic sulfur and free H_2S contents include some West Texas and Middle East crudes. Crudes containing free H_2S often require additional processing steps to remove or scavenge H_2S from the crude and its various product streams in order to meet environmental and safety regulations. Additional processing steps are also required to treat H_2S in flashed gas and in sour-water streams. Excessive levels of H_2S in upstream and midstream operations can be expected to build up in the vapor areas of tanker holds and storage tanks and affect the structural integrity of under-roof areas while producing copious amounts of iron sulfide (FeS) scale that drop into the raw crude charge. In downstream operations, H_2S containing crude presents major safety issues during operation, inspection, and maintenance activities, including the formation of pyrophoric sulfide scales during shutdowns. H_2S reacts with ammonia (NH_3) produced in the fired heater from organic nitrogen compounds to form ammonium bisulfide ($NH_3{\cdot}H_2S$) that causes sour-water corrosion and fouling at the relatively low temperatures of prefractionator, atmospheric, and vacuum column overhead systems. Combustion of petroleum products containing H_2S produces sulfuric acid (H_2SO_4) and sulfur dioxide (SO_2) both of which are highly corrosive.

1.5.3 Acidic Crude. The term "acidic crude" usually refers to crude that contains significant amounts of naphthenic acids. Acidic crudes can be either sweet or sour, depending on the amount of organic sulfur compounds and free H_2S in the crude. Most acidic crudes originate from production wells and fields in California, South America, Norway, Russia, West Africa, China, and Indonesia. Heavy crudes from geologically young formations have the highest acid content, while paraffinic crudes have the lowest. Naphthenic acids are also present in bitumen crudes derived from Canadian Athabasca oil sands and Venezuelan tar sands. Naphthenic acids attack carbon steel, low-alloy Cr-Mo steels, and 12Cr stainless steel components at temperatures above approximately 230 °C (450 °F). Naphthenic acid content is usually reported in terms of a total acid number (TAN). As a general rule, all crudes contain some amount of naphthenic acid, but typically have a TAN below 0.3. In contrast, problem acidic crudes are crudes with TAN above 0.5, often with TAN ranging up to 6 and sometimes even higher. The World Bank, in its role of monitoring crude oil markets, significantly discounts the price of crude with TAN greater than 0.5, depending on the actual TAN. For example, above 0.5 TAN, each point of excess TAN lowers the price of crude by $0.05 per dollar of Brent with the excess defined as TAN of the crude minus TAN of Brent.

1.5.4 Bitumen Derived Crude. Typical oil sands bitumen has an approximate API gravity of 8.5, 4.5% S content, and TAN of 3. Bitumen is too viscous and dense to be pumped and transported by pipeline or other forms of transportation and is diluted with water, natural gas condensate (NGC), synthetic crude oil (SCO)[12], naphtha, or kerosene in order to meet pipeline and refinery specifications. The resultant diluted bitumen crudes are primarily sour *Heavy Dilbit*[13] crude with an approximate API gravity of 20, salt content of 5-10mg/L (1.75-3.5PTB) and approximately 33% NGC, 3.3% S, with a 2.2 TAN; and sour *Medium Synbit* crude with an API gravity above 20, salt content of 5mg/L (1.75PTB) and approximately 50% SCO, 2.0% S, with a 1.5 TAN. *Light Synthetic Crude Oil (LSCO)* is usually sold to light-crude refineries, while *Heavy Synthetic Crude Oil (HSCO)* and *Heavy Dilbit* are normally processed in heavy-crude refineries. *Medium Synbit* and *SynDilBit* are usually sold to medium-crude refineries or further blended with synthetic crude for processing in a light-crude refinery.

Another type of diluted bitumen crude is *Orimulsion* made from Venezuelan Orinoco tar sands[14] bitumen by emulsifying with 30% water to facilitate flow in pipelines. *Orimulsion* can be burned in boilers as a substitute for heavy Bunker C fuel oil, but has high sulfur (S) content. Orinoco tar sands bitumen has also been diluted with light crude to make a blend sold as regular crude instead of *Orimulsion*. An example is the popular blend of Orinoco bitumen and *Venezuelan*

12. Synthetic crude is produced in an upgrader where the bitumen is heated, decoked, and combined with NGC. The vapors from the coke drums are fractionated into naphtha, kerosene, and gas oil fractions that are blended into synthetic crude oil and diesel fuel. Synthetic crude has an approximate API gravity of 32, 0.1% S content, and 0.5 TAN.

13. Not to be confused with mined bitumen emulsified with surface waters to permit transportation to an upgrader.

14. Orinoco tar sands are so named by Venezuelan authorities but are properly classified as oil sands.

Mesa crude with an approximate API gravity of 29, 0.9% S, and less than 0.3 TAN. The resultant blend is sold as *Merey* crude with an approximate API gravity of 16, 2.5% S, and 0.7 TAN.

1.5.5 Shale Crude. Shale crudes from North Dakota and South Texas are examples of high- volatility[15] light and ultralight sweet crudes. Crude volatility is expressed in terms of vapor pressure ("Reid vapor pressure") and measured in accordance to ASTM D-323[16]. While this test method normally applies to gasoline fractions, it has also been used with high-volatility crude. Volatility and vapor pressure have become important crude parameters in recent years as a result of safety and flammability issues while transporting shale crude by rail. Examples are *North Dakota Bakken* crude with a vapor pressure of 0.58-0.82bar (8.5-12psi) and *Texas Eagle Ford* crude with a vapor pressure above 0.55bar (8psi), respectively. These crudes compare to regular crudes, such as *Louisiana Light* with vapor pressure up to 0.2bar (3psi) or *Arabian Extra Light* with vapor pressure up to 0.31bar (4.5psi). In practice, high-volatility crudes require further midstream stabilization steps to remove light ends and gases prior to safe transporting to a downstream refinery.

1.5.6 Opportunity Crude. So-called opportunity crude is low-priced crude with real or perceived processing, corrosion, or fouling issues. Depending upon the know-how of the crude buyer, the experience of the unit operators, and the metallurgy of the unit, opportunity crudes can either present substantial cost savings or become a processing nightmare. Sometimes, extensive and expensive alloy upgrading is required to control unexpected corrosion and fouling issues. Since opportunity crudes, by definition become available on short notice, buying decisions have to be made on the spot, often without the benefit of up-to-date processing, corrosion, or fouling data. Opportunity crudes have, but are not necessarily limited to one or more of the following characteristics:

- Volumes are small and allow only a short processing learning curve.
- No reliable or current crude assays, distillation, corrosion, and fouling characteristics are available.
- The crude is known to be troublesome.
- High pour point, low API gravity, or high TAN make the crude difficult to desalt.
- High nitrogen (N), sulfur (S), or trace metal content make the crude difficult to process.

Sources of opportunity crude include, but are not necessarily limited to:

- Salvaged or aged cargoes that have been sitting in tankage for some time;
- Crude from small, marginal, or played-out fields;
- Heavy crudes with API gravity below 15;
- Contaminated crudes or blends containing recycled lube oils, used engine oils, or oil recovered from spills; and
- Crude from various sources with unusually high TAN.

1.6 Crude Production, Recovery, and Treating

Crude production, recovery, and treating operations can contribute directly or indirectly to corrosion issues in crude units and upgraders by contaminating crude with various chemical additives, contaminants, byproducts, and corrosion products. Examples of upstream and midstream operations for conventional crude include well stimulation and formation fracturing; pumping and extraction to move crude to the surface; and cooling, dewatering, and desalting at treating facilities. At the latter, crude is often stabilized to remove light ends and gases including free H_2S prior to transportation to a refinery. Upstream and midstream operations for bitumen include mining, hydro transportation, and diluent treating operations.

1.7 Crude Transportation

Until recently, crude was transported almost entirely by pipeline, marine tanker, or barge. Pipeline permitting issues have resulted in greater use of rail tank cars. Tank trucks are often used for short-haul transportation to pipeline access points. Various effects of accompanying corrosion and fouling issues on downstream operations should be kept in mind.

15. Volatility is the tendency of a substance to vaporize and is directly related to a substance's vapor pressure. At a given temperature, a substance with higher vapor pressure vaporizes more readily than a substance with a lower vapor pressure.
16. ASTM D-323 Standard Test Method for Vapor Pressure of Petroleum Products (Reid Method).

1.7.1 Transportation by Pipeline. Crude composition as such does not change during transportation by pipeline, but chemicals (pour point depressants, friction reducers, corrosion inhibitors) are routinely added to reduce both pumping cost and internal pipeline corrosion and can cause unpredictable downstream corrosion and fouling issues. Other potential issues include contamination of crude with stagnant water and prior cargo residuals left in pipeline sags; inadvertent or intentional contamination at pipeline access points, intermediate pipeline junctions, or terminal tanks; and contamination of crude with loose corrosion scale from internal pipeline and tank corrosion. All in all, transport by pipeline is by far the safest way to move crude from production sites to downstream refineries.

1.7.2 Transportation by Barge or Marine Tanker. Repeated transfers between ships or from ship to shore at terminals offer many opportunities for contamination and downstream fouling and corrosion issues. Cooling of crude in cargo holds and onshore tanks can cause water condensation, vapor space attack, tank bottom corrosion, and loosening of prior corrosion product scales. Similar problems are caused by stagnant seawater ballast and prior cargo residuals in cargo holds and tanks after cleanout.

1.7.3 Transportation by Railroad Tank Car or Tank Truck. Crude delivered via rail tank car or tank truck is often contaminated with rainwater or snow when access hatches were inadvertently or intentionally left open after a delivery. Unpredictable midstream stream corrosion and fouling issues occur when crude from isolated wells, marginal wells, or played-out fields is gathered by small tank-truck operators and delivered to pipelines at access points via automated, unattended LACT[17] units. Crude can also become contaminated with spent chemicals and byproducts added as a means of disposal to circumvent environmental regulations. While not corrosion and fouling issues per se, high-volatility shale crude delivered via rail tank cars has presented safety and flammability issues in recent years that need to be addressed by additional stabilization steps to remove light ends and gases including free H_2S prior to transport to downstream refineries.

1.8 Intermediate Storage of Crude

During intermediate storage, crude is likely to be subjected to many changes including atmospheric condensation due to tank "breathing" and ingress of rain or snow at defective tank roofs, seals, or open hatches that cause water or snow to enter the tanks. The principal tank corrosion issues are water condensation under roofs in the presence of H_2S causing sour-gas corrosion of roof structures and support beams; accelerated corrosion of the upper shell plates in floating-roofs tanks due to constant tank activity; and various chemical and physical changes in crude during storage that contribute to tank-bottom corrosion, such as:

- Precipitation of asphaltenes when various crudes are blended together,
- Oxidation of some components,
- Settling of formation water in cargo holds and ballast tanks,
- Separation of a brine layer when chemicals are added to lower the water content of crude prior to delivery to the refinery, and
- Settling of sediment, sand, clay, salt, mineral or biomatter as sludge.

1.9 Tank Settling of Crude in the Field

Controlled tank settling is often used in the production field to decrease the BS&W content of crude to meet pipeline specifications prior to shipment to the refinery. Unless tanks are cleaned out on a regular basis and water draws are kept open, holding and settling tanks gradually build up considerable amounts of sludge that should not be ignored. Tank-bottom sludge typically consists of:

- Brine-in-oil emulsions;
- Corrosion products, scale, mud, sand, silt, and biomatter from upstream production and transportation operations;
- Waxes, asphaltenes, organic acids, and naphthenic acid soaps extracted from some crude; and
- Entrapped puddles or pools of crude and brine.

17. A Lease Automatic Custody Transfer (LACT) unit is designed to facilitate the automatic transfer of ownership of crude or condensate between the buyer and seller.

Obvious issues caused by tank bottom sludge include among others reduced tank capacity, underdeposit corrosion of the tank bottom resulting in inventory loss, and shortened tank life. The principal causes of downstream corrosion and fouling issues are:

- Sludge pickup by the crude suction pipe causing sludge to be carried into the distribution lines at the refinery,
- Under-deposit corrosion in these lines,
- Desalter upsets when slugs of sludge are periodically and unpredictably carried into the desalter leading to salt spikes in desalted crude and accelerated corrosion in the overhead systems, and
- Fouling of the hot crude preheat exchangers (HPEs) when slugs of sludge pass through the desalter.

1.10 Chapter Summary

Origin, properties, production and transportation history, and midstream treating and storage of crude can have major consequences, as far as corrosion and fouling issues in downstream plants are concerned. API gravity, salt content, and BS&W content affect desalting parameters that, in turn, affect the severity of column overhead corrosion. Organic sulfur content and TAN affect the location, extent, and severity of high-temperature sulfidic (HTSC) and naphthenic acid corrosion (NAC). These factors are reviewed in greater detail in subsequent chapters of the *Guide*.

1.11 Read More

1. M. H. Brown et al., "Corrosion by Chlorine and by Hydrogen Chloride at High Temperatures," Ind. and Eng. *Chemistry* 39, 7 (1947).
2. H. A. Cataldi et al., "Estimation of Corrosivity of Crude Oils," *Proc. API* 33, III (1953).
3. S. Kapusta, F. van den Berg, and M.C. Place, "The Impact of Oil Field Chemicals on Refinery Corrosion Problems," paper no. 03649, CORROSION/2003 (Houston, TX: NACE, 2003).
4. K. R. Lewis et al., "Processing Corrosive Crude Oils," paper no. 377, CORROSION/99, (Houston, TX: NACE, 1999).
5. J. Gutzeit, Case Histories in *Corrosion Prevention for Practicing Engineers* by J. F. Bosich (New York, NY: Barnes & Noble, 1970).
6. G. Moins, "Stabilization Process Comparison Helps Selection," *Oil & Gas Journal*, 78, Jan. 28 (1980).
7. C. G. Munger, "Deep Pitting Corrosion in Sour Crude Oil Tankers," *Materials Processing* 15, 3, (1976).
8. W. J. Neill and Z. A. Foroulis, "Internal Corrosion in Floating Roof Gasoline Storage Tanks," *MP* 15, 9 (1976).
9. Petrolite Laboratory Staff, *Impurities in Petroleum*, (Houston, TX: Petrolite Corp., 1968).
10. C. L. Wehmeyer, "Design of Sour Crude Separation Facilities," *Petroleum Engineer*, August (1975).
11. J. McGinnis E. and E. Confrotte, "Oil Sands: Vision Resource for the Energy Industry," *Hydrocarbon Processing*, September (2007).
12. S. Golden et al., "Vacuum Unit Design for High Metals Crudes," *Process Technology Quarterly*, Q1 (2007).
13. M. Grande, "Designing Vacuum Distillation for Deep-Cut Bitumen Service," *Hydroc. Proc.*, February (2010).
14. S. Golden, "Canadian Crude Processing Challenges," *Process Technology Quarterly*, Q1 (2008).
15. R. Henderson et al., "Consider Modifying Your Refinery to Handle Heavy Opportunity Crude Oils," *Hydroc. Proc.*, September (2005).
16. B. Chambers, "Assessing Refinery Crude Corrosivity," Joint Industry Research Proposal, Honeywell Website, 2006.
17. NACE Publication 34103: *Overview of Sulfidic Corrosion in Petroleum Refining* (Houston, TX: NACE, 2004).
18. B. Messer et al., "New Theory for Naphthenic Acid Corrosivity of Athabasca Oil Sands Crudes," paper no. 04634, CORROSION/2004 (Houston, TX: NACE International, 2004).
19. H. Dettman et al., Refinery Corrosion, Organic Acid Structure, and Athabasca Bitumen, paper no. 09366, CORROSION/2009 (Houston, TX: NACE International, 2009).
20. R. Janjua, "Wastewater Treatment for Refineries Using Canadian Tar Sands SAGD Crude as Feedstock," AIChE Spring Meeting (Chicago, IL, March 2011).

2 Corrosive Compounds in Crude

In this Chapter

2.1 First Glance

The root causes of corrosion and fouling in crude units, diluent recovery units (DRUs), and vacuum distillation units (VDUs) are well known and understood. Hydrocarbon compounds that make up the bulk of crude, crude fractions, and solvents are not corrosive per se, nor do they cause corrosion prior to, during, or after heating, distillation, or condensation. This means other chemical compounds or contaminants must be present in crude that cause the corrosion and fouling issues seen on most, if not all units. These contaminants can be loosely grouped into a number of major categories listed here in order of decreasing importance:

- Organic compounds including organic sulfur compounds and various naphthenic acids cause corrosion and fouling issues in components above approximately 230 °C (450 °F). In contrast, organic nitrogen compounds do not directly cause corrosion, but can evolve ammonia (NH_3) upon heating that contributes to corrosion and fouling in upper columns and overhead systems.
- Inorganic compounds cause corrosion and fouling issues in components below approximately 230 °C (450 °F), including preheat exchangers (PHEs), upper columns, and column overhead systems. Representative compounds include inorganic chloride salts present in all crudes as suspended salt crystals and/or emulsified brine. Another troublesome inorganic compound is hydrogen sulfide present in sour crudes as free H_2S. At high enough concentrations, free H_2S can cause severe vapor space corrosion and fouling in production, treating, and transportation facilities, as well as in refineries and bitumen upgraders.

- Trace amounts of metallic mercury (Hg) and mercury salts are present in some crude from California, Asia, Africa, Russia, and Australia and have become of significant concern in recent years, because of processing, environmental, safety, and corrosion issues. Trace amounts of other metal compounds in some crude can cause corrosion, fouling, and coking issues in crude preheat exchanger trains and in fired heater tubes.
- Chemicals used during upstream operations cause corrosion and fouling issues in certain plant components. The amounts and types depend on a particular production field, but the actual concentrations in crudes delivered to the plant are usually unknown. Chemicals include primarily carbon dioxide (CO_2), strong mineral acids (HCl and H_2SO_4), organic chloride solvents, steam, and water used for opening or fracturing oil-containing formations, or elsewhere in crude recovery operations.
- Oil field chemicals (as a group) are added during midstream treating, transportation, and stabilization operations to lower the BS&W content of crude before shipment, improve sluggish flow, minimize pipeline corrosion, lower the free H_2S content of sour crude, and stabilize the scale and sand content of crude. While not necessarily corrosive per se, oil field chemicals can and often do interfere with subsequent treating, dehydrating, and desalting operations and contribute indirectly to crude corrosion and fouling issues at refineries and bitumen upgraders.
- Process additives (as a group) are used downstream at refineries and bitumen upgraders, ironically in an effort to control corrosion and fouling issues. Process additives include dilute caustic (NaOH) solutions used at the desalter and minor amounts of ammonia, liquid organic neutralizers, and filming-amine corrosion inhibitors injected into various column overhead systems for corrosion and fouling control. When overhead water condensate is used as desalter water or injected as wash water, these chemicals end up entrained in crude going to the fired charge heater.
- Cooling water treatment chemicals (as a group) enter from water-cooled overhead condensers as a result of leaking tubes. They too are recycled when overhead water condensate is used as desalter water or injected as wash water.

2.2 Corrosive Organic Compounds in Crude

The overwhelming bulk of corrosion and fouling issues in crude units, diluent recovery units (DRU), and vacuum distillation units (VDU) can be attributed to organic sulfur and acid compounds present in crude and its various fractions; the remainder is usually caused by one or more of the other categories of chemical compounds listed above.

2.2.1 Organic Sulfur Compounds. Organic sulfur compounds are present in crude at concentrations of 0.1-6.5% S and sometimes higher and can cause serious corrosion in the high-temperature circuits above approximately 230 °C (450 °F). Corrosion is caused by certain reactive sulfur (RS) compounds including aliphatic sulfides (C_n–S–C_m), aliphatic disulfides (C_n–S–S–C_m), mercaptans (C_n–SH), and polysulfides (C_n–S_n–S–C_m). In addition to reactive sulfur compounds, crude contains nonreactive sulfur (NRS) compounds that are not corrosive. These include primarily thiophenes (aromatic 5-membered heterocyclic compounds with one sulfur atom) and cyclic sulfur compounds.

2.2.2 Organic Acid Compounds. Organic acid compounds are present in various amounts in all crudes. However, only crudes with a TAN above 0.5 from certain regions and countries (California, Venezuela, China, Indonesia, Russia, Nigeria, Norway, and Canada, among others) contain sufficient amounts of organic acids to cause serious corrosion issues in the high-temperature circuits above approximately 230 °C (450 °F). Organic acid compounds can partially break down in the fired heaters of the atmospheric and vacuum sections to form short-chain organic acids, such as acetic acid (CH_3-COOH), formic acid (H-COOH), or proprionic acid (CHÐCHÐ-COOH) that can contribute to overhead corrosion below approximately 230 °C (450 °F).

Organic acid content is usually expressed in terms of total acid number (TAN) that represents the amount potassium hydroxide (KOH) in milligram (mg) needed to neutralize all acids other than strong mineral acids in a 1-gram sample of crude or its fractions. TAN of crudes typically ranges from less than 0.3 to more than 6, corresponding to approximately 0.1 to 2.5% naphthenic acids [N-$(CH_2)_n$-COOH]. Besides naphthenic acids, TAN can also include certain other organic acids and their derivatives, including other carboxylic acids, hydrolysable esters, acidic nitrogen compounds, and calcium naphthenates found in some crude. Certain types of neutralizers, corrosion inhibitors, and other additives formulated with naphthenic acid derivatives and used during upstream and midstream operations also affect TAN. As a general rule, TAN differs from crude to crude, and from fraction to fraction of a given crude; increases with the boiling point (BP) of crude fractions; varies with the type and structure of the acids; and is affected by compounds other than organic acid derivatives (for example, by some mercaptans).

2.2.3 Organic Nitrogen Compounds. Organic nitrogen compounds are present in naphthene-base crudes from certain regions, including California and Wyoming. Potentially corrosive organic nitrogen compounds include alkylquinolines (heterocyclic compounds with two aromatic rings and one nitrogen atom); pyrrols (aromatic 5-membered heterocyclic compounds with one nitrogen atom); and amides (R–[C=O]–NH–CH_3). The nitrogen content ranges from 0.2-2.5% N and increases with the BP of crude fractions. A portion of organic nitrogen compounds breaks down in the atmospheric fired heater to form ammonia (NH_3) that contributes to corrosion and fouling in the atmospheric distillation columns and overhead systems by reacting with hydrogen chloride (HCl) to form the hydrochloride salt ($NH_3 \cdot HCl$) and with hydrogen sulfide (H_2S) to form the bisulfide salt ($NH_4 \cdot HS$).

Other organic nitrogen compounds stay with reduced crude and are converted to ammonia (NH_3) at the higher temperatures of the vacuum fired heater. Ammonia then travels up the vacuum column to form an ammonium chloride ($NH_3 \cdot HCl$) mist when exposed to HCl vapors. The mist particles sublime and form corrosive deposits in the upper vacuum column and dome, vacuum overhead line, and vacuum overhead condensers upstream of the steam ejectors. Ammonium chloride mist also forms immediately downstream of the injection point when ammonia is injected intentionally into the upper vacuum column or vacuum overhead vapor line for pH control. Ammonium chloride is soluble in overhead water condensate and leaves at the first downstream separator drum.

2.3 Corrosive Inorganic Compounds in Crude

Another important group of corrosive or potentially corrosive compounds in crude includes certain inorganic salts, gases, metals, and other naturally occurring compounds. Examples are inorganic chloride salts, hydrogen sulfide (H_2S), elemental sulfur (S_{xn}), metal sulfides, oxidized sulfur compounds (oxides and sulfates), mercury (Hg) and mercury salts, and various trace metals.

2.3.1 Inorganic Chloride Salts. Inorganic chloride salts are present in all crudes as suspended salt crystals and emulsified brine. Depending on prior history, crudes typically contain 100-400mg/L or 30–120PTB chloride salt (as Cl), but preferably less than 100mg/L or 30PTB as delivered to the refinery or upgrader. The main sources of inorganic chloride salts are formation water produced with the crude, heavy brines used during some crude recovery operations, water floods used during enhanced recovery operations, surface waters used during bitumen mining operations, brackish or seawater used to wash down and recover residual crude (heel) from cargo holds of barges and marine tankers, and brackish or seawater used as ballast when barges or marine tankers run empty during the return voyage. In the absence of actual and up-to-date salt analyses, inorganic chloride salts in raw crude or diluted bitumen are considered to consist of approximately 75% sodium chloride (NaCl), 15% magnesium chloride ($MgCl_2$), and 10% calcium chloride ($CaCl_2$).

Inorganic chloride salts can cause serious corrosion by water condensate collecting in gathering lines, field treating facilities, transfer tanks, barges, marine tankers, pipelines, and receiving tanks and distribution lines at the refinery and upgrader. The worst corrosion damage occurs, however, in the atmospheric section of crude units, the diluent recovery section of bitumen upgraders, and the respective vacuum sections. When feed to these sections is heated prior to distillation, $MgCl_2$ and $CaCl_2$ salts are partially hydrolyzed to form hydrogen chloride (HCl) vapor. The vapor enters various distillation columns, goes overhead, and dissolves in overhead water condensate to form dilute hydrochloric acid (HCl). To minimize overhead corrosion, the salt content of incoming crude or diluted bitumen must be reduced to below 3-10mg/L or 1-3PTB by tank settling, desalting, and/or caustic addition. In most overhead systems, reaction of HCl vapor with ammonia (NH_3) also causes corrosion under ammonium hydrochloride (NH_3-HCl) deposits at temperatures below approximately 230 °C (450 °F).

Hydrogen chloride also causes corrosion problems in vacuum overhead systems. Overhead water condensate upstream of the ejectors can contain 20-25PPM or more HCl at pH values below 4 or 5 and account for most corrosion problems in the precondensers. As noted in Chapters 14 and 15 of the *Guide*, approximately 90% of the $MgCl_2$ and 10% of the $CaCl_2$ remaining in desalted crude is hydrolyzed to HCl at typical preheat exchanger and crude heater temperatures. The remaining 10% of the $MgCl_2$ and 90% of the $CaCl_2$ end up in reduced crude going to the vacuum section. Here the remaining $MgCl_2$ and an additional 5% of the $CaCl_2$ react with steam in the vacuum heater to form HCl vapor that rises in the vacuum column. Another source of chlorides in vacuum overhead systems is entrained seawater in reduced crude shipped by tankers and added to the vacuum column charge. Sometimes, certain types of waste products containing organic chlorides are also added to reduced crude as a means of disposal.

2.3.2 Hydrogen Sulfide. Hydrogen sulfide is present in some sour crude as dissolved free H_2S. Hydrogen sulfide as such does not contribute to HTSC. At high enough concentrations, however, H_2S can cause severe vapor space corrosion and fouling in gathering lines, field treating facilities, terminal tanks, barges, marine tankers, rail tank cars, and pipelines. It can also cause vapor space corrosion and fouling in the crude receiving tanks, distribution lines, and preheat exchangers (PEs) at the refinery or upgrader. Hydrogen sulfide that enters the DRU, preflash, atmospheric, and vacuum columns goes overhead and partially dissolves in light naphtha and water condensate. Approximately four or five times as much H_2S dissolves in light naphtha compared to water condensate. Depending on the water pH, hydrogen sulfide is present as dissolved molecular hydrogen sulfide (H_2S) at pH below 7 and as bisulfide[1] ions (HS^-) at pH above 7. The actual amounts depend on the H_2S partial pressure.

Reaction with ammonia (NH_3) produces ammonium bisulfide ($NH_4{\cdot}HS$) and can lead to severe sour water corrosion and subsequent fouling by iron sulfide (Fe_xS_y) deposits in all overhead systems. When overhead water condensate is recycled to the desalter as desalter water, H_2S and ($NH_4{\cdot}HS$) will redissolve in the desalted crude and return once again to the downstream columns. In most overhead systems, $NH_4{\cdot}HS$ can be expected to deposit (sublime) at temperatures below approximately 100 °C (210 °F). H_2S in the vacuum section originates with reactive sulfur (RS) compounds in reduced crude (atmospheric resid) that are decomposed at the higher temperatures of vacuum heaters compared to crude heaters. H_2S travels up the vacuum column and in combination with steam condensate can corrode the vacuum column dome, precondenser, steam ejectors, intercondensers, vacuum receiver, and connection piping. However, in the absence of significant amounts of HCl, vacuum overhead condensate water has pH values above 5 and corrosion is relatively minor.

2.3.3 Elemental Sulfur. Elemental sulfur (S_{xn}) is present in some types of crude. It can also form in situ when oxygen (air) enters with water floods; surface waters; at leaking pump seals and water-cooled heat exchanger tubes; with injected dilution, stripping, or motive steam; and with wash water. Like H_2S, elemental sulfur does not contribute significantly to corrosion in the high-temperature circuits of crude units or bitumen upgraders. It can cause, however, fouling of upstream and midstream treating facilities, terminal tanks, marine barges, tankers, rail tank cars, and pipelines. Sulfur deposits can also cause fouling at the downstream refinery and upgrader, typically in crude receiving tanks, distribution lines, preheat exchangers (PHEs), and overhead condensers. The presence of elemental sulfur in deposits is readily apparent from the bright-yellow color when a component is opened to the atmosphere for inspection during shutdown.

2.3.4 Oxidized Sulfur Compounds. Oxidized sulfur compounds, including various oxides and sulfates (SO_2, SO_3^{-2}, and SO_4^{-2}) are formed along with elemental sulfur when oxygen (air) enters a unit. Like H_2S or elemental sulfur, sulfur oxides and sulfates do not contribute significantly to corrosion in the high-temperature circuits of crude units and upgraders, but can cause corrosion in the low-temperature circuits including the overhead systems of various distillation columns.

2.3.5 Mercury and Mercury Salts. The metallic mercury content in mg/kg Hg of certain crudes from California, Asia, Africa, Russia, and Australia has become an issue in recent years. As a general rule, the higher the mercury content, the less desirable and expensive crude is, because mercury and mercury compounds present operating, safety, product quality, and environmental issues. When mercury-containing crude is heated prior to distillation, mercury vaporizes and recondenses in the upper preflash and atmospheric columns and their overhead systems. Mercury not only reduces the quality of refined products, but can also poison catalysts in downstream units. Charging mercury-laden crude can present various operating, safety (during inspection and maintenance operations), and environmental issues. Wastewater and solid waste streams can have mercury concentrations that exceed regulatory limits. Localized corrosion of carbon steel components and stress-corrosion cracking of copper and aluminum-base tubulars have been attributed to mercury deposits that originated in crudes.

The total mercury content of crudes processed in U.S. refineries in 2004 was reported to have averaged 7.3μg/kg (0.007ppm) Hg, while the volume-weighted mean including all species in both dissolved and suspended forms was 3.5±0.6μg/kg (0.0035±0.0006ppm) Hg. Measured concentrations ranged from below the analytical detection limit of 0.5μg/kg (0.0005ppm) to approximately 600μg/kg (0.6ppm) Hg[2].

2.3.6 Other Trace Metals. Iron (Fe), nickel (Ni), vanadium (V), zinc (Zn), tin (Sn), selenium (Se), or molybdenum (Mo) content (in mg/Kg or ppm) in certain crudes has been an issue for many years. As a general rule, the higher the metals content, the less desirable and expensive crude is, because trace metals can present operating, corrosion, and fouling issues. Metals may have to be removed at some stage of the refining process, because burning heavy Bunker C fuel oil

1. The correct term is disulfide but bisulfide is commonly used in the corrosion literature.
2. S. Mark Wilhelm et al., "Mercury in Crude Processed in the United States" *Environmental Science and Technology*, American Chemical Society, (2004), published on the Web, June, 2007.

in fired heaters and boilers can leave deposits of vanadium and nickel oxides in fire boxes, flue ducts, and on heater tubes that lead to fireside corrosion. It is also desirable to remove trace amounts of arsenic, vanadium, and nickel prior to further processing to avoid poisoning certain catalysts.

2.4 Problem Chemicals Added in Upstream and Midstream Operations

A variety of corrosive or potentially corrosive compounds can be present in crude, depending on prior production and treating history. These include primarily carbon dioxide (CO_2) added during enhanced recovery operations, surface waters added during enhanced recovery operations, sulfuric acid (H_2SO_4) and hydrochloric acid (HCl) used for formation fracturing operations, organic chloride solvents used in fracturing and wellbore cleanup operations, oxygen (air) introduced with injected water floods and at leaking pump seals, and oilfield chemicals.

2.4.1 Carbon Dioxide. Carbon dioxide (CO_2) is used during enhanced crude recovery operations and has been identified in overhead water condensate of atmospheric columns. Usually, it cannot be shown for sure that carbon dioxide entered with raw crude, dilution (velocity) steam at the fired heater, or stripping steam at the atmospheric column and actually contributed to overhead corrosion. Carbon dioxide may also form by decomposition of bicarbonates added to crude or steam prior to distillation.

2.4.2 Water. Excessive residual water is normally removed from crude prior to transportation to minimize downstream corrosion issues and meet pipeline and refinery specifications. Typically, refinery specifications call for crude receipts with no more than 1-5% BS&W depending on API gravity to avoid having to pay for hauling and removing excess water in crude.

2.4.3 Sulfuric Acid. Minor amounts of sulfuric acid (H_2SO_4) have been identified in some overhead water condensates. Aside from contaminating raw crude during upstream fracturing operations, sulfuric acid is sometimes used to lower the desalter pH. Again, it cannot be shown convincingly why significant amounts of sulfuric acid remained with desalted crude and contributed to overhead corrosion instead of leaving the desalter with the effluent brine. Trace amounts of sulfuric acid can also form when oxygen (air) enters at leaking pump seals, at leaking water-cooled heat exchanger tubes, or with injected steam and wash water and then react with hydrogen sulfide (H_2S).

2.4.4 Organic Chlorides. Various types of organic chlorides (in mg/L Cl) found in certain crude have been an issue for many years. Organic chlorides (also called nonhydrolysable chlorides) rarely originate with the crude itself, but are still used periodically to dissolve waxy deposits during wellbore cleanup operations despite stringent regulations against this practice. When contaminated crude is heated prior to distillation, a fraction of the organic chlorides is hydrolyzed in the presence of water to form hydrogen chloride (HCl) vapor. The vapors enter the preflash, atmospheric, and vacuum columns, go overhead, and dissolve in overhead water condensate to form dilute hydrochloric acid (HCl). The remainder of the organic chlorides dissolves as such in side streams and overhead light naphtha condensate and can interfere with downstream reformer operations. As a general rule, crude with organic chloride content greater than 1mg/L Cl is unacceptable, as far as overhead corrosion is concerned.[3] Contaminated crude should be rejected, unless it can be trickle fed to the regular crude charge. Additional information on organic chlorides is covered in Chapter 11 of the *Guide*.

2.4.5 Oxygen (Air). As noted above, oxygen (air) reacts with hydrogen sulfide and forms yellow elemental sulfur (S_x) and inorganic polysulfide deposits that can plug preheat and overhead exchanger bundles. However, often it is not clear if the oxygen originated with raw crude, entered as a result of leaking pump seals, or with cooling water leaks in water-cooled overhead condensers. Oxygen can accelerate overhead corrosion in the form of pitting when collecting in air pockets in condenser shells or by reacting with dissolved ammonium bisulfide ($NH_3{\cdot}H_2S$) to form various polysulfide ($NH_3{\cdot}H_2S{\cdot}S_X$) solutions. Oxygen can cause severe corrosion problems in the vacuum section as a result of air entry at leaking flanges, valve stem seals, pump seals, and sight-glass cover seals. Other sources of oxygen are air entry during shutdowns and cooling water leaks at precondenser tube bundles upstream of the ejectors. Poorly deaerated steam can be a significant source of oxygen entry downstream of the ejectors. Air leakage into the vacuum section deteriorates ejector performance and can "break" the ejector operation.

3. Test per ASTM D 4929 (latest revision), *Standard Test Methods for Determination of Organic Chloride Content in Crude Oil* (West Conshohocken, PA: ASTM).

2.4.6 Oilfield Chemicals. Crude can contain a group of so-called "oilfield chemicals" that can interfere with treating, dehydrating, and desalting of midstream and downstream operations. Oilfield chemicals include:

- Paraffin inhibitors, dispersants, and solvents;
- Asphaltene inhibitors, dispersants, and solvents;
- Foamers and defoamers;
- Biocides;
- Corrosion and scale inhibitors;
- Flow improvers; and
- Hydrogen sulfide (H_2S) scavengers.

Oilfield chemicals are often used to aid in removing produced water from crude. They can also be used in combination with heater treating, tank settling, crude stabilizing, and desalting in midstream operations. Depending on the specific compounds and conditions involved, oilfield chemicals can partition into the water phase, hydrocarbon phase, or both. Typical issues at the refinery and upgrader attributed to oilfield chemicals have included poor crude settling and dehydration, poor desalting, corrosion and fouling of preheat exchangers and fired heater tubes, and unexpected column overhead corrosion. Another problem is fouling and underdeposit corrosion in atmospheric columns that is caused by certain H_2S scavengers added to crude. For example, the most widely used scavenger is MEA-triazine [hexahydro-1,3,5-tris (hydroxyethyl)-s-triazine] or HHTT. Upon reacting with H_2S, MEA-triazine produces dithiazine and monoethanolamine (MEA). The latter reacts with hydrogen chloride (HCl) to form MEA-chloride deposits that have caused severe fouling and underdeposit corrosion in atmospheric columns and elsewhere.

2.5 Problem Chemicals Added in Downstream Operations

Corrosive or potentially corrosive chemicals are often added to crude at the refinery, including dilute caustic (NaOH) solutions added at the desalter to increase the desalter water pH. Caustic solutions may also be added to desalted crude downstream of the desalter at concentrations of up to 14kg/Mm^3 (5PTB) NaOH to decrease the amount of hydrogen chloride (HCl) evolved in the fired crude heater by converting $MgCl_2$ and $CaCl_2$ to the harmless NaCl. Dilution or velocity steam used at the fired heater and stripping steam used at the atmospheric column can become contaminated with oxygen and carbon dioxide (CO_2) from ineffective boiler feed water (BFW) treatment. Steam can also introduce various types of mineral scale and chemicals added to reduce steam condensate corrosion.

Other corrosive or potentially corrosive chemicals can end up in desalted crude when overhead water condensate is used as desalter water. Examples are ammonia (NH_3) or organic neutralizers, ammonium bisulfide ($NH_3{\cdot}H_2S$) and hydrogen sulfide (H_2S), corrosion products, organic chloride solvents from shop cleaning wastes, spent catalyst streams added intentionally or unintentionally to slop oil sent to the API Separator as a convenient means for disposal, and water treatment chemicals entering as a result of cooling water leaks in water-cooled overhead condensers. In the vacuum section, residual organic nitrogen compounds in reduced crude are converted to NH_3 at the higher temperatures of the vacuum heater relative to the crude heater. Ammonia travels up the vacuum column and reacts with HCl vapors to form an ammonium chloride ($NH_3{\cdot}HCl$) mist. The mist particles sublime and form corrosive deposits in the upper vacuum column and dome, vacuum vapor line, and vacuum overhead condensers upstream of the ejectors. $NH_3{\cdot}HCl$ mist also forms immediately downstream of the injection point when ammonia is injected intentionally into the upper vacuum column or vapor line for pH control. $NH_3{\cdot}HCl$ is soluble in overhead water condensate and leaves at the first downstream separator drum.

2.6 Chapter Summary

Origin, properties, upstream production history, and midstream treating of crude have significant consequences, as far as refinery and upgrader corrosion and fouling are concerned. The hydrocarbon compounds that make up crude are not corrosive per se, nor do they cause corrosion or fouling prior to, during, or after heating, distillation, or condensation. Instead, various chemical compounds or contaminants present in, or added to, raw crude are the root cause of corrosion and fouling issues seen on most units. These compounds include primarily organic sulfur and acid compounds present in all crude; inorganic chloride salts present in all crude; chemicals added during enhanced recovery operations; oilfield chemicals added to improve crude properties during upstream production and midstream treating, transportation, and storage operations; and chemicals added downstream at the refinery and upgrader to improve desalting and to control corrosion and fouling.

Sulfur content and TAN affect the location, extent, and severity of high-temperature sulfidic corrosion (HTSC) and naphthenic acid corrosion (NAC) at temperatures above approximately 230 °C (450 °F). API gravity and salt content affect the desalting parameters which, in turn, affect the severity of corrosion and fouling in the upper distillation columns and column overhead systems at temperatures below approximately 150 °C (300 °F). H_2S and mercury (Hg) in crude can cause operating, safety, environmental, corrosion, and product quality issues. These factors and various ways of controlling, or removing problem chemicals are reviewed in greater detail in subsequent chapters of the *Guide.*

2.7 Read More

1. H. A. Cataldi et al., "Estimation of Corrosivity of Crudes," *Proc. API* 33, III (1953).
2. L. Engel, "Korrosion in Mineralolraffinerien," *Erdoel umd Kohle* 27, 6 (1974).
3. J. Gutzeit, "Refinery Corrosion Overview" in *Process Industries Corrosion - The Theory and Practice*, B. J. Moniz, W. I. Pollock, eds. (Houston, TX: NACE, 1986).
4. J. Gutzeit, R. D. Merrick, and L. R. Scharfstein, "Corrosion in Petroleum Refining and Petrochemical Operations" in *Volume 13 - CORROSION*, ASM Metals Handbook, 9th ed. (Metals Park, OH: ASM, 1987).
5. J. Gutzeit, R. D. Merrick, and L. R. Scharfstein, "Corrosion in Petroleum Refining and Petrochemical Operations" in *Corrosion in the Petrochemical Industry*, ASM Monograph, (Metals Park, OH: ASM International, 1994).
6. S. Kapusta et al., "The Impact of Oil Field Chemicals on Refinery Corrosion Problems," paper no. 03649, CORROSION/2003 (Houston, TX: NACE, 2003).
7. K. R. Lewis et al., Processing Corrosive Crudes, paper no. 377, CORROSION/99 (Houston, TX: NACE, 1999).
8. D. McIntyre, J. English, and G. Kobrin, "Mercury Attack of Ethylene Plant Alloys," paper no. 106, CORROSION/89 (Houston, TX: NACE, 1989).
9. C. G. Munger, "Deep Pitting Corrosion in Sour Crude Tankers," *MP* 15, 3 (1976).
10. W. J. Neill and Z. A. Foroulis, "Internal Corrosion in Floating Roof Gasoline Storage Tanks," *MP* 15, 9, (1976).
11. Petrolite Laboratory Staff, *Impurities in Petroleum* (Houston, TX: Petrolite Corp., 1968).
12. R. L. Turissini, et al., "Mercury Liquid Metal Embrittlement Causes Aluminum Plate Heat Exchanger Failure," *MP* 37, 6 (1998).
13. C. L. Wehmeyer, "Design of Sour Crude Separation Facilities," *Petroleum Engineer*, August (1975).
14. S. M. Wilhelm et al., "Mercury in Crude Processed in the United States" *Environmental Science and Technology*, American Chemical Society (2004). Published on the Web, June, 2007.
15. S. M. Wilhelm, "Impact of Mercury in Crude on Refining," *World Refining*, January/February (2003).
16. S. Wongkasemjit and A. Wasantakorn, "Laboratory Study of Corrosion Effect of Dimethyl- Mercury on Natural Gas Processing Equipment," *J. Corr. Sci. and Eng.* 1, 12 (1999).

3 Plant Overview

In this Chapter

3.1 First Glance

This chapter provides a brief overview of metallurgy, corrosion, and fouling issues in major components of crude units and bitumen upgraders that are reviewed in greater details in subsequent chapters of the *Guide*. Distillation is the first process in the refining of crude in modern petroleum refineries and usually consists of atmospheric distillation and vacuum distillation. Depending on the designer and engineering company; age, size, and complexity of the refinery; and types of crude processed, there can be significant differences in crude unit layouts. For example, on some units no vacuum distillation is provided nor is it required to meet the needs of various localities and jurisdictions. Some units are simple topping units that provide LPG, diesel, fuel oil, kerosene, atmospheric gas oil (AGO), and asphalt near the source of production. Diesel units are topping units that concentrate specifically on diesel-range products without the full range of other products available with an atmospheric distillation column.

Bitumen upgraders are designed to convert mined bitumen into crude that meets pipeline specifications for shipping to refinery crude units. Some upgraders also produce a range of other products for local consumption, such as LPG, diesel, fuel oil, or gas oil (AGO). Depending on the process, mined bitumen is first crushed, mixed with warm water, and moved by pipeline to an extraction plant where a number of process steps are used to separate the bitumen from tailings. After adding water, natural gas condensate (NGC), or naphtha, the diluted bitumen (dilbit) is sent by pipeline to an upgrader for further processing. At the upgrader, the bulk of the diluent is removed in the diluent recovery unit (DRU) for reuse. The dilbit is then sent to the vacuum distillation unit (VDU) of the upgrader. Both units are similar to the atmospheric and vacuum distillation sections of a crude unit and present similar corrosion and fouling issues.

3.2 Crude Unit Overview

Crude distillation consists of two steps: atmospheric distillation to produce light naphtha, heavy naphtha, kerosene, fuel oil, atmospheric gas oil, and reduced crude (atmospheric resid) followed by vacuum distillation to separate the reduced crude into heavy distillates (vacuum gas oils) and asphalts (vacuum resid). Distillation is an energy intensive process that requires a number of preheat exchangers (PHEs) and a fired heater to bring the crude up to distillation temperature. Crude dewatering and desalting steps are often necessary to meet product specifications and protect the unit from excessive corrosion and fouling.

3.2.1 Atmospheric Section. A schematic of the atmospheric section of a typical crude unit is shown in Figure 3-1. The processing of crude involves major changes in temperature and pressure that dictate the choice of materials of construction (metallurgy). Crude is first heated to approximately 315-343 °C (600-650 °F) in a series of cold preheat exchangers (CPEs), hot preheat exchangers (HPEs), and a fired heater. The resultant crude liquid/vapor mixture is then sent via a transfer line to the flash zone of the atmospheric column where hydrocarbon liquids are separated from vapors by distillation. The CPEs often use overhead vapors while the HPEs use various product sidestreams as the heating medium. A mix valve combined with a single or double desalter often follows the CPE train and reduces the crude salt content to acceptable levels before the crude enters the HPE train and fired heater. Inline mixers are sometimes added just downstream of the mix valve to improve contact between crude and desalter water. Velocity or dilution steam is often added at the fired heater to optimize flow through the heater and minimize coke build-up in the heater coils. Distillation steam is added at the bottom of the atmospheric column to aid in the distillation process.

As crude is heated, the pressure is raised by the main charge pumps from near-ambient to approximately 12-15bar (180-225psi) to overcome the pressure drop through the CPE train and desalter. The desalted crude exits at 95-150 °C (200-300 °F) and is repressurized with the desalted crude pumps (not shown in Figure 3-1) to approximately 17-20bar (260-300psi) to pass through the HPE train and fired heater. The hot crude/vapor mixture then enters the flash zone of the atmospheric distillation column at approximately 315-343 °C (600-650 °F). The column is designed to separate the various crude fractions by boiling point range, as follows:

- Steam and light crude fractions (fuel gas and light naphtha) are taken off as mixed vapors at the top of the column where temperatures can range from 93 °C (200 °F) to 180 °C (350 °F). The actual column top temperature depends on the type of crude charged, the required product slate, and the specific design and operating parameters of the crude unit. The vapors are then cooled and condensed in the crude column overhead system.
- Intermediate crude fractions including heavy naphtha, kerosene, fuel oil, diesel, and light atmospheric gas oil (LAGO) are taken off as sidestreams at various levels and temperatures of the column, and sent to separate stripping columns.
- Heavy crude fractions including heavy atmospheric gas oil (HAGO) are taken off near the bottom of the column at temperatures near 300 °C (580 °F) and sent to a separate stripping column.
- The remainder is reduced crude (atmospheric resid) and leaves the very bottom of the column, usually at temperatures near 315 °C (600 °F), and goes to the vacuum section for further processing.

Many crude units use a preflash column upstream of the fired heater. Preflashing reduces the total volume of liquid and vapor that passes through the fired heater and atmospheric column and, therefore, reduces size, construction costs, and operating costs of both the heater and the column. Crude usually enters the bottom of the preflash column at temperatures near 200 °C (400 °F). Steam and light crude fractions (fuel gas and light naphtha) are taken off the top of the preflash column at 93–150 °C (200-300 °F) temperature and condensed and cooled in the preflash column overhead system. Water and light naphtha condensate from the preflash column overhead system are then combined with the corresponding streams from the atmospheric column overhead system.

FIGURE 3-1 SCHEMATIC OF ATMOSPHERIC SECTION OF A TYPICAL CRUDE UNIT

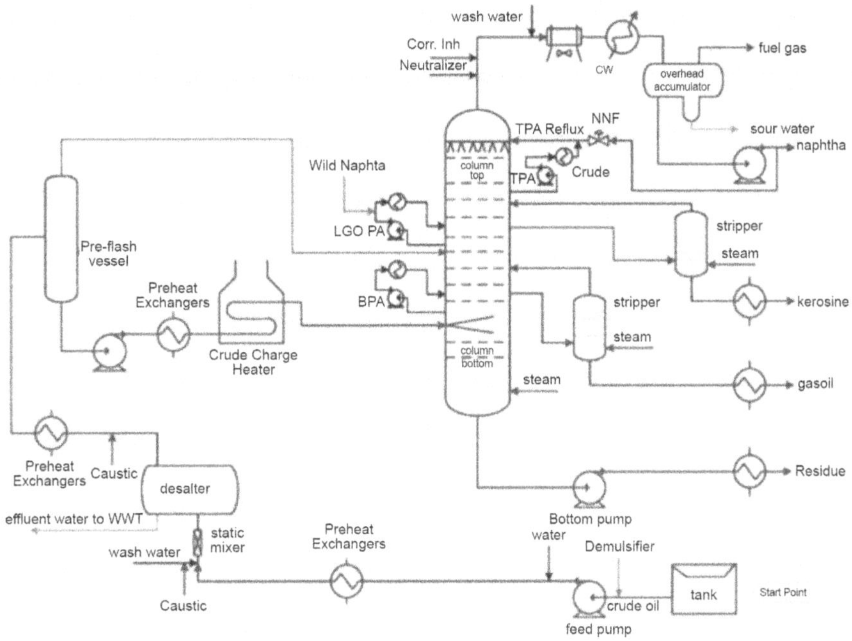

Adapted from Saad Al-Dhafiri et al., Synergistic Effect of Erosion-Corrosion And Galvanic Corrosion in Crude Distillation Tower Top Section. (No other information available.)

After preflashing at 2.5-3.5bar (40-50psi), the preflash crude pumps again raise the pressure to approximately 16-17bar (240-260psi) upstream of the hottest PE. On some crude units, a preflash drum is used instead of the preflash column to remove a portion of the steam, fuel gas, and light naphtha from the preheated crude before it enters the fired heater. Using a preflash drum instead of a preflash column avoids the corrosion issues seen in preflash column overhead systems. Steam, fuel gas, and light naphtha vapors taken off the top of the preflash drum are sent directly to the outlet of the fired heater or to the top of the crude column.

3.2.2 Vacuum Section. Simplified schematics of a vacuum distillation column and 3-stage vacuum ejector system are shown in Figure 3-2 and Figure 3-3, respectively. Actual designs can vary significantly, depending on whether the goal is to obtain fuels or lube stocks. Steam is usually added at the bottom of the vacuum column to aid distillation. Vacuum columns tend to be quite large with diameters of 12m (40ft) and more to handle the large vapor volumes produced at the low operating pressures. Pump-around circuits are used to provide liquid to trays, as needed. A number of side-stream strippers are used on lube units. In the vacuum section, reduced crude (atmospheric resid) at temperatures near 315 °C (600 °F) is reheated in the fired vacuum heater to approximately 350-370 °C (670-700 °F) before entering the flash zone of the vacuum column. Vacuum distillation at absolute pressures as low as 10-40mm Hg is used to keep temperatures below 370 °C (700 °F) to prevent premature thermal cracking and coking that can plug furnace tubes and transfer lines. Vacuum columns operate well above the dewpoint and no water condensate is formed. The required amount of ejector stages depends on the level of vacuum required. The vacuum column typically operates at an absolute flash zone pressure of 0.07-0.2bar (50-150mm Hg) depending on the type of unit and number of ejector stages. The column is designed to separate reduced crude (atmospheric resid) into various fractions by boiling range, as follows:

- Steam and the lightest fractions (vacuum overhead) are taken off the top of the column where temperatures can range anywhere from 38 °C (100 °F) to 93° C (200 °F), depending on the reduced crude, and the specific design and operating parameters of the column. The vapors are piped to a barometric precondenser (where used) and two or three steam-ejector stages. All condensate goes to an accumulator (seal drum or sump).

- An intermediate fraction, light vacuum gas oil (LVGO), is taken off as a sidestream near the top of the column at temperatures ranging from 120 °C (250 °F) to 150 °C (300 °F).
- A heavy fraction, heavy vacuum gas oil (HVGO) is taken off near the middle of the column, usually at temperatures from 230 °C (450 °F) to 260 °C (500 °F).
- Vacuum column bottoms (vacuum resid) are taken off the very bottom of the column, usually at temperatures near 340 °C (650 °F) to 390 °C (725 °F) and sent to a delayed coker for further processing or to storage for marketing as asphalt.

FIGURE 3-2 SCHEMATIC OF VACUUM DISTILLATION COLUMN IN PETROLEUM REFINERIES

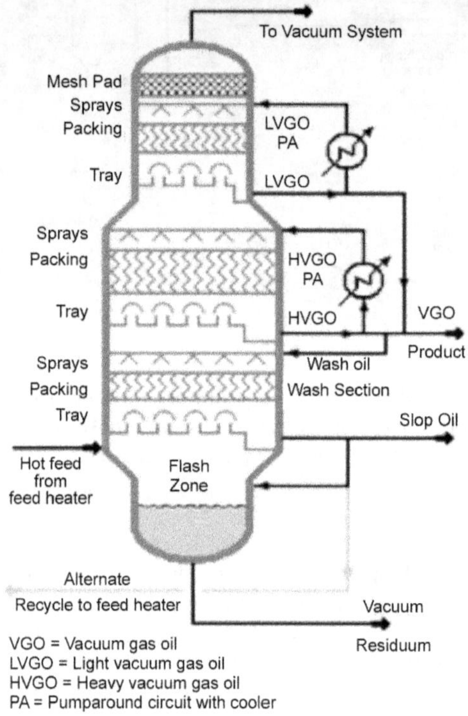

Adapted from M. Beychok, "Vacuum Distillation," *Wikipedia*, Feb. 29, 2008.

FIGURE 3-3 THREE-STAGE VACUUM EJECTOR SYSTEM

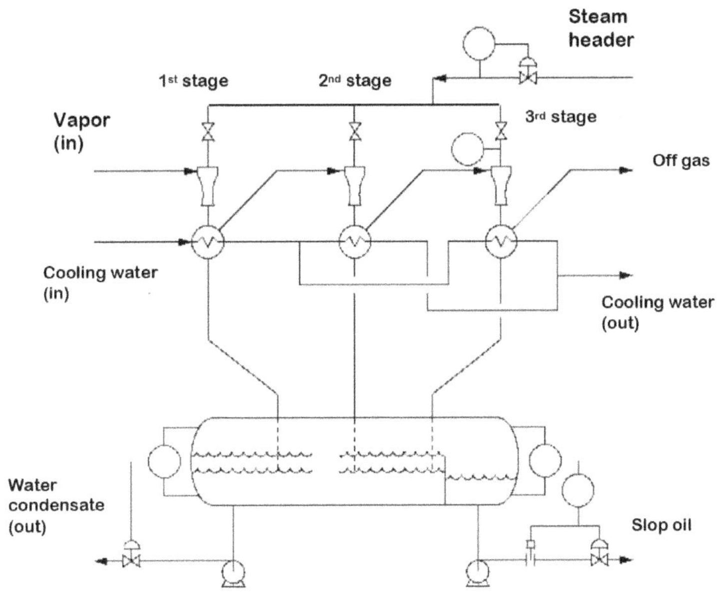

Adapted from G. R. Martin, "Understand Real-World Problems of Vacuum Ejector Performance," *Hydroc. Proc.* November 1997.

3.3 Upgrader Overview

A bitumen upgrader converts diluted heavy bitumen (DB) into synthetic crude that can be sent to a refinery crude unit. For practical reasons, bitumen upgraders are located as close as practical to the bitumen mining operations and are often designed by the same engineering companies that design crude units. As a result, both tend to be quite similar as far as major components and metallurgy are concerned. The initial step of upgrading is a simple distillation process in the diluent recovery unit (DRU) to remove and recover the naphtha diluent and water that was used to separate bitumen froth from the tailings and allow transport to the upgrader proper. The resultant diluted bitumen, or dilbit, should not be confused with diluted bitumen crude sold as bitumen crude to refineries as heavy *Sour Dilbit* crude. The recovered naphtha is recycled for reuse in the froth treatment process. Along with the DRU, a vacuum distillation unit (VDU) is used in the conversion process. Since mined bitumen is too viscous and dense to be pumped and transported by pipeline or most other forms of transportation, it must be diluted with natural gas condensate (NGC), synthetic crude oil (SCO)[1], naphtha, or kerosene in order to meet pipeline and refinery specifications.

3.3.1 DRU Section. Dilbit (DB) from tankage is heated in a series of preheat exchangers (PEs) and send to the 1st stage of a 3-stage diluent stripper. The diluent stripper is similar to a preflash column on a refinery crude unit. The feed enters the diluent stripper at approximately 150 °C (300 °F). Because of several heated pump-around circuits, the stripper operates above the dew point. No aqueous corrosion occurs during normal operation and no water is present in any of the sidestreams. Stripper overhead vapors leave at 130 °C (270 °F) and consists primarily of water vapor and

1. Synthetic crude is produced in an upgrader where the bitumen is heated, decoked, and combined with NGC. The vapors from the coke drums are fractionated into naphtha, kerosene, and gas oil fractions that are blended into synthetic crude oil and diesel fuel. Synthetic crude has an approximate API gravity of 32, 0.1% S content, and 0.5 TAN.

hydrocarbon diluent along with traces of H_2S, HCl, and NH_3. This overhead stream is cooled by heat exchange with dilbit feed in the feed PEs and piped to the diluent accumulator where the vapors are separated from recovered diluent and water condensate and sent off-site for further processing. The recovered diluent is sent to tankage. The stripper bottoms leave at 168 °C (335 °F), are heated by heat exchange with hot gas oil sidestreams in a series of heat exchangers, and then sent to a fired heater for further heating to 315-343 °C (600-650 °F). The heater effluent is piped to the lower section of the diluent (DR) column where the stream is separated by steam stripping into reduced bitumen (the bottoms fluid), kerosene, and overhead vapors. The reduced bitumen is the primary feed to the vacuum feed stripper of the VDU. The overhead vapors exit the diluent column at 180 °C (360 °F) and are cooled in several DB preheat exchangers and also piped to the aforementioned diluent accumulator.

3.3.2 VDU Section. The VDU section is practically identical to the vacuum distillation column and ejector system of a typical crude unit, shown above in Figure 3-2 and Figure 3-3, respectively. In the VDU, DR bottoms (reduced bitumen) are sent to a vacuum feed stripper and heated in the fired vacuum heater to approximately 350 °C (670 °F) to 370 °C (700 °F) before entering the flash zone of the vacuum column. Vacuum columns typically operate at an absolute flash zone pressure of 0.07-0.2kg/cm² (50-150mm Hg) depending on the type of vacuum unit and ejector system. The column is designed to separate hot DR bottoms into HVGO, LVGO, and vacuum resid. The latter is usually sent to a delayed coker for further processing. Vacuum column overhead vapors consist of steam, light hydrocarbons, HCl, H_2S, and NH_3 and are piped to a barometric precondenser (where used) and two or three steam-ejector stages. The condensate is separated in a vacuum accumulator (seal drum or sump).

3.4 Preheat Exchangers

Preheat exchangers heat the incoming crude charge by heat exchange with hot overhead vapors or product side streams. The cold preheat exchangers (CPEs) are upstream of the desalter and operate typically below 150 °C (300 °F), hence carbon or low-alloy Cr-Mo steel tubes can be used depending on the temperature of the hot side. The hot side can experience corrosion problems depending on the heating medium, usually overhead vapors or light naphtha. The cold side can experience fouling issues due to sludge carryover from upstream tankage and piping. The hot preheat exchangers (HPEs) are downstream of the desalter and operate typically between 150 °C (300 °F) and 230 °C (450 °F). They 5Cr or 9Cr steel tubes depending on the temperature of the hot side. Either side can experience corrosion problems depending on the temperatures of the specific heating medium, usually HAGO, reduced crude, or bitumen.

3.5 Fired Heaters and Transfer Lines

Fired heaters (furnaces) are the principal heat source for high-temperature processes and follow the HPEs. Crude units include at least one fired heater in the atmospheric section to heat the desalted crude charge to the main atmospheric column and at least one fired heater in the vacuum section to heat the reduced crude (atmospheric resid) charge to the vacuum column. The same applies to the DRU and VDU of bitumen upgraders. Depending on past revamps and upgrades, there may be additional fired heaters to increase unit throughput. Most fired heaters contain steam coils to supply low or medium-pressure process steam. All fired heaters use either gas or oil as fuel. Heater tubes, return bends, and headers represent the most expensive components of a fired heater. Since they operate with metal temperatures well above 230 °C (450 °F), low-alloy Cr-Mo steels up to and including 9Cr-1Mo are required for acceptable life against HTSC and NAC in the atmospheric section. Type 316 stainless steel tubes, return bends, and headers are required for fired heaters of the vacuum section to control NAC.

Transfer lines connect the heater outlet manifold to a downstream distillation column. Since transfer lines also operate with wall temperatures well above 230 °C (450 °F), alloying against HTSC is required, and low-alloy Cr-Mo steel up to and including 9Cr-1Mo is required for acceptable tube life. Unless used as a cladding, 12Cr stainless steel is not recommended, because of the possibility of 475 °C (885 °F) embrittlement. Solid or clad Type 316 and Type 317 stainless steel lines are often required when naphthenic acids are present.

3.6 Distillation Columns

The top four trays, corresponding shell areas, and the top head of most columns require alloying against acid corrosion and under-deposit pitting. The lower trays and corresponding shell areas operating at temperatures above 230 °C (450 °F) require alloying against HTSC. The flash zone also requires alloying against NAC when acidic crudes are charged. Corrosion issues in the preflash and diluent stripper overhead systems are similar to those seen in single-stage crude column overhead systems. Because the preflash column charge bypasses the fired heater, the extent of hydrolysis of inorganic salts to yield hydrochloric acid (HCl) and the thermal decomposition of naphthenic acids to yield acetic acid (CH_3-COOH), formic acid (H-COOH), or proprionic acid (CHÐCHÐ-COOH) is reduced. Most preflash columns and drums operate at temperatures below 230 °C (450 °F) and do not require alloying against HTSC or NAC. Components in vacuum columns operating at temperatures above 230 °C (450 °F) require alloying with 12Cr stainless steel against HTSC. With acidic crudes, austenitic Type 316 or Type 317 stainless steels are required to control NAC.

3.7 Overhead Condensing Systems

Overhead coolers and condensers are chosen with the goal of maximizing heat transfer and minimizing energy and maintenance costs. Several designs and configurations are used on crude units and upgraders, depending on type of feed, type of column, specific design, and operating parameters. Each has its advantages and disadvantages, as discussed below.

3.7.1 Single-Stage Type 1 Atmospheric Condensing Systems. With a single-stage (single-drum) Type I condensing system, overhead vapors are cooled and condensed in a set of coolers and condensers and then piped to an overhead receiver drum where fuel gas and hydrocarbon condensate are separated from steam condensate. A part of the hydrocarbon condensate is returned to the column as reflux, the remainder is sent on as product for further processing or storage. Steam condensate is withdrawn from an attached or remote water boot for use as desalter water, injection as overhead wash water, or both. Any excess is sent to a sour water stripper.

3.7.2 Two-Stage Type 2 and Type 3 Atmospheric Condensing Systems. With two-stage (two-drum) Type 2 and Type 3 condensing system, only a portion of the overhead vapors is cooled and condensed in the first stage and separated in the first-stage receiver drum. The uncondensed portion is sent to the second stage where it is further cooled, condensed, and then separated in the second-stage receiver drum. Two types of two-stage condensing systems are used, depending on column top temperatures. Type 2 systems are used with "hot columns" having top temperatures in the 150-180 °C (300-350 °F) range. No water normally condenses in the first stage and the first-stage receiver simply operates as a reflux drum. The first stage is said to operate "dry." Type 3 systems are used with "cold columns" having column top temperatures in the 105-120 °C (220-250 °F) range. Water condenses in both stages and is withdrawn from both receiver drums. The first stage is said to operate "wet." Note that some cold columns can have top temperatures below 93 °C (200 °F) resulting in wet conditions in the top of the column.

From a corrosion point of view, the difference between dry and wet overhead operation is quite important: Corrosion in dry areas and dry stages of overhead condensing systems is caused primarily by ammonium hydrochloride (NH_3·HCl) and neutralizer hydrochloride (R-NH_2·HCl) salt deposits, while corrosion in wet areas and stages is caused by dilute hydrochloric acid (HCl). With acidic crudes, acetic acid (CH_3-COOH), formic acid (H-COOH), and proprionic acid (CHÐCHÐ-COOH) may also contribute to overhead corrosion.

3.7.3 Multistage Atmospheric Condensing Systems. Single and two-stage overhead condensing systems are used with most preflash and crude distillation columns and corresponding upgrader columns, depending on column top temperatures. Under special conditions, however, overhead systems may consist of three or more condensing stages. While there are probably certain advantages from a process point of view, the operating parameters and corrosion rates in multistage condensing systems are more difficult to monitor and control than those in single or two-stage systems.

3.7.4 Vacuum Condensing Systems. The overhead condensing system of vacuum columns in crude units and VDUs of upgraders is similar to a single-stage overhead system used in the atmospheric section, but with steam jet ejectors and intercondensers tacked onto the vapor line. The steam condensate from each condenser shell drains directly into a large receiver (seal drum, hot well, or vacuum sump). Noncondensables are sent to the fuel-gas system. Steam ejector/condenser stages often are imposing attachments to vacuum columns, as shown in Figure 3-6. During startup, all ejectors are operational, but during steady operation fewer ejectors need to be engaged. Ejectors normally operate with superheated, medium-pressure steam (motive steam). Back pressure is reduced by steam condensation and the static head of a condensate-filled standpipe. As a general rule, corrosion issues are similar to those seen in preflash and atmospheric overhead systems, but usually less severe because the bulk of HCl vapors has ended up in those two columns.

3.7.5 Overhead Condensing Issues. All overhead systems are subject to several types of low-temperature process and waterside corrosion problems. Collectively referred to as overhead corrosion, these continue to be a major and costly maintenance problem on most units. Subsequent chapters in the *Guide* review in some detail the accepted methods for controlling corrosion in atmospheric overhead systems, including selective alloy upgrading, neutralizer injection, corrosion inhibitor injection, and water washing.

Several points should be noted regarding corrosion in both atmospheric and vacuum overhead systems:

1. Similar corrosion and fouling issues are often, but not always seen in parallel banks of a given cooler or condenser;
2. More severe corrosion is seen in the upper (hotter) upstream units than in the lower (cooler) downstream units of any given bank of coolers or condensers; and
3. Ammonium hydrochloride ($NH_3 \cdot HCl$) and neutralizer hydrochloride ($R\text{-}NH_2 \cdot HCl$) salts can deposit on and in all types of tubes and the resultant underdeposit corrosion can lead to early tube failures, even with titanium tubes. Overhead water injection is often required to control underdeposit corrosion.

In vacuum overhead systems, corrosion and fouling can occur in various components at temperatures between approximately 120 °C (250 °F) and 40 °C (100 °F). The affected components include primarily the upper shell areas and dome of the vacuum column and various components of the vacuum overhead system. Specific examples are the vacuum overhead vapor lines; any vacuum overhead precondensers upstream of the steam ejectors; the steam ejectors; the steam surface and other condensers downstream of the ejectors; and the hot well and the vacuum gas KO drums. In most cases, corrosion and fouling mechanisms are similar to those seen in atmospheric overhead systems of crude units, as discussed in Chapter 14 of the *Guide*.

FIGURE 3-6 FIRST-STAGE EJECTOR SYSTEM ON CRUDE UNIT VACUUM COLUMN

Adapted from J.R. Lines, "Understanding Ejector Systems Necessary to Troubleshoot Vacuum Distillation," *Hydroc. Proc.* 73, 10 (1994).

3.8 Heat Exchangers, Coolers, and Condensers

Depending on heat transfer, flow rate, flow velocity, and pressure drop constraints, heat exchangers, coolers, and condensers are configured as single or parallel banks of one or more units in series. Feed/overhead shell-and-tube heat exchangers are often used as the first set of coolers and condensers in single-stage and two-stage overhead systems for preflash, crude, and DRU columns. These are the hottest heat exchangers of an overhead system and allow feed to be preheated with what otherwise would be waste heat. By using cold feed as the cooling medium, waterside corrosion and fouling issues seen with water-cooled coolers or condensers in this service are avoided. With proper design and corrosion controls in place, crude/versus overhead heat exchangers can normally use carbon steel tubes and last at least 10 years.

Air-cooled fin-fan coolers or condensers are an alternative to shell-and-tube heat exchangers. They too avoid the corrosion and fouling issues seen with water-cooled coolers or condensers, but overhead heat is "wasted" to the atmosphere instead of being used to heat an incoming feed or a product stream. Air-cooled coolers or condensers normally use aluminum-finned carbon steel tubes that last at least 10 years.

Water-cooled coolers or condensers often follow feed/overhead heat exchangers or air-cooled coolers and condensers. Even when used as trim coolers at relatively low overhead temperatures, water-cooled heat exchangers are often subject to corrosion and fouling on both the overhead and watersides. With water-cooled coolers or condensers, overhead heat is "wasted" to the cooling water system. Naval brass tubesheets and Admiralty brass, copper nickel 70-30, or titanium tubes in combination with treated cooling water are usually required for optimum tube life. Organic coatings and sacrificial Zn/Al anodes are needed to protect the carbon steel channel boxes and tubesheets where used. Tube bundles in water-cooled coolers and condensers often last only 1-3 years with carbon steel tubes, 5-10 years with Admiralty brass tubes, 7-12 years with copper nickel 70-30 tubes, but at least 10 years with titanium tubes.

3.9 Chapter Summary

Crude unit and upgrader designs vary depending on age, size, available feedstocks, required product slate, and complexity of the plant. Many crude units consist of an atmospheric section and a vacuum section. The corresponding upgrader units consist of a diluent recovery unit (DRU) and vacuum distillation unit (VDU). Temperature and pressure variations during the processing of crude and dilbit greatly affect the choice of metallurgy. Elevated temperatures increase the severity of corrosion. Elevated pressures increase the strength requirements of components operating at elevated temperatures (for example, fired heater tubes).

Several configurations of overhead condensing systems are used. With single-stage (Type 1) condensing systems, the upstream condensers usually operate "dry" (no water present), while the downstream condensers operate "wet." With two-stage two-drum (Type 2) systems no water condenses in the first stage and the first-stage receiver drum simply operates as a reflux drum. With two-stage two-drum (Type 3) systems, water can condense in both stages and is withdrawn from both receiver drums. As a general rule, single-stage condensing systems are used on the DR columns of upgraders. The difference between dry and wet operation is important: Corrosion in dry areas or stages of overhead condensing systems is caused primarily by ammonium hydrochloride ($NH_3{\cdot}HCl$) and neutralizer hydrochloride ($R\text{-}NH_2{\cdot}HCl$) deposits, while corrosion in wet areas or stages is caused primarily by dilute hydrochloric acid (HCl).

Overhead condensing systems of vacuum columns in crude units and in VDUs of upgraders are of an entirely different design. While similar to single-stage condensing systems, they are followed by steam jet ejectors and intercondensers to facilitate vacuum distillation at absolute pressures as low as 10-40mmHg to keep temperatures below 370 °C (700 °F) and prevent premature thermal cracking and coking that can plug furnace tubes and transfer lines.

3.10 Read More

1. API Division of Refining, "Guide for Inspection of Refinery Equipment," Chapter II, *Conditions Causing Deterioration or Failures*, 2nd ed. (Washington, DC: API, 1973).
2. R. J. Hengstebeck, *Petroleum Processing – Principles and Applications* (New York, Toronto, London: McGraw-Hill, 1959).
3. W.L. Nelson, *Petroleum Refinery Engineering*, 4th ed. (New York, Toronto, London: McGraw-Hill, 1958).
4. J. R. Lines, "Understanding Ejector Systems Necessary to Troubleshoot Vacuum Distillation," *Hydroc. Proc.* 73, 10 (1994).
5. D. A. Hansen and R. B. Puyear, *Materials Selection for Hydrocarbon and Chemical Plants* (New York, NY: Marcel Dekker, 1996).
6. E. L. Hildebrand, "Materials Selection for Petroleum Refineries and Petrochemical Plants," *MP* 11, 7 (1972).
7. M. J. Humphries and G. Sorell, "Corrosion Control in Crude Oil Distillation Units," *MP* 15, 2 (1976).
8. R. I. Jacobs and J. A. McMaster, "Titanium Tubing: Economical Solution to Heat Exchanger Corrosion," *MP* 11, 7 (1972).
9. T. M. Krebs, "Selection of Furnace Tubing - Part1: Mechanical Considerations, Types of Tubing Used," *Petroleum Engineer* 30, 2 (1958).
10. D. M. McCue, "Design Considerations for Titanium Heat Exchangers," paper no. 60, CORROSION/81 (Houston, TX: NACE, 1981).
11. J. A, McMaster, "Selection of Titanium for Petroleum Refinery Components, *MP* 18, 4 (1979).
12. G. E. Moller, I. A, Franson, and T. J. Nichol, "Experience with Ferritic Stainless Steel in Petroleum Refinery Heat Exchangers," *MP* 20, 4 (1981).
13. NACE RP0170-2004, "Protection of Austenitic Stainless Steels and Other Austenitic Alloys from Polythionic Acid Stress Corrosion Cracking During Shutdown of Refinery Equipment" (Houston, TX: NACE, 2004).
14. W. J. Neill, "Experience with Titanium Tubing in Oil Refinery Heat Exchangers," *MP* 19, 9 (1980).
15. R. W. Schutz and L. C. Covington, "Effect of Oxide Films on the Corrosion Resistance of Titanium," paper no. 61,CORROSION/81 (Houston, TX: NACE, 1981).
16. L. M. Wyatt et al., *An Atlas of Corrosion and Related Failures,* MTI publication no. 18 (St. Louis, MO: Materials Technology Institute of the Chemical Process Industries, 1987).
17. A. Garg, "Improve Vacuum Heater Reliability," *Hydroc. Proc.* 78, 3 (1999).
18. 18. L. Engel, "Korrosion in Mineralolraffinerien," *Erdöl und Kohle* 27, 6 (1974).
19. 19. G. R. Martin, "Understand Real-World Problems of Vacuum Ejector Performance," *Hydroc. Proc.* (November 1997).

4 Materials of Construction (Metallurgy)

In this Chapter

4.1 First Glance

The principal material of construction (metallurgy) of crude units and upgraders is and has always been carbon steel. When annual or biennial turnarounds were the norm, regular replacement of corroded carbon steel components "in kind" presented no particular issues or restrictions on plant operations[1]. Compared to most other alloys, carbon steel is cheap, readily available, easily fabricated, and sufficiently strong for many components. However, changing economics during the past 30 to 40 years have dictated increasingly longer unit runs. Today, the goal of most plants is run 5 to 10 years or more between turnarounds. If necessary, individual components are bypassed, repaired, or replaced on the run. Since feed slates have become more aggressive and less predictable, low-alloy Cr-Mo steels and stainless alloys are increasingly substituted for carbon steel to extend the service life of critical components in many types of services. It should be recognized that there is no such thing as a unit free of corrosion issues, unless all components are gold-plated; obviously such units will never be built.

1. Some jurisdictions still require annual or biannual boiler and pressure vessel inspections which often mean a complete unit shutdown.

4.2 Materials Guidelines, Recommended Practices, and Specifications

The selection of construction materials for refineries and upgraders is based on a combination of mandatory construction and engineering guidelines (Codes), nonmandatory recommended practices (RPs), and standard materials specifications (MSs).

4.2.1. Mandatory Construction and Engineering Guidelines (Codes). Mandatory construction and engineering guidelines (Codes) provide rules and procedures for the mechanical design, materials, fabrication, repair, and inspection of major plant components, including pressure vessels, piping, pumps, and tanks. Codes are jointly developed by the industry and certain Code-issuing organizations, primarily in the United States, Great Britain, Germany, Japan, China, and Russia. Code-issuing organizations in the United States include ASME[2], ANSI[3], and API. Code compliance is mandatory in all states. Outside the United States, Code compliance is often mandatory, but statutory authorities are allowing petroleum companies increasing freedom to prioritize inspections and integrity assessments by means of risk-based inspection (RBI) or risk-based inspection management (RBIM) programs. In general, Codes provide procedures for establishing minimum and maximum design temperatures and pressures. Codes also provide advisory information on potential corrosion issues, but normally do not specify materials and corrosion rates for specific services. Codes do not cover special heat treatments, hardness limitations, or other factors, including methods of corrosion control, nor do Codes address the effects of process upsets on design criteria. Codes do not specify corrosion allowances, but provide rules for including corrosion allowances when calculating required wall thicknesses. Consequently, the refining industry developed its own criteria for selecting suitable materials of construction. For example, in the case of high-temperature components, potential corrosion issues expected with various crudes tend to be based on most criteria listed in Table 4-1.

TABLE 4-1 CRITERIA USED BY REFINING INDUSTRY TO PREDICT THE CORROSIVENESS OF CRUDES (1996 SURVEY)

Basis	Users, %
Prior experience with crude	85
Sulfur content of crude fractions	70
TAN of crude fractions	70
TAN of whole crude	62
Sulfur content of whole crude	39
Naphthenic acid content of crude and crude fractions	38
Sulfur compounds analyses of crude and crude fractions	23

4.2.2 Nonmandatory Standard Procedures. Nonmandatory standard procedures provide rules and procedures for the inspection, repair, and corrosion control of major plant components. Nonmandatory standard procedures are issued by API and NACE in the form of recommended practices (RPs). While recommended practices are nonmandatory, they are often treated as mandatory, because of safety and potential litigation concerns.

4.2.3 Standard Materials Specifications. Standard materials specifications for refinery steels and other materials are issued in the United States by ASTM[4] and AISI[5]. Standard materials specifications for refinery steels and other materials are also issued in a number of other countries, including, but not limited to, Great Britain, Germany, Japan, Sweden, China, and Russia. Compliance with standard materials specifications for steels and other materials is mandatory in most states and in some countries outside the United States.

2. ASME International.
3. American National Standards Institute.
4. ASTM International, formerly the American Society for Testing and Materials.
5. American Iron and Steel Institute.

4.3 Covered Components

As written, mandatory construction and engineering guidelines (Codes) apply only to new construction of major components, including pressure vessels, piping, and tanks. In practice, however, Codes are also followed for "repairs of major components" and for "new construction and repairs of other than major components," such as pumps.

4.3.1 Unfired Pressure Vessels. Unfired pressure vessels must comply with the ANSI/ASME Boiler and Pressure Vessel Code, Section VIII. Acceptable vessel steels and allowable stress values must comply with the ANSI/ASME Boiler and Pressure Vessel Code, Sections IIA and IIB. Welded pressure vessels must comply with the ANSI/ASME Boiler and Pressure Vessel Code, Section IX. In most cases, Codes list acceptable methods for calculating minimum wall thicknesses of vessels based on design temperatures and pressures, qualifying welding processes, heat treatment during fabrication, and inspection following construction. In the United States, materials compositions must comply with the ASTM Standard Materials specifications.

4.3.2 Process Piping. Piping must comply with the ANSI/ASME Boiler and Pressure Vessel Code, Section IX. This Code lists acceptable methods for calculating minimum wall thicknesses based on design temperatures and pressures, qualifying welding processes, heat treatment during fabrication, and inspection following construction. In the United States, materials compositions must comply with the ASTM Standard Materials Specifications.

4.3.3 Tanks. Tanks must comply with API publications API Std. 620 and API Std. 650. Tanks must also comply with relevant API publications dealing with cathodic protection (API RP 651), linings (API RP 652), inspection and repairs (API Std. 653), as well as certain other API publications. In the United States, materials compositions must comply with the ASTM Standard Materials Specifications. In Europe, the applicable tank design codes and guidelines include BS[6]2654 (manufacture), DIN[7]4119 (design standards), EEMUA[8]154 (demolition), and EEMUA 159 (maintenance & inspection).

4.3.4 Pumps. Pumps must comply with API publication API Std. 610. In the United States, materials compositions must comply with the ASTM Standard Materials Specifications.

4.4 ASTM Specifications for Common Steels

Common ferrous and nonferrous alloys for refinery and upgrader services are described below in approximate order of increasing corrosion resistance, decreasing availability, and increasing cost. Additional relevant materials information is provided in the chapters on various corrosion types. Except for unusual circumstances, minimum quantities, long lead times, and handling surcharges apply to all but the most common materials. As a result, some companies have stockpiled special alloys to ensure adequate supplies for future maintenance operations.[9] Table 4-2 lists current and past ASTM specifications for a number of steels. Many specifications cover several steel compositions that are further identified by a grade suffix (for example: A182-F11 designates a 1.25Cr-0.5Mo forging specification, while A182-F22 a 2.25Cr-1Mo forging specification). ASME specifications are similar except that SA-XXX is used instead of A-XXX as a prefix for identification. While Table 4-2 includes steel castings specifications, plant use of steel castings is often limited to certain components (pump impellers, valve bodies, etc.) in nonhazardous applications. Steel castings are rarely used for pressure-containing components in hydrocarbon services, because casting defects are generally unpredictable. Carbon steel and alloy steel bolts and nuts are covered by specifications A193, A194, A320, A354, A449, A453, A540, and A563.

6. British Standards produced by the BSI Group.
7. Deutsches Institut für Normung.
8. Engineering Equipment and Materials Users Association.
9. One major company maintained the largest supply of titanium heat exchanger tubes in the United States to ensure availability during periods of shortage.

TABLE 4-2 ASTM SPECIFICATIONS FOR COMMON REFINERY STEELS

Steel Type	Pipes (P) & Tubes (T)	Plates (Gr)	Castings	Forgings (F)
Carbon steel	A53, A106, A120, A134, A135, A139, A178, A179, A192, A210, A211, A214, A226, A333, A334, A369 A381*, A524, A587. A671, A672, A691	A283, A285, A299, A442, A455, A515, A516, A537, A570, A573*	A27*, A216, A352	A105, A181, A234, A268, A350, A372, A420, A508, A541
C0.5Mo	A161*, A209, A250, A335, A369, A426, A672, A691	A204, A302, A517, A533	A217, A352, A487	A182,A234,A336, A508, A541
1.25Cr0.5Mo (-X11)	A199, A200*, A213, A335,	A387, A389*, A517 A369, A426, A691	A217, A389*	A182, A234, A336, A541
2.25Cr1Mo (-X22)	A199, A213, A335, A369,	A387, A542 A426, A691	A217, A487	A182, A234, A336, A541, A542
3CrlMo (-X21)	A199, A200*, A213, A335 A369, A426, A691	A387		A182, A336
5Cr0.5Mo (-X5)	A199, A200*, A213, A335, A369, A426, A691	A387	A217	A182, A234, A336
7Cr0.5Mo (-X7)	A199, A200*, A213, A335, A369, A426	A387		A182, A234
9Cr1Mo (-X9)	A199, A200*, A213, A335, A369, A426	A387	A217	A182, A234, A336
Stainless Steel (All types) A511*	A213, A249, A268, A269 A271*, A312, A358, A376,	A167, A176*, A240, A412, A457	A297*, A351, A447*	A182, A336, A403, A473*

*These specifications are not approved by the ANSI/ASME Boiler and Pressure Vessel Code or by the ANSI/ASME Code for Pressure Piping B31.3.

4.5 Common Materials of Construction (Metallurgy)

4.5.1 Carbon Steel. The principal material of construction (metallurgy) of crude units is and has always been carbon steel (low-carbon steel, plain carbon steel, or mild carbon steel)[10]. Compared to other materials, carbon steel is cheap, readily and often locally available, easily fabricated by conventional means, and sufficiently strong for most plant components.[11] Carbon steel is typically used for piping, pumps, valves, and fittings; support structures, tanks and desalter vessels; heat exchangers, coolers, and condensers; and distillation columns and overhead receiver drums. Carbon steel is sufficiently corrosion resistant for most high-temperature hydrocarbon services up to approximately 230 °C (450 °F), many low-temperature hydrocarbon services including overhead systems, and many cooling water services. But depending on the specific process conditions, carbon steel components may require certain corrosion control measures, such as neutralizer or corrosion inhibitor injection. When carbon steel is used in cooling water services, some type of cooling water treatment is usually required and some components may have to be protected with organic coatings and sacrificial anodes.

4.5.2 Silicon-Killed Carbon Steel. Over 15% of HTSC failures seen in refinery piping have been attributed to ASTM A53 and API 5L carbon steel pipe containing less than 0.15% Si. In comparison, ASTM A106 pipe deoxidized with silicon contains 0.15-0.3% Si and has shown superior resistance to long-term HTSC. As a general rule, all carbon steel pipe and pipe fittings with suspect Si content in HTSC services above 240 °C (450 °F) should be inspected for thinning;

10. In the *Guide*, the term carbon steel is preferred.
11. Carbon steel can be welded without postweld heat treatment (PWHT) provided the carbon equivalent (CE) is less than 0.35.

locations with higher than average corrosion should be regularly reinspected every five years or less. Additional information on the effect of silicon in steels on HTSC is covered in Chapter 6 of the *Guide.*

4.5.3 Low-Alloy Cr-Mo Steels. The 5Cr-0.5Mo to 9Cr-1Mo steels are used primarily instead of carbon steel for piping, furnace tubes, furnace headers, vessels, and heat exchanger components in HTSC services above 240 °C (450 °F). These steels are much more resistant than carbon steel, and an upgrade to 5Cr steel is often considered when corrosion rates in carbon steel piping and heater tubes exceed 0.25mm/y (10mpy). Steels containing 9% Cr are even more resistant than grades containing less chromium and are often selected as a first upgrade for piping and heater tubes operating at high temperatures and flow velocities. These steels may require postweld heat treating (PWHT) to prevent hydrogen stress cracking ("sulfide cracking") at hard welds in some aqueous corrosion applications. As a general rule, resistance toward HTSC increases with chromium content of the steel. Steels containing less than 5% Cr such as 2.25Cr-1.0Mo are only marginally better than carbon steel. They should no longer be specified for new components, but are still found on many existing units. Like carbon steel, 2.25Cr and 5Cr pipe and pipe fittings systems with Si content below 0.15% Si in HTSC services should be inspected for thinning; locations with higher than average corrosion should be regularly reinspected every 5 years or less. The cutoff temperature for 5Cr steel in HTSC services is 340 °C (650 °F) instead of 240 °C (450 °F). Additional information on the effect of Cr in steels on HTSC is covered in Chapter 6 of the *Guide.*

4.5.4 Admiralty Brass. Admiralty brass (69Cu-30Zn-1Sn), originally named "Admiralty metal" for its marine applications, is used for heat exchanger tubes instead of carbon steel to provide improved resistance against cooling water corrosion. The alloy is classified among a subfamily known as alpha brasses[12]. Brass tubes have also reasonably good resistance against process corrosion below 200 °C (400 °F) because they form protective zinc sulfide films in the presence of hydrogen sulfide (H_2S). As a general rule, brass has limited strength at temperatures above 230 °C (450 °F) and limited resistance against HTSC and should never be used under these conditions, even if the strength requirements are met. Admiralty brass like all brasses is subject to stress-corrosion cracking ("season cracking") in the presence of residual ammonia, moisture, and air when brass tube bundles are opened to the atmosphere prior to cleaning during shutdown. A dilute sulfuric acid wash has been used to prevent cracking. While all currently available Admiralty brass tubes are inhibited against preferential dezincification through the addition of 0.05% As (arsenic), uninhibited tubes may still be found in older plants.

4.5.5 Naval Brass. Naval brass (59Cu-40Zn-1Sn) is used for heat exchanger tube sheets instead of carbon steel for improved resistance against cooling water corrosion when a high-strength, corrosion resistant material is required. The alloy is classified among a subfamily known as duplex or alpha-beta brasses[13]. Like Admiralty brass, it has reasonably good resistance against process corrosion below 200 °C (400 °F) because protective zinc sulfide films are formed in the presence of hydrogen sulfide (H_2S). All brasses have limited strength at elevated temperatures above 230 °C (450 °F) and limited resistance against HTSC and should never be used under these conditions, even if the strength requirements are met. Naval brass like Admiralty brass is subject to stress-corrosion cracking in the presence of residual ammonia, moisture, and air when tube bundles are opened to the atmosphere prior to cleaning during shutdown. A dilute sulfuric acid wash has been used to prevent cracking. While all currently available brass production is inhibited against preferential dezincification through the addition of 0.05% As (arsenic), uninhibited naval brass tubesheets may still be found in older plants.

4.5.6 Aluminum Alloys. Aluminum alloys are used primarily for improved resistance of internals (demister pads, packed beds, pall rings) against HTSC and corrosion by organic acids including naphthenic acids in the upper vacuum column. As a general rule, aluminum alloys have good resistance toward organic sulfur compounds and many organic acids, but have limited strength at temperatures above approximately 230 °C (450 °F). Aluminum coated or clad steel has been used where increased strength is required. Aluminum spray metallized coatings have been used in various applications, but often fail due to spalling. Alonizing™ is a proprietary aluminum diffusion treatment at 900 °C (1,650 °F) for forming a thin layer of an iron-aluminum alloy measuring less than 1.25mm (0.05in.) on steel surfaces. Alonized™ steel is expensive, and upgrading to low-alloy Cr-Mo steel is often more cost-effective. The main benefit of alonizing™ is a significant reduction in the amount of sulfide scale formed on all steels, including stainless steels at elevated temperatures. Alonized™ carbon steel heat exchanger tubes can become distorted due to the high-temperature diffusion treatment and are prone to pit at discontinuities and cracks in the alonized surface layer. This can be a problem under chloride-containing deposits in low-temperature hydrocarbon services when a wet overhead stream is used to cool hot crude.

4.5.7 Ferritic 12Cr Stainless Steels. AISI Types 405 and 410S stainless steels are used primarily instead of low-ally Cr-Mo steels for piping, pressure vessel cladding and internals, and heat exchanger shells to provide improved resistance against HTSC. 12Cr stainless steel is the default material for column trays and structured packing. As a general rule, 12Cr stainless

12. Alpha brasses are malleable, can be worked cold, and are used for tube drawing, pressing, or similar applications.
13. Duplex brasses can be strengthened by hot working.

steels has only limited resistance against naphthenic acid corrosion (NAC), and then only in certain applications. AISI Type 410S is the most frequently used grade, but AISI Type 405 is also used with good results. 12Cr stainless steels will pit in chloride-containing cooling water services and under chloride-containing deposits in hydrocarbon services below 200 °C (400 °F), such as crude unit overhead systems. 12Cr stainless steel has occasionally become subject to metallurgical changes at temperatures above approximately 400 °C (750 °F). Commonly known as 475 °C (885 °F) embrittlement, affected steels become brittle upon cooling to ambient temperatures below 95 °C (200 °F). While embrittlement can be reversed by annealing, some plants avoid using 12Cr stainless steel for heater tubes, return bends, and headers. Welds require postweld heat treatment to prevent hard welds and possible hydrogen stress cracking (sulfide cracking) in some applications.

4.5.8 Martensitic 12Cr Stainless Steel. AISI Type 410 stainless steel is a hardenable version of 12Cr stainless steel and is used for limited applications where elevated strength and hardness are required.

4.5.9 Copper Nickel (Cupronickel) Alloys. The 70Cu-30Ni and 90Cu-10Ni alloys are used primarily as a substitute for Admiralty brass for improved resistance of cooler and condenser tubes against cooling water corrosion. Copper nickel alloys, in contrast to the brasses, are not subject to stress corrosion cracking by residual ammonia, moisture, and air when tube bundles are opened to the atmosphere prior to cleaning during shutdown. Copper nickel 90-10 has poor resistance against wet hydrogen sulfide (H_2S), and copper nickel 70-30 is preferred for overhead corrosion applications. Most copper alloys including copper nickel alloys have limited resistance against HTSC. They should not be used above 70 °C (160 °F) even if their strength requirements are met.

4.5.10 Austenitic Stainless Steels. AISI Types 316 and 317 austenitic Cr-Ni stainless steels are used primarily instead of 12Cr stainless steel for improved resistance of piping, furnace tubes, furnace headers, pressure vessel cladding and internals, and heat exchanger shells against high-temperature corrosion by organic acids. These alloys are also preferred for pump housings and internals, instrument and flush lines, and vacuum ejectors in the presence of turbulent flow conditions. Austenitic stainless steels may be subject to crevice corrosion and pitting under wet chloride-containing deposits in low-temperature hydrocarbon services including overhead systems.

Chloride stress corrosion cracking (CSCC) can occur at temperatures above 65 °C (150 °F). Both CSCC and pitting can also occur under wet exterior insulation and in chloride-containing cooling water services during shutdowns unless components are protected in accordance with NACE Standard RP0170[14]. Austenitic stainless steel cladding on the inside of piping and vessels is under compressive stresses and less susceptible, but not immune to CSCC compared to cladding on the outside of piping and vessels. Stainless steel clad components are also less susceptible to CSCC compared to components fabricated from solid stainless steel. When used in crude and upgrader units, austenitic stainless steels usually are not subject to polythionic acid stress corrosion cracking (PASCC), because of the type of sulfide scale formed and sensitization is usually of no great concern.

4.5.11 Duplex Stainless Steels. Duplex stainless steels such as Alloy 2205 (22Cr-5.5Ni-3Mo) or Sandvik 3RE60 (18Cr-5Ni-2.5Mo) are austenitic-ferritic stainless steels used primarily for improved corrosion resistance against chlorides in overhead condensers, vacuum ejector condensers, desalter brine coolers, desalter feed versus brine exchangers, and gas oil versus crude exchangers. These steels have overall corrosion resistance similar to that of common austenitic stainless steels, but improved resistance to pitting and CSCC in chloride-containing cooling water services and under chloride-containing deposits in low-temperature hydrocarbon services, including overhead systems. In some services, however, duplex stainless steels can be sensitive to hydrogen embrittlement and cracking at weldments and have performed poorly in some low-temperature hydrocarbon services, including overhead systems.

4.5.12 Stainless Steel 904L. Stainless steel 904L (20Cr-25Ni-4.5Mo-1.5Cu) is a high-alloy austenitic stainless steel that is not chemically stabilized[15]. The addition of copper to this alloy gives it greatly improved resistance toward strong reducing acids, particularly hydrochloric (HCl) acid and sulfuric (H_2SO_4) acid. It is also highly resistant to pitting, crevice corrosion, and SCC. Stainless steel 904L contains expensive alloying elements and many applications in which only this grade has previously performed well can now be handled at lower cost by duplex stainless steels, such as Alloy 2205. As a result, stainless steel 904L is less often specified than it was in the past. In many environments, stainless steel 904L has a corrosion resistance intermediate between Type 316L stainless steel and the highly alloyed 6Mo and similar "super austenitic" grades.

14. While NACE Standard RP0170 addresses primarily protection against polythionic acid stress corrosion cracking (PASCC), shutdown procedures such as nitrogen blanketing and alkaline washing can also be used to protect against chloride stress corrosion cracking (CSCC).
15. 904L stainless steel is subject to sensitization (carbide precipitation at grain boundaries) upon exposure to 550 - 850 °C (1000 – 1550 °F) that leaves the grain boundaries susceptible to preferential attack.

4.5.13 Titanium. Titanium (Ti) tubes are used instead of carbon steel or Admiralty brass tubes in water-cooled coolers and condensers for improved resistance against cooling water corrosion. Several alloys are used depending on the severity of the process conditions:

- The standard, unalloyed, grades (primarily Grade 2) are often used in overhead systems. For cost reasons, welded Grade 2 tubing with wall thickness of BWG18[16] average instead of minimum is preferred over seamless tubing. Tube wall thicknesses less than BWG18 (for example, BWG20 or 22) should be avoided, because these thin-walled tubes are often damaged when bundles are pulled for cleaning and maintenance.
- The molybdenum and nickel bearing grades (primarily Grade 12 with 0.3% Mo and 0.8% Ni) have improved strength and somewhat better corrosion resistance than the standard unalloyed grades, but are more expensive.
- The palladium-bearing grades (primarily Grade 7 with 0.2% Pd), perform even better under reducing conditions than the standard grades, but are far more expensive.

Titanium tubes are required for any and all cooling water services involving brackish and/or seawater. They are often substituted for Admiralty brass tubes to avoid ammonia SCC and low-temperature process corrosion, including overhead corrosion. However, Ti tubes can pit and hydride (embrittle) under chloride-containing deposits, and tube bundles may have to be kept free of deposits by regular water washing. They can pit and hydride when used with steel tube sheets or baffles. But the most important issue is vibration damage because of the thin tube walls specified for titanium, unless the usual baffle spacing for brass or steel tubes is decreased.[17]

4.5.14 Alloy 400. Alloy 400 (66Ni-32Cu) was formerly known as Monel 400 and is often used instead of carbon steel for crude column top trays, top cladding, and selective cladding of overhead condenser shells and separator drums to prevent low-temperature process corrosion. As a general rule, Alloy 400 should not be used in these applications at temperatures above approximately 120 °C (250 °F) because the alloy has limited resistance against HTSC. The alloy is resistant to HCl corrosion and is fully resistant to stress-corrosion cracking, but may be attacked in the presence of H_2S at temperatures above 250 °F, as shown by the data in Table 4-3.

TABLE 4-3 CORROSION IN STIRRED NAPHTHA/WATER BLENDS AT 350F AND PH 4.3

Alloy	Corrosion Rate, mm/y (mpy)	
	H_2S Saturated	H2S Saturated + 1,500 mg/L Cl
Carbon Steel	0.37 (15)	0.7 (28)
Alloy 400	2.8 (112)	2.48 (99)
12Cr	0.275 (11)	0.025 (1)
Type 304 SS	0.1 (4)	0.05 (2)

4.5.15 Alloy 625. Alloy 625 (60Ni-21Cr-9Mo-5Fe) was formerly known as Inconel 625 and is used primarily as a substitute for carbon steel or Admiralty brass to provide improved resistance of heat exchanger tubes against overhead corrosion. As a general rule, Alloy 625 should only be used in low-temperature hydrocarbon services including overhead systems at temperatures up to approximately 200 °C (400 °F) when steam, hydrogen chloride (HCl), and hydrogen sulfide (H_2S) are present. Alloy 625 is fully resistant to stress corrosion cracking (SCC), but not to pitting under chloride-containing deposits in most hydrocarbon services. Therefore, heat exchanger tube bundles must be kept free of deposits by water washing.

4.5.16 Alloy 20. Alloy 20 (36Ni-20Cr-3.5Mo-2.5Cu) was formerly known as Carpenter 20 stainless steel and was developed for applications requiring superior resistance toward HCl corrosion and pitting. While the alloy is actually a high-Ni alloy, it is often selected to address SCC issues seen with Type 316L stainless steel. The copper content helps protect it from sulfuric acid corrosion in certain alkylation processes.

16. Birmingham Wire Gauge.
17. The author was present when a brand-new titanium condenser bundle failed within two hours after having been placed in service as a replacement for a corroded steel bundle. First blamed on corrosion by the unit operators, the refinery metallurgist hollered the iconic reply into his phone: "Nothing corrodes out in two hours!" As it turned out, the titanium tubes actually failed due to a combination of baffle-hole vibration and mid-span tube collision damage, because the baffle spacing had not been decreased to account for the reduced rigidity of thin-walled titanium tubes.

4.5.17 Alloy C-276. Alloy C-276 (58Ni-15Cr-16Mo-6Fe-2Co-3W) was formerly known as Hastelloy C-276 and is used primarily instead of carbon steel, Alloy 400, and Alloy 625 for the upper crude column trays, top cladding, vapor lines, and selective cladding of condenser shells and separator drums against overhead corrosion. Alloy C-276 is also used for injection nozzles, static mixer elements, and pipe fittings to prevent impingement attack. Currently, two variations of C-type alloys are available: Alloy C-276 and Alloy C-22. A third variation, Alloy C-4, has been discontinued, but may still be listed in refinery specifications. All C-type alloys have excellent resistance against HTSC and NAC. The high Mo content of Alloy C-276 is definitely beneficial, as far as corrosion-erosion and pitting attack are concerned. However, the cost of Alloy C-276 for heat exchanger tubes may be hard to justify and the tubes may be hard to obtain in small quantities.

Despite a few isolated claims to the contrary, Alloy C-276 has excellent resistance against pitting and impingement corrosion under the reducing conditions encountered in most overhead systems, but not necessarily so under oxidizing conditions. The results of standard crevice corrosion tests[18] in 10% $FeCl_3$ solution confirm the superior resistance of Alloy C-276, as shown in Table 4-4. These tests are useful for comparing the resistance of stainless steels and related alloys to the initiation of pitting and crevice corrosion. Ferric chloride solutions are used for the tests, because ferric chlorides are found within pits or crevices on ferrous and nickel alloys in chloride solutions.

TABLE 4-4 RESULTS OF ASTM STANDARD CREVICE CORROSION TESTS (100HR)

Material	Average Corrosion Rate, mm/y (mpy)		
	25 °C (77 °F)	50 °C (122 °F)	75 °C (167 °F)
Alloy C-276	0.005 (0.2)	0.005 (0.2)	0.035 (1.4)
Alloy 625	0.037 (1.5)	3.1 (124)	12.7 (510)
Alloy 20	5.12 (205)	9.5 (380)	17.5 (700)
Type 316L SS	7.8 (312)	11.5 (460)	19.5 (780)

4.6 Temperature Limits for Common Steels

Temperature limits apply above and beyond those dictated by corrosion considerations when carbon steel, low-alloy Cr-Mo steels, and stainless steels are used at various process temperatures. Recommended target ranges are shown in Table 4-6.

TABLE 4-6 TEMPERATURE LIMITS FOR COMMON STEELS[19]

Temperature Limit °C	°F	Materials Limitations
1100-1200	2000-2200	Refractory lining required
>930	>1700	AISI 310 required
>650	>1200	Austenitic stainless steels required
>540	>1000	1.25Cr-0.5Mo to 9Cr-1Mo steels
>450	>840	Minimum of 1.25Cr-0.5Mo required
>350	>660	"475 °C (885 °F) embrittlement" of 12Cr and 17Cr
> -20	> -4	Impact testing of CS per Code.
> -50	> -58	SA-516 steel (all grades) preferred.
> -60	> -75	2.25Ni and 3.5Ni steels per SA-203 and impact testing.

18. ASTM G48–11 *Standard Test Methods for Pitting and Crevice Corrosion Resistance of Stainless Steels and Related Alloys by Use of Ferric Chloride Solution.*
19. Beyond those dictated by corrosion considerations.

4.7 Composition of Common Stainless Steels and Nickel Alloys

AISI designations and chemical composition of a number of common stainless steels and nickel alloys used in refineries and upgraders are listed for convenience in Table 4-7. Note that with improved analytical methods and fabrication processes, the alloy content in current products may barely meet the low end of the required range for a given material. This can be a problem, when materials are selected based on corrosion and fouling data obtained with earlier compositions. For example, current production of Type 316 stainless steel may no longer have the maximum 3% Mo content on which past corrosion tests were based. While Type 316 stainless steel with 2% Mo still falls into the required chemistry range, it is not resistant to naphthenic acid corrosion. Unless Type 316 can be obtained with 3% Mo, Type 317 containing 3-4% Mo should be ordered instead.

TABLE 4-7 COMMON STAINLESS STEELS AND NICKEL ALLOYS

	Elemental Composition, %						
Designation	**C_{max}**	**Cr**	**Ni**	**Cu**	**Mo**	**Fe**	**Other**
Ferritic Stainless Steels							
AISI 405	0.08	11.5-14.5	-	-	-	bal	Al: 0.1-0.3
AISI 410S	0.08	11.5-13.5	0.6	-	-	bal	Nb+Ta: opt.
AISI 430	0.12	14-18	0.5	-	-	bal	
Martensitic Stainless Steels							
AISI 410	0.15	11.5-13.5	-	-	-	bal	
AISI 431	0.2	15-17	1.25-2.5	-	-	bal	
Austenitic Stainless Steels							
AISI 304	0.08	18-20	8-12	-	-	bal	Si: 1
AISI304L	0.03	18-20	8-12	-	-	bal	Si: 1
AISI 310	0.25	24-26	19-22	-	-	bal	Si: 1.5
AISI 316	0.08	16-18	10-14	-	2-3	bal	
AISI 316L	0.03	16-18	10-14	-	2-3	bal	
AISI 317	0.08	18-20	11-14	-	3-4	bal	
AISI 317L	0.03	18-20	11-14	-	3-4	bal	
AISI 321	0.08	17-19	8-11	-	-	bal	Ti: 4xC
AISI 347	0.08	17-19	9-13	-	-	bal	Nb+Ta: 10xC
Age-Hardenable Stainless Steels							
17-4 PH	0.05	15.5-17.5	3-5	3-5	-	bal	Nb: 0.15-0.45
17-7 PH	0.07	16-18	6.5-7.75	-	-	bal	Al: 0.75-1.5
Duplex Stainless Steels							
2205	0.03	21-23	4.5-6.5	-	2.5-3.5	bal	Si: 0.8, N: 0.08-0.2
3RE60	0.03	17-19	4-6	-	2-3	bal	
High Nickel Alloys							
904L	0.02	19-23	23-28	1-2	4-5	bal	
Alloy 20	0.05	19-21	32-38	3-4	2-3	bal	Nb: 1 max
Alloy 400	0.15	-	63-70	bal	-	2.5 max	
Alloy 600	0.08	14-17	72min	0.5max	-	6-10	
Alloy 625	0.10	20-23	60	-8-10	5max	Nb+Ta:	3.15-4.15
Alloy 800	0.04	19-23	30-35	0.75max	-	bal	
Alloy 825	0.03	19.5-23.5	48-46	1.5-3	2.5-3.5	bal	Al: 0.2max
Ti: 0.6-1.2							
Alloy B-2	0.02	1	bal	-	26-30	2	Co: 1
Alloy C-276	0.02	14.5-16.5	bal	-	15-17	4-7	Co: 2.5,W: 3-4.5
Alloy C-22	0.01	20-22.5	bal	-	12.5-14.5	2-6	Co: 2.5 W: 2.5-3.5

4.8 Metallurgy of Selected Components

The metallurgy of major components of crude units and upgraders depends primarily on the sulfur and naphthenic acid content (TAN) of the crude charged and is summarized below for three principal operating modes:

- Units processing primarily sweet crude with less than 0.6% S and TAN below 0.3,
- Units processing primarily sour crude with more than 0.6% S and TAN below 0.3, and
- Units processing primarily acidic and/or opportunity crude with TAN above 0.3 regardless of organic sulfur content.

Tables 4-8 and 4-9 provide specific recommendations for the metallurgy of major components in the atmospheric and vacuum sections of these units. Corrosion mechanisms, examples of corroded components, and methods for controlling corrosion are further discussed in subsequent chapters of the *Guide*.

4.8.1 Units Processing Sweet Crude. Units processing primarily sweet crude with less than 0.6% S, TAN below 0.3, and no free H_2S have relatively few corrosion and fouling issues in the high-temperature circuits. As listed in Table 4-8 (left column), these units are generally constructed from carbon steel. Fired heater tubes and transfer lines in the atmospheric and vacuum sections are often made from carbon steel or 2.25Cr-1Mo steel. Columns operating at temperatures above 230 °C (450 °F) are internally clad with 12Cr stainless steel to prevent impingement corrosion under highly turbulent flow conditions encountered in the flash zone and at tray downcomers. Valve trays are also made from 12Cr stainless steel. Even low corrosion rates can destroy the valves in carbon steel trays.

4.8.2 Units Processing Sour Crude. Units processing primarily sour crude with more than 0.6% S and TAN below 0.3 require additional alloy protection against HTSC above 230 °C (450 °F). As listed in Table 4-8 (right column), the extent of alloying depends on the design and operating practices on a given unit. Based on plant experience, flow velocities in heater tubes, headers, U-bends, and transfer lines should be maintained below 60m/sec (200fps). A minimum of 5Cr-0.5Mo steel is usually required for crude heater tubes, headers, and U-bends, but 9Cr-1Mo or 12Cr stainless steel tubes are used for longer tube life. Vacuum heaters use 9Cr-1Mo steel for tubes, headers, and U-bends, while transfer lines are often made from 5Cr-0.5Mo steel. When corrosion issues arise, upgrading to the next higher Cr-Mo steel is common practice. Various columns operating at temperatures above 230 °C (450 °F) are internally clad with 12Cr stainless steel. Most bubble cap and valve trays are also made from 12Cr stainless steel.

4.8.3 Units Processing Acidic or Opportunity Crude. Crude units processing naphthenic acid crude with TAN above 0.3 regardless of organic sulfur content require additional upgrading at areas of high velocity or turbulence, because naphthenic acids form no protective corrosion scales. Since opportunity crude may or may not contain significant amounts of naphthenic acids, alloy requirements are identical to those of crude units processing naphthenic acid crude. As shown in Table 4-9, austenitic stainless steels are often required for corrosion control since upgrading with low-alloy Cr-Mo steels or 12Cr stainless steel is usually ineffective. Only AISI Type 316 stainless steel containing 2-3% Mo or AISI Type 317 stainless steel containing 3-4% Mo is resistant to most, but not all forms of NAC. Based on plant experience, flow velocities in heater tubes, headers, U-bends, and transfer lines should be maintained below 50m/sec (175fps). Lines for gas oil, reduced crude, and vacuum resid have relatively low flow velocities and can usually be fabricated from 9Cr-1Mo steel. With few exceptions (for example, flash zones), the lower shell of the atmospheric column and various sidestream columns require no special alloying against NAC. They do require, however, alloying against HTSC at temperatures above 230 °C (450 °F). Vacuum column trays and shell areas at temperatures between 290 °C (550 °F) and 340 °C (650 °F) and certain other components may have to be made of or clad with AISI Type 316 or AISI Type 317 stainless steel.

TABLE 4-8 METALLURGY FOR SELECTED COMPONENTS

	Low-Sulfur Crude (S<0.6% and TAN<0.3)	High-Sulfur Crude (S>0.6% and TAN<0.3)
ATMOSPHERIC SECTION		
Inlet Piping	CS	CS
Charge Pumps	CS/12Cr trim (API S-8)	same as low S
Desalter	CS	CS
Preheat Exchangers	CS	CS
Preflash Drum/Column	CS	CS
Fired Heater Tubes	CS or 2.25Cr	5Cr or 9Cr
Heater Return Bends	CS or 2.25Cr	5Cr or 9Cr
Transfer Line	CS	5Cr or 9Cr
Atm. Column Shell	CS clad w/12Cr from bottom up to 230 °C (450 °F) level	same as low S
	CS clad w/ Alloy 400 or Alloy C-276 from top down to 4th tray.	same as low S
Atm. Column Trays	12Cr, except Alloy 400 or Alloy C-276 for upper 4 trays.	same as low S
Sidestream Column Shells	CS clad w/12Cr from bottom up to 230 °C (450 F) level	same as low S
Sidestream Column Trays	12Cr	same as low S
Overhead Vapor Line	CS	same as low S
	CS w/ Alloy C-276 overlay at elbows and tees	same as low S
Overhead Exchangers	CS	same as low S
Overhead Condensers	CS w/ Adm. brass or titanium tubes	same as low S
Overhead Separator Drums	CS w/ Alloy C-276 overlay at high-corrosion areas	same as low S
Overhead Rundown Piping	CS	same as low S
Overhead Reflux Line	CS	same as low S
Overhead Reflux Pump	316 /316 SS trim (API A-8)	same as low S
HAGO Lines above 315 °C (600 °F)	CS or 2.25Cr	5Cr
HAGO Stripper	CS /12Cr internals	CS /12Cr clad 12Cr internals
HAGO Pump	CS /CI trim (API S-1)	316/316 SS trim (API A-8)
LAGO Lines above 500 °F 260 °C)	CS	5Cr
LAGO Stripper	CS/12Cr internals	same as low S
LAGO Pump	CS/CI trim (API S-1)	12Cr/12Cr trim (API C-8)
Reduced Crude Line	CS	5Cr
Reduced Crude Pump	12Cr/12Cr trim (API C-8)	316/316 SS trim (API A-8)
VACUUM SECTION		
Vacuum Heater Tubes	2.25Cr	5Cr or 9C
Heater Return Bends	2.25Cr	5Cr or 9Cr
Transfer Line	2.25Cr	5Cr or 9Cr
Vacuum Column Shell	CS w/12Cr clad from bottom up to 230 °C (450 °F) level	same as low S
Vacuum Column Trays	CS	12Cr
Overhead Vapor Line	CS	CS
WC Overhead Condensers	CS w/ Adm. brass tubes	same as low S
Steam Ejectors	Cast iron	Cast iron, Cast 316
Overhead Receiver (Hot Well)	CS	same as low S
Overhead Rundown Piping	CS	same as low S
HVGO Lines above 315 °C (600 °F)	CS	5Cr
HVGO Pump	12Cr w/12Cr trim (API C-8)	same as low S
LVGO Lines above 260 °C (500 °F)	CS	5Cr
LVGO Pump	12Cr w/12Cr trim (API C-8)	same as low S
Vacuum Resid Line	CS	5Cr

TABLE 4-9 METALLURGY FOR SELECTED COMPONENTS WITH ACIDIC CRUDE (TAN>0.3) AND OPPORTUNITY CRUDE

ATMOSPHERIC SECTION	
Crude Pumps	CS w/12Cr trim (API S-8)
Crude Heater Tubes	9Cr
Heater Return Bends	9Cr
Transfer Line	9Cr or CS w/316L clad
Crude Column Shell	CS w/12Cr clad from bottom up to 230 °C (450 °F) level
	CS w/ Alloy 400 or Alloy C-276 clad from top head down to 4th tray
Crude Column Trays	12Cr, except Alloy 400
or Alloy C-276 for upper 4 trays	
Sidestream Column Shell	CS w/12Cr clad from bottom up to 230 °C (450 °F) level
Sidestream Column Trays	12Cr
Overhead Vapor Line	CS
CS w/ Alloy C-276 overlay for elbows	and tees
Overhead/Crude Exchangers	CS
WC Overhead Condensers	CS w/Adm. brass or titanium tubes
Overhead Separator Drums	CS w/Alloy C-276 cladding at high corrosion areas
Overhead Reflux Pump	316 w/316 SS trim (API A-8)
HAGO Lines >315C (600F)	9Cr
HAGO Stripper	CS w/12Cr clad and 12Cr internals
HAGO Pump	316 w/316 SS trim (API A-8)
LAGO Lines >260 °C (500 °F)	5Cr
LAGO Stripper	CS w/12Cr clad and 12Cr internals
LAGO Pump	12Crw/12Cr trim (API C-8)
Reduced Crude Line	9Cr
Reduced Crude Pump	316 w/ 316 SS trim (API A-8)

VACUUM SECTION	
Vacuum Heater Tubes	316L SS
Heater Return Bends	316L SS
Transfer Line	CS /316L SS clad
Vacuum Column Shell	CS /316L or 317 SS clad to top P.A.R.
	CS /12Cr clad from top P.A.R. to top of column
Vacuum Column Trays	316L or 317 SS
Overhead Vapor Line	CS
WC Overhead Condensers	CS w/ Adm. brass tubes
Steam Ejectors	Cast iron, cast 316 SS
Overhead Receiver (Hot Well)	CS
Overhead Rundown Piping	CS
HVGO Lines >315 °C (600 °F)	9Cr
HVGO Return Lines	5Cr
HVGO Surge Drum	CS w/ 316L SS clad
HVGO Pump	316/316 trim (API A-8)
LVGO Lines >260 °C (500 °F)	5Cr
LVGO Surge Drum	CS w/12Cr clad
LVGO Pump	12Cr w/12Cr trim (API C-8)
Vacuum Resid Line	9Cr
Vacuum Resid Pump	316/316 SS trim (API A-8)

4.8.4 Pump Specification. Pumps are specified in accordance to ANSI/API Std. 610 Centrifugal Pumps for Petroleum, Petrochemical and Natural Gas Industries. The four principal types are API S-1, API S-6, API C-8, and API A-8 listed in order of increasing degree of alloying and cost in Table 4-10. These types cover most if not all applications in crude unit hydrocarbon services. Reflux pumps usually require AISI Type 316 stainless steel housings and internals because of high flow turbulence.

TABLE 4-10 CENTRIFUGAL PUMPS FOR REFINERY SERVICES

Class	Material Selection		
	Case	Impeller	Trim
API S-1	Carbon steel	Cast iron	Cast iron
API S-8	Carbon steel	12Cr	12Cr
API C-8	12Cr	12Cr	12Cr
API A-8	Type 316 SS	Type 316 SS	Type 316 SS

4.8.5 Overhead Systems. Column overhead systems are generally made from carbon steel. The principal exceptions are tubes in water-cooled condensers that tend to corrode primarily on the water side, and Admiralty brass or titanium tubes are often used in this service. Alloy 625 tubes are used to control process side corrosion. Alloy C-276 may be required at elbows and tees of the overhead vapor line to control impingement attack by HCl. Corrosion of condenser tubes continues to be a significant cost factor on many units and initial costs rapidly increase with alloy upgrading. However, overall costs also depend on such factors as available shop facilities, labor costs, workload, or current materials stock lists, and up-to-date price quotes should be obtained in all cases.

4.8.6 Acceptable Corrosion Allowances. Mandatory construction and engineering Codes provide rules and procedures for the design, fabrication, repair, and inspection of major plant components, but do not specify corrosion allowances as such.[20] This means that suitable corrosion allowances in terms of extra wall thickness be specified for most components. Acceptable corrosion allowances are listed in Table 4-11 and depend primarily on inspection history, cost of materials, properties of the process stream, and type of component involved. With low-cost materials and thick sections (such as cast steel pump impellers), a higher corrosion rate can be tolerated than with expensive materials and thin sections (such as alloy condenser tubes). With relatively benign streams (such as crude at near-ambient temperatures), a higher corrosion rate can be tolerated than with potentially hazardous streams (such as flammable liquids or explosive hydrocarbon vapors at elevated temperatures).

TABLE 4-11 REQUIRED CORROSION ALLOWANCE BASED ON EXPECTED CORROSION RATE

Expected Corrosion mm/y	(mpy)	Required Corrosion Allowance, mm (in.) For 10-Year Life	For 15-Year Life	Rating
0.02- 0.1	(1-5)	2.0 (1/16)	3.0 (1/8)	Excellent
0.1-0.5	(5-20)	5.0 (3/16)	8.0 (5/16)	Good
0.5-1	(20-50)	12.0 (1/2)	18.0 (3/4)	Fair
1-5	(50-200)	Alloy upgrade	Alloy upgrade	Poor

With easily accessible and repairable components (such as above-ground process piping) a higher corrosion rate can be tolerated than with inaccessible components that can only be repaired or replaced during shutdown (such as fired heater tubes). As a general rule, higher corrosion rates can also be tolerated with piping, because compared to pressure vessel repairs, piping repairs are less constrained by Code requirements, less expensive, and can often be done on the run (for example, by clamping). Also, to meet Code thickness requirements, the depth of localized corrosion and pitting in piping can be discounted for limited spot size. In contrast, the maximum depth of any defect in pressure vessels applies to the entire thickness requirements. Finally, corrosion is easier to predict, measure, and inspect in piping than in pressure vessels walls.

20. Codes do provide, however, rules for including corrosion allowances when calculating required wall thicknesses.

4.9 Chapter Summary

The selection of materials of construction (metallurgy) for crude units and upgraders is based on a combination of mandatory construction and engineering Codes, nonmandatory recommended practices, and standard materials specifications. Codes apply only to new construction of major components, but are also followed for repairs of major components and for other than major components because of safety and potential litigation concerns. Recommended practices are nonmandatory, but are often treated as mandatory because of safety and potential litigation concerns. Compliance with standard materials specifications for steels and other materials is mandatory in most states and in some countries outside the United States.

Crude unit and upgrader metallurgy varies, depending primarily on the type of crude charged, component involved, and operating conditions. Units running sweet crudes containing less than 0.6% S and with a TAN below 0.3 have relatively few corrosion issues in the high-temperature circuits. These circuits were historically constructed of carbon steel except for vessel internals made from 12Cr steel. Units running sour crudes containing more than 0.6% S but with a TAN below 0.3 require alloy protection against HTSC. Usually, these units require a minimum of 5Cr-0.5Mo steel for the fired heater tubes, headers, U-bends, transfer lines, and some piping circuits to ensure that protective sulfide scales are formed. Units running acidic crudes with a TAN above 0.3 regardless of sulfur content require additional upgrading at areas of high velocity and/or turbulence because no protective scales are formed. Upgrading with low alloy Cr-Mo steels and 12Cr stainless steel is usually ineffective and austenitic 300 Series stainless steels such as Type 316 SS or Type 317 SS are required for reasonable long-term protection. Units running opportunity crudes have identical requirements as units running acidic crudes, because opportunity crudes often contain unknown amounts of naphthenic acids. Overhead condensing systems on various columns are constructed of carbon steel except areas where accelerated corrosion due to high HCl content, high vapor velocities, chloride salt deposition, or cooling water corrosion occurs. These areas may require selective alloying with Admiralty brass, copper nickel alloys, high-nickel alloys, or titanium.

4.10 Read More

1. API Refining Department, *Controlling Weld Hardness of Carbon Steel Refining Equipment to Prevent Environmental Cracking*, publication 942, 2nd ed. (Washington, DC: API, 1982).
2. J. Gutzeit, "Refinery Corrosion Overview" in *Process Industries Corrosion - The Theory and Practice*, B. J. Moniz, W.I. Pollock, eds. (Houston, TX: NACE, 1986).
3. J. Gutzeit, R. D. Merrick, and L. R. Scharfstein, "Corrosion in Petroleum Refining and Petrochemical Operations," in *Volume 13 - Corrosion, ASM Metals Handbook*, 9th ed. (Metals Park, OH: ASM, 1987).
4. J. Gutzeit, R. D. Merrick, and L. R. Scharfstein, "Corrosion in Petroleum Refining and Petrochemical Operations," in *Corrosion in the Petrochemical Industry*, ASM Monograph (Metals Park, OH: ASM International, 1994).
5. D. A. Hansen and R. B. Puyear, *Materials Selection for Hydrocarbon and Chemical Plants* (New York, NY: Marcel Dekker, 1996).
6. E. L. Hildebrand, "Materials Selection for Petroleum Refineries and Petrochemical Plants," *MP* 11, 7 (1972).
7. M. J. Humphries and G. Sorell, "Corrosion Control in Crude Oil Distillation Units," *MP* 15, 2 (1976).
8. R. I. Jacobs and J. A. McMaster, "Titanium Tubing: Economical Solution to Heat Exchanger Corrosion," *MP* 11, 7 (1972).
9. T. M. Krebs, "Selection of Furnace Tubing – Part 1: Mechanical Considerations, Types of Tubing Used," *Petroleum Engineer* 30, 2 (1958).
10. D. M. McCue, "Design Considerations for Titanium Heat Exchangers," paper no. 60, CORROSION/81 (Houston, TX: NACE, 1981).
11. W. A. McGill and M. J. Weinbaum, "Aluminum Diffused Steels Last Longer," *Oil & Gas J.* 70 (Oct. 9, 1972).
12. ASM Committee on Petroleum Refinery Applications, "Selection of Steel for High-Temperature Service in Petroleum Refinery Applications," *Metals Handbook*, 8th ed. Vol. I (Metals Park, OH: ASM, 1961).
13. W. A. McGill and M. J. Weinbaum, "Alonized Heat Exchanger Tubes Give Good High Temperature Service," *MP* 17, 1 (1978).
14. W. A. McGill and M. J. Weinbaum, "The Selection, Application and Fabrication of Alonized Systems in the Refinery Environment," *Proc. API* 54, III (1975).
15. B. B. Morton, "Metallurgical Methods for Combating Corrosion and Abrasion in the Petroleum Industry," *J. of the Institute of Petroleum* 34, 289 (1948).

16. J. A. McMaster, "Selection of Titanium for Petroleum Refinery Components," *MP* 18, 4 (1979).
17. G. E. Moller, I. A. Franson, and T. J. Nichol, "Experience with Ferritic Stainless Steel in Petroleum Refinery Heat Exchangers," *MP* 20, 4 (1981).
18. R. A. Mueller, "Corrosion Allowance – Not Always a Simple Concept," *MP* 45, 3 (2006).
19. NACE RP 0170-2004, "Protection of Austenitic Stainless Steels and Other Austenitic Alloys from Polythionic Acid Stress Corrosion Cracking During Shutdown of Refinery Equipment" (Houston, TX: NACE, 2004).
20. W. J. Neill, "Experience with Titanium Tubing in Oil Refinery Heat Exchangers," *MP* 19, 9 (1980).
21. C. A. Robertson and H. L. Myers, "Application and Use of Aluminum Coatings in Oil Refinery Processes," *MP* 6, 9 (1967).
22. C. H. Samans, "Which Steel for Refinery Service? Part 1: Steels for Service below 650 °F," *Hydroc. Proc.* 42, 10 (1963).
23. C. H. Samans, "Which Steel for Refinery Service? Part 2: Steels for Service above 650 °F," *Hydroc. Proc.* 42, 11 (1963).
24. D. B. Bird, "Titanium Exchanger Tubing Works Well in Refinery Service," *Hydroc. Proc.* 55, 5 (1976).
25. R. J. Scanlan and R. M. Boothman, "Corrosion Monitoring Experience in the Refinery Industry Using the FSM Technique," paper no. 03655, CORROSION/2003 (Houston, TX: NACE, 2003).
26. R. W. Schutz and L. C. Covington, "Effect of Oxide Films on the Corrosion Resistance of Titanium," paper no. 61, CORROSION/81 (Houston, TX: NACE, 1981).
27. L.M. Wyatt et al., *An Atlas of Corrosion and Related Failures,* publication no. 18 (St. Louis, MO: Materials Technology Institute, 1987).
28. R. A. White and E. F. Ehmke, *Materials Selection for Refineries and Associated Facilities* (Houston, TX: NACE, 1991).
29. D. Kumar et al., "Material Selection for Designing SAGD Plants," paper no. 1716, CORROSION /2012 (Houston, TX: NACE, 2012).
30. A. J. Brophy, "Stress Corrosion Cracking of Austenitic Stainless Steels in Refinery Environments," *MP* 13, 5 (1974).
31. A. S. Couper and H. F. McConomy, "Stress Corrosion Cracking of Austenitic Stainless Steels in Refineries," *Proc. API* 46, III (1966).
32. A. S. Couper, "Minimizing Stress Corrosion Cracking - Testing Austenitic Stainless Steels for Modern Refinery Applications," *MP* 8, 10 (1969).
33. W. A. Derungs, "Corrosion in Distilling and Vacuum-Distilling Plants for the Processing of Middle East Crude Oils," *Proc. API* 36, III (1956).
34. Z. A. Foroulis, "Corrosion and Corrosion Inhibition in the Petroleum Industry," *Werkstoffe und Korrosion* 33, 2 (1982).
35. I. A. Franson and L. C. Covington, "Application of Titanium to Oil Refinery Environments," *Proc. API* 56, III (1977).

5 Corrosion Monitoring

In this Chapter

5.1 First Glance

With the arrival of API's risk-based inspection (RBI) and various proprietary risk-based inspection management (RBIM) and corrosion prediction programs, actual online corrosion monitoring seems to have become a thing of the past. While these programs have certainly lead to a greater awareness and emphasis of the critical components that should be monitored, they often rely on historical corrosion data in inspection files that may or may no longer be relevant. Hence, up-to-date corrosion data are more necessary than ever for establishing realistic IOW[1] and AOL[2] parameters.

Direct monitoring tools include visual inspection, ultrasonic thickness (UT) readings, electrical resistance probes, coupon weight-loss probes, hydrogen probes, and linear polarization probes, among others. Each type is useful for certain applications provided its limitations are kept in mind. Some direct monitoring tools have been modified to take condensation, salt deposition, and fouling profiles into consideration. These so-called "cold finger probes" featuring an internal cooling medium with or without temperature control have been used as electrical resistance, coupon weight loss, and hydrogen activity probes. Elaborate bypass condensers have been used to simulate initial condensate corrosion on probe sensing elements. Indirect monitoring tools include iron counts and pH measurements for monitoring

1. IOW: Integrity Operating Window is a key process variable, which, if exceeded, will compromise the design life of the asset and may require a fitness-for-service review.
2. AOL: Allowable Operating Limit is a key process variable, which, if exceeded, will reduce the expected remaining life of the asset and may pose a threat to the unit's or refinery's safety.

corrosion in aqueous overhead condensate, and parallel-tube heat exchangers for evaluating the loss in heat transfer due to corrosion and fouling in heat exchangers.

In recent years, various new types of corrosion monitoring tools have been proposed as a result of extensive research in the United Kingdom and Norway to find reliable methods for continuously monitoring the integrity of subsea pipelines. They include among others the CEION™, FSM™, and nanoCorr™ analyzers. All are expensive and usually require special sensors, precision electronics, software programs, and trained personnel for interpretation of the data obtained. Regardless of what tools and methods are used, corrosion monitoring should be an ongoing effort and not be discontinued even when no changes are seen. Once monitoring is discontinued, historical data are missing and no conclusions can be drawn when corrosion issues occur at a later date.

5.2 Definition of Corrosion

Corrosion has been defined as the process and rate of weakening, wearing away, or destroying a metal by chemical action. In practice, the term is also applied to nonmetals and other than chemical action. The two major types of corrosion seen in crude units, bitumen upgraders, and elsewhere in petroleum plants include, 1. high-temperature corrosion above approximately 230 °C (450 °F) caused primarily by organic sulfur and acid compounds in the absence of free water, and 2. low-temperature corrosion below approximately 230 °C (450 °F) caused primarily by inorganic acids and salts in the presence of free water. Both types have been studied extensively in the laboratory and can be monitored online and offline in the field.

From a practical view point, corrosion is best expressed in terms of penetration rates (mm/y or mpy [3]) instead of a variety of other ways (for example, % weight loss, $mg/cm^2/day$, or $g/in^2/hr$) often seen in the scientific and early corrosion literature. Penetration rates are preferred because they can be related directly to the rate of wall thickness loss and, therefore, the expected life of a component in a process unit. A penetration rate of 1mm/y is equivalent to 40mpy, while a penetration rate of 1mpy is equivalent to 0.025mm/y.

5.3 Recommended Monitoring Locations

For ongoing monitoring of general corrosion in pressure vessels, retractable Corrosometer™, weight-loss coupon, or hydrogen activity probes are simply installed in the vessel wall using available access nozzles or mounted on manhole covers. Special probes and nozzles are needed for monitoring tray corrosion. For monitoring general corrosion in piping systems, probes are installed in horizontal and/or vertical piping. In overhead vapor lines, the preferred location is the horizontal segment between the column and the first set of heat exchangers or condensers. Probes with water-cooled tubular sensing elements should be used here to force shock condensation on the probe elements. For monitoring general corrosion in overhead condensate lines, probes are installed in vertical rundown lines from the condensers to the receiver drum.

The preferred monitoring location for monitoring general corrosion in vacuum overhead systems is upstream of the ejectors in the vertical rundown piping from the water-cooled precondenser to the seal drum (hot well or vacuum sump). On some units, corrosion is also monitored downstream of the ejectors in the vertical rundown piping between the water-cooled ejector condensers and the seal drum. For long probe life, probes with wire elements are preferred over probes with tubular elements. For best probe sensitivity, carbon steel elements are used regardless of metallurgy. A desirable goal is to control carbon steel corrosion to less than 0.1mm/y (4MPY).

Many plants rely primarily on ultrasonic thickness (UT) measurements to monitor the wall thickness of piping and equipment offline at regular inspections. UT measurements are also done online as needed to monitor suspect areas. The main disadvantages are necessary sensor calibration for each type of metal, good contact with the metal surface, and interpretation of the measurements by an experienced inspector. While it is easy to convert UT measurements to corrosion rates in terms of thickness loss with time and then predict a likely remaining service life, the measurements often lack accuracy and repeatability. Even minor deviations in test points can produce thickness measurements that for various reasons increase rather than decrease with time.

3. Mpy: Mils per year, where 1 mil = 0.001in.

UT inspection points in vapor lines are usually located at high risk areas (elbows, tees, and other fittings) and are often "cherry-picked" to represent the worst-case scenarios for corrosion in these lines. The lines obviously remain prime areas for potential leaks and warrant future monitoring on an ongoing basis. As a general rule, the measured corrosion rates should be considered with caution for several reasons: 1. Any baseline readings that fall completely outside the nominal wall thickness range for the piping should be suspect; 2. Follow-up readings often add to the confusion; and 3. Calculating and reporting corrosion rates to more than one significant figure is unrealistic, unless used to identify specific inspection points on piping schematics.

5.4 Offline Corrosion Monitoring Tools

Other than visual examination, offline corrosion monitoring tools that have been used over many years include primarily UT readings and radiography. With few exceptions, both can also be used online and are well suited for most units and components. UT readings at selected locations are useful for quantifying corrosion damage while a unit is offline or online and have the advantage over other monitoring tools in that corrosion is easily expressed in terms of metal thickness loss. Radiography is a special inspection technique that is used to examine piping and vessels walls for thickness loss due to corrosion. As shown in Figure 5-1, the method involves visual evaluation of radiographic films by qualified inspectors, but the results are subject to human interpretation. Radiography allows examination of pipe and vessel walls without having to remove exterior insulation.

FIGURE 5-1 GENERAL SETUP FOR RADIOGRAPHIC INSPECTION OF INSULATED PIPE

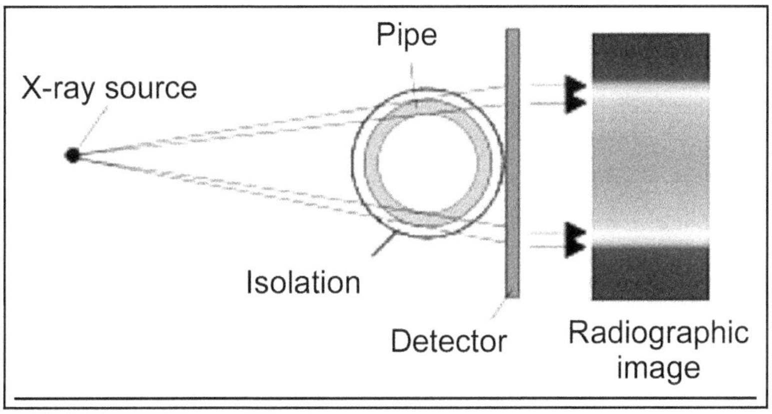

Adapted from Y. Onel et al., "Radiographic Wall Thickness Measurement of Pipes by a New Tomographic Algorithm," 15th World Conference on Non-Destructive Testing, Rome, Italy, 2000.

5.5 Online Corrosion Monitoring Tools

The principal difficulty with monitoring corrosion on line is deciding on the best locations for measuring wall thickness or placing some kind of monitoring probe. As discussed in later chapters, corrosion and fouling are likely to occur almost anywhere in a given component or group of components and can constantly shift in response to process changes. Based on past experience, the worst corrosion and fouling is often, but not always, found at elbows, tees, flow reducers, and similar fittings of pipe segments. Areas where water can condense and collect or where salt deposits can form are also prime locations that should be considered for regular monitoring. The following questions may have to be addressed when selecting a monitoring location:

- Should probes be inserted into the process stream proper or into a slip or bypass stream for better sensitivity or temperature control?
- Are relative rather than absolute changes in corrosion or fouling rate acceptable?

- Should probes be placed in pipe bends (sags), vessel boots, dead end segments, or vessel internals?
- Should the inner (short) or outer (long) radius of pipe bends be monitored?
- What about horizontal versus vertical pipe segments?
- What about areas with or without external insulation?
- What about general accessibility? Despite best intentions, areas that are difficult to reach will not be monitored regularly in the dead of winter.

5.5.1 Direct Online Monitoring Tools

Direct corrosion monitoring tools that have been used over many years include UT readings, electrical resistance probes, coupon weight-loss probes, hydrogen activity probes, and linear polarization probes, among others. With few exceptions, all are well suited for most units, components, and processes. Suitable instruments are relatively cheap to purchase or easily modified if necessary.

5.5.1.1 Ultrasonic Thickness Readings. While differing from other monitoring tools, periodic UT readings at a specific location are one of the basic methods for measuring corrosion damage while a unit is either on or off stream. Aside from accessibility issues, use of UT sensors is sometimes restricted by elevated metal wall temperatures and special standoffs may be required. UT readings have an advantage over other monitoring tools in that corrosion is easily expressed in terms of metal thickness loss.

5.5.1.2 Electrical Resistance Probes. Electrical resistance or Corrosometer™ probes can be permanently mounted in a pipe or vessel wall or inserted and periodically removed through a valved access nozzle to monitor corrosion onstream. Readings can be obtained manually several times a day or week with a portable instrument, like the one shown in Figure 5-2. With electrical resistance probes, a metal[4] wire, tube, or strip sensor element is exposed to the process stream. As the cross-section of the sensor element is reduced by corrosion, the electrical resistance increases and produces a proportional change in the probe reading. Electrical resistance probes usually require several days of readings for meaningful data to accumulate. Corrosion is generally proportional to the slope of consecutive probe readings versus time. The probe readings can be easily converted to conventional corrosion rates using Equation 5-1 or Equation 5-2:

To determine corrosion rate in metric units:

$$CR_{mm/y} = \frac{0.365\,(R_{final} - R_{initial})\,P_{mm}}{t} \qquad \text{(Eq5-1)}$$

where:

$CR_{mm/y}$ = Corrosion rate, mm/y
$R_{initial}$ = Initial instrument reading, divisions
R_{final} = Current instrument reading, divisions
P_{mm} = Probe span, mm
t = Time between measurements, day

To determine corrosion rate in customary units:

$$CR_{mpy} = \frac{0.365\,(R_{final} - R_{initial})\,P_{mils}}{t} \qquad \text{(Eq.5-2)}$$

where:

CR_{mpy} = Corrosion rate, mpy
$R_{initial}$ = Initial instrument reading, divisions
R_{final} = Current instrument reading, divisions
P_{mil} = Probe span, mil
t = Time between measurements, day

4. For greater sensitivity, carbon steel is generally preferred over other metals as a basis of comparison.

FIGURE 5-2 PORTABLE CORROSOMETER™ INSTRUMENT FOR MEASURING CORROSION IN THE SHELLSIDE INLET OF AN HEAT EXCHANGER

Retractable electrical resistance probes and retractable weight-loss coupon probes can be used interchangeably at the same locations. Both types can also be mounted on the same holder at the same time to allow for direct comparison of the results obtained with the two methods. The accuracy of the data obtained with either method depends on probe location relative to active corrosion sites. A wide selection of available probe assemblies is commercially available. Probes with flush-end elements, shown in Figure 5-3, can be installed through a nozzle and placed flush with the pipe or vessel wall to minimize the effects of flow turbulence. In addition, retractable probes have been modified to permit sampling and readings of the liquid and accumulated deposits on column trays.

FIGURE 5-3 FLUSH-END CORROSION PROBE FOR MINIMIZING HIGH READINGS DUE TO FLOW TURBULENCE

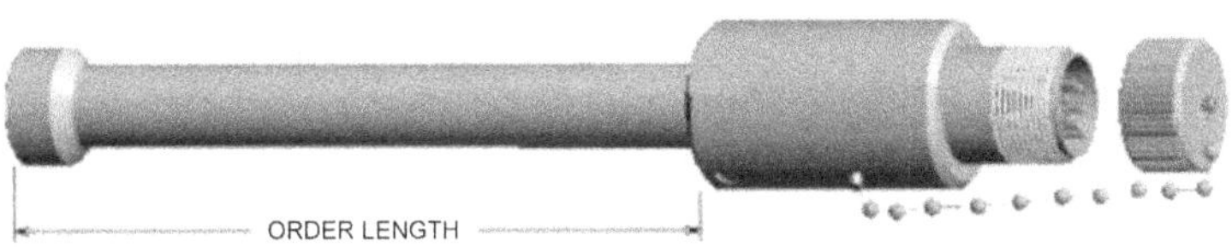

Courtesy of Rohrback Cosasco Systems, Inc., Santa Fe Springs, CA.

Standard sensor elements are available in various metals, shapes, and thicknesses to accommodate a wide range of corrosive conditions. There is a tradeoff between probe sensitivity and probe life: highly sensitive probe elements have a short useful life and require more frequent replacement, while less sensitive probes have a long useful life. Strip and tube loop elements have greater surface areas and are more sensitive than wire loop or cylindrical elements. The latter are preferred for most refinery and upgrader applications, because they are more resistant to handling and flow-induced damage. In practice, probe elements are available with specified probe spans ranging from 4 to 80 mils, as shown in Figure 5-4. The resultant probe life ranges from days to years depending on the corrosion rate, as shown in Table 5-1.

FIGURE 5-4 AVAILABLE PROBE ELEMENTS IN STRIP, TUBE, AND WIRE SHAPES AND PROBE SPANS RANGING FROM 4 TO 80 MILS

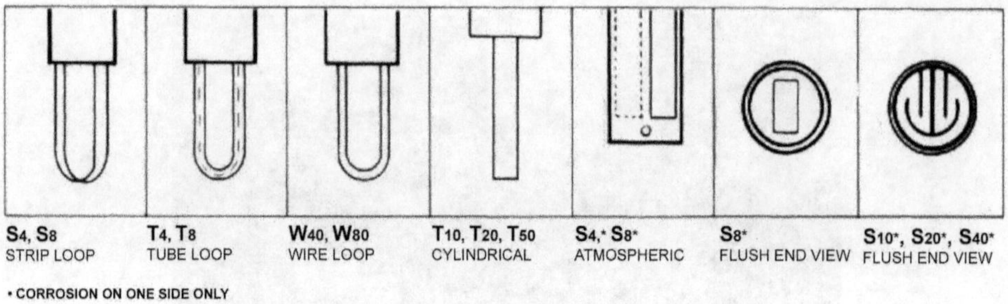

Courtesy of Rohrback Cosasco Systems, Inc., Santa Fe Springs, CA.

TABLE 5-1 PROBE LIFE AS A FUNCTION OF CORROSION RATE AND PROBE SPAN

Corrosion Rate, mm/y (mpy)	Probe Life For Various Probe Spans, mm (mil)*		
	0.1 (4)	0.25 (10)	0.5 (20)
0.025 (1)	2 yr	5 yr	10 yr
0.25 (10)	2.4 mo	0.5 yr	1 yr
1.25 (50)	12 d	1 mo	2 mo
2.50 (100)	6 d	2 wk	1 mo

* Assumes probe element does not break off.

The minimum time between probe readings depends on the sensitivity of different probe elements in terms of probe span and on expected corrosion rates, as shown in Table 5-2. At high corrosion rates, readings taken with highly sensitive probe elements at short intervals (hours) can provide meaningful data. However, while offering improved response times compared to conventional probes, the entire probe span is rapidly used up if unanticipated corrosion excursions occur. In most cases, sufficiently long intervals are required to compensate for faulty readings due to sudden temperature spikes or other variable factors. If necessary, retractable probes allow changing probe elements while the unit is on stream.

TABLE 5-2 MINIMUM TIME BETWEEN PROBE READINGS AS A FUNCTION OF CORROSION RATE AND PROBE SPAN

Corrosion Rate, mm/y (mpy)	Minimum Times Between Probe Readings For Various Probe Spans, mm (mil)				
	0.1 (4)	0.25 (10)	0.5 (20)	1.0 (40)	2.0 (80)
0.0125 (0.5)	29d	73d	6mo	1y	2y
0.025 (1)	15d	36d	73d	4mo	8mo
0.125 (5)	3d	7d	15d	1mo	2mo
0.250 (10)	35hr	4 d	7d	14d	1mo
0.625 (25)	14hr	35hr	3d	6d	12d
1.250 (50)	7hr	18hr	35hr	3d	6d

Unless calculated automatically by the instrument, consecutive probe readings (R) can be used directly to determine changes in corrosion rate with exposure time. When the instantaneous probe readings (R) are plotted against time, the final slope of the probe reading versus time curve (i.e., $\Delta R/t$) represents a reasonably accurate long-term corrosion rate. However, when the instantaneous probe readings are first converted to corrosion rates that are plotted against time, the final slope of the corrosion rate versus time curve (i.e., $\Delta CR/t$) represents the expected change in corrosion rate with time and not the long-term corrosion rate.

The principal advantage of electrical resistance probes over weight-loss coupon probes is their ability to provide readings remotely and continuously as part of an integrated corrosion monitoring system with readouts on the operators' panel board. An example of a remote measuring probe is shown in Figure 5-5. Several precautions apply, however, when considering the installation of an integrated corrosion monitoring system. All instrumentation must meet the refinery's fire codes and must be explosion proof. Any interconnections must be hard-wired in place and the electronic transmission network must be approved. When operational issues occur, they cannot be addressed by vendor personnel, but must be turned over to a refinery electrician or instrument mechanic who may or may not be available because of other overriding priorities. This means that an integrated corrosion monitoring system may well be out of order for some period of time.

FIGURE 5-5 REMOTE ELECTRIC RESISTANCE PROBE MOUNTED ON VERTICAL PIPE WALL

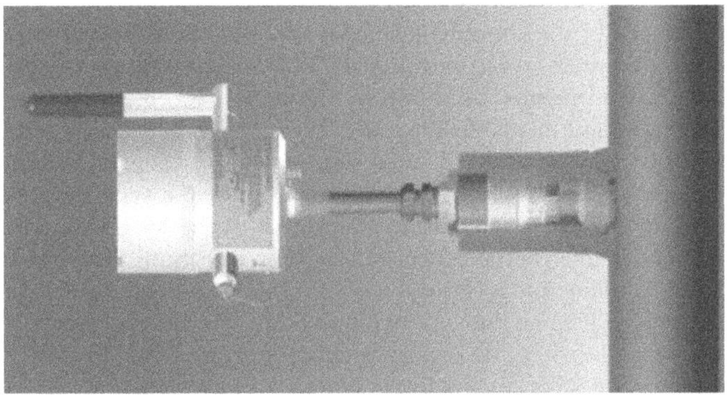

Courtesy of Rohrback Cosasco Systems, Inc., Santa Fe Springs, CA.

5.5.1.3 Coupon Weight-Loss Probes. Corrosion coupons have been used for many years prior to the development of other corrosion monitoring devices. Coupons are typically mounted in duplicate on some sort of rack that is either placed directly inside a component or inside a sidestream loop for better access, flow control, or temperature control. Special coupon racks can be placed on the top and other trays of the various columns to measure corrosion in accumulated condensate or under deposits. Aside from allowing the measurement of average corrosion rate, coupons can be visually inspected and chemically analyzed to determine the type, composition, amount, and thickness of any scale formed. Coupons can also reveal any tendency toward pitting, stress corrosion cracking (SCC), and other forms of localized corrosion, such as velocity-accelerated corrosion or impingement attack. Compared to more sophisticated measuring techniques that rely on expert interpretation of the results, coupons reveal to the naked eye at once what is going on, as far as corrosion and fouling are concerned. When coupons are used, corrosion rate in metric units is calculated from the measured weight loss as follows:

$$CR_{mm/y} = \frac{86.1 \cdot WL}{d \cdot A \cdot t} \qquad \text{(Eq.5-3)}$$

where:

$CR_{mm/y}$ = Corrosion rate, mm/y
WL = Weight loss of coupon, mg
d = Density of metal, g/cm^3
A = Coupon area, cm^2
t = Exposure time, hr

In the United States, corrosion rate in mpy is usually calculated by using a combination of metric and customary units:

$$CR_{mpy} = \frac{534 \cdot WL}{d \cdot A \cdot t} \qquad \text{(Eq.5-4)}$$

where:

CR_{mpy} = Corrosion rate, mpy
WL = Weight loss of coupon, mg
d = Density of metal, g/cm^3
A = Coupon area, in^2
t = Exposure time, hr

Semipermanent coupon racks are only practical when a unit or component comes down for repairs or maintenance at regular intervals. With the current trend of extending runs to 5-10 years and longer, coupon racks are often forgotten and left in place. Even when they are eventually recovered during a turnaround, the data obtained are usually no longer pertinent. Retractable coupon holders, similar to those used for electrical resistance probes, can also be used to determine changes in corrosion rate with time. This is easily done by mounting one or more metal coupons (test specimens) on the coupon holder and removing and replacing coupons at various time intervals. The tee-handle of the coupon holder is used to indicate which side of the coupons faces the flow direction. As in the case of retractable electrical resistance probes, the coupon holder must be secured in place with a special bracket to prevent it from being pushed through the packing gland by process pressure, as shown in Figure 5-6. To obtain long-term corrosion rates with retractable coupon holders, coupon weight loss (WL) instead of corrosion rate (CR) should be plotted against time (t). The final slope of the WL versus t curve (i.e., $\Delta WL/t$) then represents a reasonably accurate long-term corrosion rate. When corrosion rate (CR) is plotted against time (t), the final slope of the CR versus t curve (i.e., $\Delta CR/t$ or $\Delta WL/t/t$) represents the expected change in corrosion rate with time instead of the long-term corrosion rate. The minimum interval between coupon retrieval depends on the expected corrosion rate. Sufficient exposure time should be allowed to compensate for errors in the handling and weighing of coupons. As a general rule, the minimum recommended exposure time (t_{min}) in hours is 2000/CR when the expected corrosion rate CR is in mpy. The minimum recommended exposure time (t_{min}) in hours is 50/CR, when the expected corrosion rate CR is in mm/y. However, corrosion rates obtained at shorter-than-recommended intervals are always better than no data at all.

FIGURE 5-6 RETRACTABLE COUPON HOLDER FOR ONSTREAM REMOVAL AND REPLACEMENT OF COUPONS

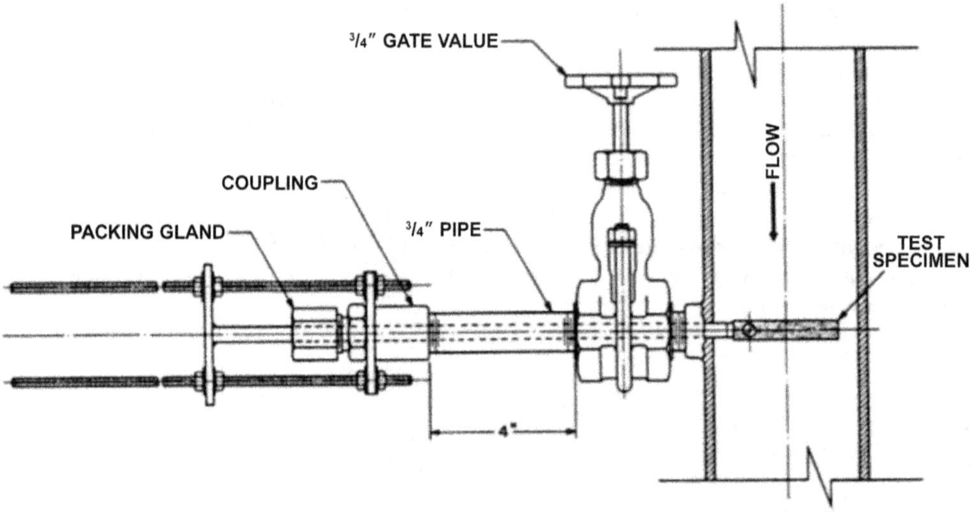

Courtesy of Betz Process Chemicals, The Woodlands, TX.

Retractable coupon holders can also be used to evaluate different materials of construction at the same time. Coupon holders can be designed to pass through the same valved access nozzle used for electrical resistance probes by using smaller coupons. Shown in Figure 5-7 are the front and back sides of four different alloys evaluated in parallel over the same time span, including carbon steel (top coupon), two different low-alloy Cr-Mo steel coupons, and a 12Cr stainless steel (bottom coupon). The purpose of the test was to determine if NAC was occurring in a HVGO stream. Since all coupons other than the carbon steel coupon corroded at similar rates in this test, NAC was judged to be a problem in the affected line. Accelerated corrosion damage due to velocity effects is clearly seen on the carbon steel coupon.

Some monitoring tools have been modified to take condensation, salt deposition, and fouling profiles into consideration. So-called "cold finger probes" featuring an internal cooling medium and temperature control have been adapted as coupon-weight loss probes. Others have used tubular specimens that can be weighed before and after exposure to obtain a corrosion rate as shown in Figure 5-8. Inspection departments are sometimes tempted to build coupon probes in-house to avoid having to purchase Corrosometer™ probes. Experience has shown, however, that the overall costs of building, purchasing, and evaluating various types of retractable probes are comparable. Moreover, wireless transmission of data from purchased remote electrical resistance probes can reduce their installation costs significantly.

FIGURE 5-7 RETRACTABLE COUPON HOLDER FOR EVALUATING FOUR DIFFERENT MATERIALS AT THE SAME TIME

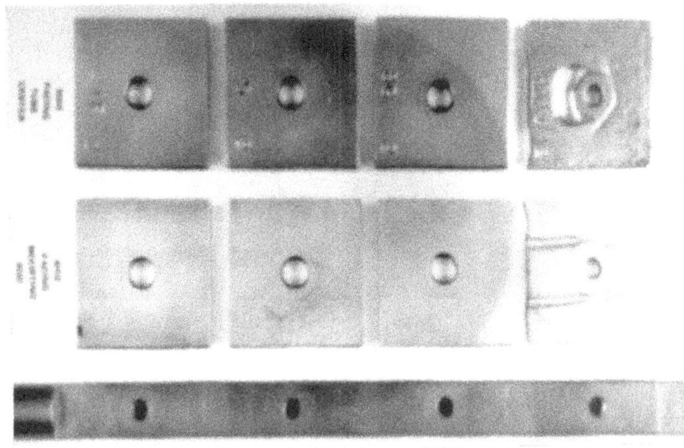

FIGURE 5-8 WATER-COOLED TEST LOOP FOR EVALUATING CORROSION UNDER CONDENSING CONDITIONS

5.5.1.4 Hydrogen Activity Probes. Hydrogen activity probes are excellent tools for measuring hydrogen blistering and cracking in steel exposed to corrosive condensate. The cathodic portion of many corrosion reactions evolves atomic hydrogen (H_0) on the outer surface of the probe wall, a portion of which then diffuses through the probe wall and combines to form molecular hydrogen (H_2) on the inside of the probe wall. Molecular hydrogen is trapped and a pressure gage is used to measure the resultant increase in hydrogen pressure inside the probe body. An example of a simple finger-type hydrogen probe is shown in Figure 5-9. This type of probe can be remotely monitored by using stainless steel capillary tubing to connect the probe body to the pressure gage and bleed-off valve at ground level.

In overhead systems, hydrogen activity probes are useful for measuring aqueous HCl corrosion in overhead water condensate at pH below 5. They are not useful at higher pH values unless cyanides are present to dissolve the protective iron sulfide films on the metal surface. Despite some claims to the contrary, hydrogen activity probes are useless for measuring high-temperature corrosion in hydrocarbon streams of crude units. The reason for this is: 1. Hydrogen evolution requires enough corrosion to provide a high-enough H_0 potential for hydrogen entry, and 2. The presence of a free water phase on an essentially bare metal surface. An oil layer would interfere with hydrogen evolution and entry.

That does not mean that considerable effort has not been expended over the years to find a way around this dilemma. Aside from finger-type probes, external patch-type probes have been used for measuring the hydrogen flux directly through a pipe or vessel wall. These probes use various methods for hydrogen detection, including pressure gages, electrochemical cells, and vacuum pumps. Most rigs are expensive and require special sensors, precision electronics, dedicated software, and trained personnel for interpreting the data obtained. In most cases, they are difficult to use with complex geometries, and unexpected results tend to be discounted or ignored. Hence, they are not recommended for most plant applications.

FIGURE 5-9 FINGER-TYPE HYDROGEN PROBE AND PRESSURE GAGE

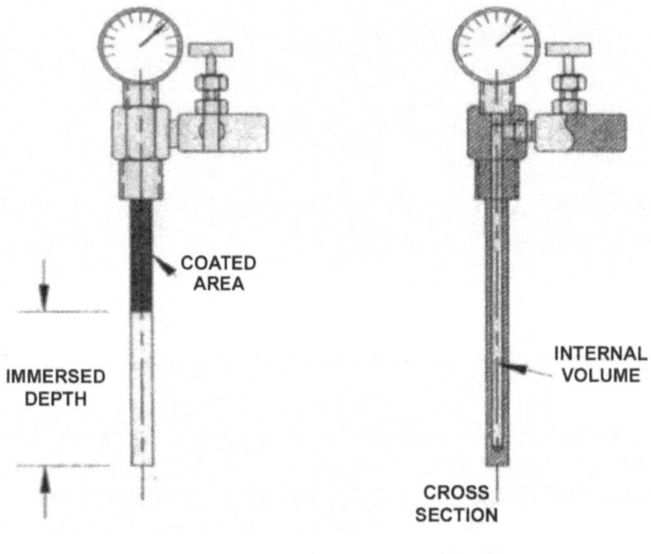

Adapted from NACE Publication IC 184, "Hydrogen Permeation Measurement and Monitoring Technology" (Houston, TX: NACE).

5.5.1.5 Polarization Probes. Two- and three-prong polarization or Corrator™ probes are not practical for monitoring corrosion in most nonaqueous hydrocarbon streams. They are not recommended despite claims that various proprietary improvements, such as harmonic distortion analysis (HDA) and electrochemical noise (EN) analysis, can overcome the issues with hydrocarbon streams. They can be used, however, in cooling water services or in components where a more-or-less continuous water phase is always present.

5.5.2 Indirect Corrosion Monitoring Tools

Indirect corrosion monitoring tools that have been used over many years include primarily iron counts and pH measurements. With few exceptions, iron counts are suited for most processes and components while pH measurements are useful for monitoring corrosion in overhead water condensate. Parallel-tube heat exchangers have been used for evaluating heat transfer loss due to corrosion and fouling in preheat exchangers. Except for the latter, indirect corrosion monitoring tools are relatively easy to use, and test equipment can be purchased or modified at low cost.

5.5.2.1 Iron Counts. Monitoring corrosion with iron counts (actually "ion counts") can be done easily, inexpensively, and quickly in the field, but is not always the best indicator of corrosion. This is especially true for stainless steels and nickel alloys that have corrosion rates 1-2 orders of magnitude lower than those of carbon steel. Pitting corrosion and localized attack are likely to be missed entirely by iron counts, because the affected areas are small compared to the overall surface areas involved. While a long list of dissolved ion species in water samples may look impressive at first glance, they tend to be useless for tracking corrosion in the field. Iron counts, unlike corrosion rates obtained by other methods, indicate corrosion upstream of and not at the sampling point. Iron counts are usually reported as mg/L or mg/kg in various process streams and have been relied upon mostly to detect corrosion during upstream, midstream, transportation, and storage operations. As a general rule, however, iron counts are unreliable indicators of corrosion in downstream applications unless:

- Precipitation of insoluble sulfides and oxides is taken into account.
- The same ion species are determined at a given sampling point in a system.
- Other test methods are also deployed to determine detection limits, accuracy, precision, and possible interferences.
- The size of the corroding surface areas is taken into account.
- The flow rate at the time of sampling is taken into account and the results are reported as an iron production rate (for example, weight/d).

However in the case of general or uniform corrosion, the "iron production rate" can be quite useful, since it can be converted into a quasi corrosion rate (CR) if the size of the corroded area (A) is known:

$$\text{CR (mpy)} = \text{gm/day/m}^2 \bullet 1.8 \text{ for the most common metals} \quad \text{(Eq. 5-6)}$$

Using metric units:

$$\text{CR (in mm/y)} = \text{IC (in kg/d)} / 2.2 \bullet 10^{-2} \bullet \text{A (in m}^2\text{)} \quad \text{(Eq. 5-7)}$$

Using customary units:

$$\text{CR (in mpy)} = \text{IC (in lb/d)} / 1.1 \bullet 10^{-4} \bullet \text{A (in ft}^2\text{)} \quad \text{(Eq. 5-8)}$$

5.5.2.2 PH Readings. In the absence of added neutralizers, pH readings are an excellent indicator of the total HCl content of water condensate and acid corrosion in overhead systems (for obvious reasons, pH readings are useless for hydrocarbon streams) [5]. Since HCl dissociates completely in water, condensate with pH4 contains 10^{-4}gmoles/L or 3.5mg/L HCl and corrodes CS at approximately 100mpy in laboratory tests. Once a neutralizer is added, a portion of the HCl is tied up as the dissolved neutralizer hydrochloride salt (R-NH2·HCl) and pH readings now reflect only the free HCl content. Special chloride ion electrodes can be placed upstream of the neutralizer injection point in order to accurately track chloride intrusions, but are rarely used. There is also no need for monitoring residual amine other than the daily neutralizer consumption rate monitored by the process additives supplier. In practice, pH readings should not be obtained with test papers or colorimeters. Neither method is reliable nor does it lend itself to providing a continuous readout.

Stagnant water condensate in water boots or sample connections does not give representative readings regardless of the method used. PH data from once-a-shift grab samples are also pretty useless, because neutralizer-pump settings are often "tweaked" by the operators on subsequent shifts to maintain the required pH. The result can be a typical sinusoidal pH curve with the pH drifting between a maximum and minimum value from shift to shift. Weekly or monthly follow-up pH data for "well-aged" samples are even worse: even if accurate, the results are too late to permit any kind of corrective measures to be implemented. PH spikes due to sudden chloride intrusions or neutralizer interruptions can

5. While the pH of entrained water in hydrocarbon streams can be measured by collecting a mixed sample under pressure, the procedure often is too cumbersome.

only be determined accurately by monitoring the pH of a flowing sample stream continuously and using a calibrated electronic pH meter with the output recorded on the operators' instrument panel. This is particularly important in cold climates, where obtaining grab samples once a shift under snowy and icy conditions is highly inconvenient at best and dangerous at worst and often avoided by the unit operators.

In most overhead systems, the preferred location for a pH probe is the water condensate collected in the accumulator drum. In refinery crude units, accumulator drums are typically fitted with a water boot that allows installing probes for continuous pH monitoring. When an overhead stream contains a relatively large volume of water condensate, the accumulator boot should be outfitted with a level control to allow continuous drainage and replenishment of the boot content. The opposite concern is that only a small amount of water condensate available for pH measurements at some locations. Since automatic pH and chloride control is used in numerous process operations, pH equipment manufacturers should be asked to recommend equipment suitable for petroleum applications. Other concerns are hydrocarbon contaminants in water condensate, for example oily or "black water," that interfere with accurate pH readings. These may require special sampling cells, filters, or techniques in addition to special pH electrodes.

5.6 Other Types of Monitoring Tools

A number of other types of corrosion probes have been proposed as a result of extensive research to find reliable methods for continuously monitoring subsea pipelines in the offshore industry. Most are expensive and require special sensors, electronics, software, and trained personnel for interpreting the data obtained. These tools are neither needed nor recommended for crude unit or upgrader applications. In practice, unexpected results obtained with these tools tend to be discounted or ignored and conventional methods become the fallback position. Typical examples of tools offered in the past by various inspection companies and some process additives suppliers include (among others) the CEION™, FSM™, nanoCorr™, Shell pulsed eddy current (PEC), and long-range guided-wave UT methods listed below.

5.6.1 CEION™ Method. The CEION™ method detects corrosion based on changes in current density due to surface defects, conductive deposits, or solutions that influence an electrically conductive path through a metallic sample. No major plant installations have been reported.

5.6.2 FSM™ Method. The FSM™ (field signature method) measures corrosion by feeding an electric current over several meters (ft) of an actual piping structure and sensing the resulting electric field pattern. One major refinery installation (in the United Kingdom) has been described in some detail[6]. As can be seen, the readout of the 3-dimensional corrosion pattern shown in Figure 5-8 can be fairly complex and difficult to interpret in practical terms.

FIGURE 5-8 EXAMPLE OF FSM READOUT FOR CRUDE UNIT OVERHEAD LINE

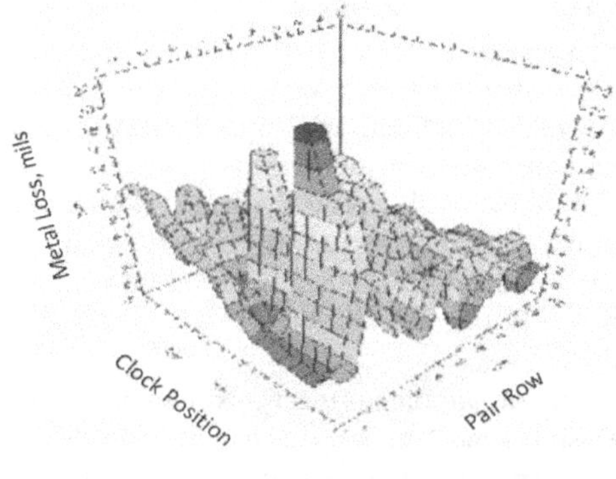

Courtesy of CorrOcean, Houston, TX.

6. K. Wold and G. Sirnes, "FSM Technology -16 Years of Field History –Experience, Status and Further Developments" paper no.07331, CORROSION/2007 (Houston, TX: NACE, 2007).

5.6.3 NanoCorr™ Method. The nanoCorr™ analyzer uses a multi-electrode, polarization type probe for real-time corrosion measurements. All polarization probes are intended, however, for aqueous systems and are ineffective for nonaqueous and/or high temperature applications.

5.6.4 Shell Pulsed Eddy Current (PEC) Probe. The pulsed eddy current probe "sees" a metal wall profile through overlying corrosion scale and insulation and has the advantages of being noninvasive, requiring no surface preparation, and being able to resist temperatures ranging from ambient to approximately 425 °C (800 °F). PEC has been used for periodic thickness measurements to monitor corrosion at fixed locations of overhead piping. While it can detect large areas of localized attack, it tends to miss isolated pitting. PEC can be used, however, at relatively high temperatures because it requires no direct contact with the metal wall. The principal limitations are that wall thickness is averaged under the probe's footprint, thickness changes rather than absolute thickness values are measured, and the probe is difficult to use with complex geometries.

5.6.5 Long-Range Guided-Wave UT. Long-range guided-wave UT is a nondestructive technique for measuring wall thickness of long pipe runs without the need to remove the pipe insulation. A set of transducers is clamped around the circumference of the pipe at a single location and a burst of ultrasonic guided waves is sent down the pipe. Areas of corrosion damage reflect the signals back to the transducers for analysis. Normally, other UT techniques must be deployed to verify the findings obtained with long-range guided-wave UT.

5.6.6 Corrosion Data Management Systems. While not a corrosion monitoring tool as such, UltraPIPE[7] is a popular inspection data management system (IDMS) for process facilities. This program enables operators to input, organize, analyze, and document thickness inspection data and is useful for monitoring corrosion and predicting the remaining life of fixed equipment. An example of a monitoring screen for the UltraPIPE program is shown in Figure 5-9.

FIGURE 5-9 EXAMPLE OF MONITORING SCREEN FROM ULTRAPIPE® PROGRAM

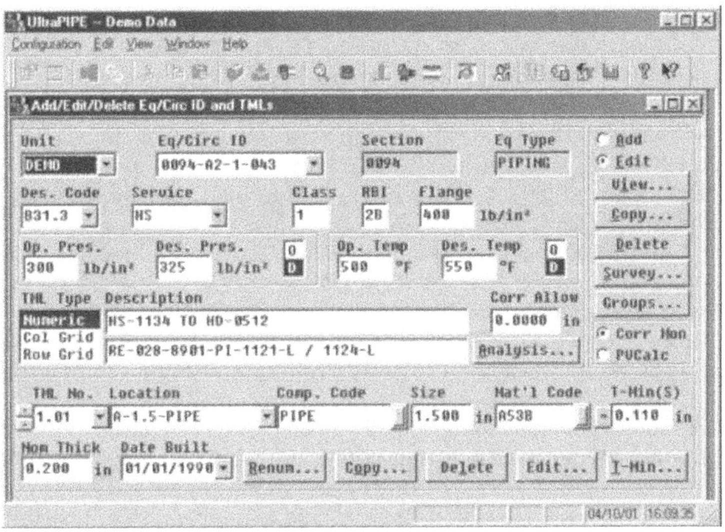

Courtesy of Siemens Oil & Gas Consulting.

5.7 Chapter Summary

While risk-based inspection (RBI) systems have lead to greater awareness of the critical components that should be monitored, they still require up-to-date corrosion data for establishing realistic IOW and AOL parameters. Common tools for monitoring various forms of corrosion online or offline include ultrasonic thickness readings, electrical resistance probes, coupon weight-loss probes, and hydrogen probes. Each type is useful for certain applications provided its limitations are kept in mind.

7. UltraPIPE is a proprietary program from Siemens Oil & Gas Consulting.

Some monitoring tools have been modified to take condensation, salt deposition, pitting, and stress corrosion cracking (SCC) into consideration. New types of corrosion monitoring tools have been proposed in recent years, but most are expensive and require special sensors, precision electronics, software programs, and trained personnel for interpretation of the data obtained. They cannot be justified for routine use in most plants.

5.8 Read More

1. M. W. Jooston et al., "Internal Corrosion Monitoring of Subsea Production Flowlines – Probe Design and Testing," paper no. 77, CORROSION/98 (Houston, TX: NACE, 1998).
2. M. W. Jooston et al., "Internal Corrosion Monitoring of Subsea Oil and Gas Production Equipment," *MP* 34, 4 (1995).
3. R. J. Scanlan and R.M. Boothman, "Corrosion Monitoring Experience in the Refinery Industry Using the FSM Technique," paper no. 03655, CORROSION/2003 (Houston, TX: NACE, 2003).
4. B. Jack and C. White, "Sulfidation of Low Silicon Steels – An Inspection Challenge," presentation at API 2013 Inspection Summit (Galveston, TX: API, 2013).
5. A. S. Couper, "Minimizing Stress Corrosion Cracking - Testing Austenitic Stainless Steels for Modern Refinery Applications," *MP* 8, 10 (1969).
6. K. Wold and G. Sirnes, "FSM Technology -16 Years of Field History –Experience, Status and Further Developments," paper no. 07331, CORROSION/2007 (Houston, TX: NACE, 2007).
7. NACE Publication 1C184, "Hydrogen Permeation Measurement and Monitoring Technology" (Houston, TX: NACE).
8. R. Strommen et al., "The FSM Technology--Operational Experience and Improvements in Local Corrosion Analysis," paper no. 338, CORROSION/96 (Houston, TX: NACE, 1996).
9. S. R. Williams, "A New Non-Invasive Corrosion Monitoring System," paper no. 260, CORROSION/97 (Houston, TX: NACE, 1997).
10. A. F. Denzine and M.S. Reading, "An Improved, Rapid Corrosion Rate Measurement Technique For All Process Environments," paper no. 287, CORROSION/97 (Houston, TX: NACE, 1997).
11. A. F. Denzine and M.S. Reading, "A Critical Comparison Of Corrosion Monitoring Techniques Used In Industrial Applications," paper no. 288, CORROSION/97 (Houston, TX: NACE, 1997).
12. NACE Standard RP0497-2004, "Field Corrosion Evaluation Using Metallic Test Specimens," (Houston, TX: NACE, 2004).
13. D. Abayarathna and Ali Naraghi, "Overview of Hydrogen Permeation Measurement and Monitoring Techniques," paper no. 396, CORROSION/98 (Houston, TX: NACE, 1998).
14. NACE International Publication 3T199, "Techniques for Monitoring Corrosion and Related Parameters in Field Applications," (Houston, TX: NACE).
15. A. Jarragh et al., "The Value of Internal Corrosion Monitoring Activities in Asset Integrity Management," paper no. 2223, CORROSION/2013 (Houston, TX: NACE, 2013).
16. Lietai Yang et al., "Real-Time Monitoring of Carbon Steel Corrosion in Crude Oil and Brine Mixtures Using Coupled Multielectrode Sensors," paper no. 05293, CORROSION/2005 (Houston, TX: NACE, 2005).
17. NACE Standard TM0172-2001, "Determining Corrosive Properties of Cargoes in Petroleum Product Pipelines" (Houston, TX: NACE, 2001).
18. NACE Standard RP0189-2002, "On-Line Monitoring of Cooling Waters" (Houston, TX: NACE, 2002).
19. NACE Standard RP0192-98, "Monitoring Corrosion in Oil and Gas Production with Iron Counts (Houston, TX: NACE, 1998).
20. Jose Vera, "Applications of Hydrogen Permeation for Corrosion Assessment," paper no. 00461, CORROSION/2000 (Houston, TX: NACE, 2000).
21. R. D. Tems, F. W. H. Dean, "Field Application of a New, Portable, Non-Intrusive Hydrogen Monitor for Sour Service," paper no.00471, CORROSION/2000 (Houston, TX: NACE, 2000).
22. R. D. Tems, A. L. Lewis & A. I. Abdulhadi, "Field And Laboratory Measurements With Hydrogen Permeation Measurement Devices," paper no. 02345, CORROSION/2002 (Houston, TX: NACE, 2002).
23. O.Yepez, J.R.Vera, and R.Calarotti, "A Comparison Between an Electrochemical and a Vacuum Loss Technique as Hydrogen Probes For Corrosion Monitoring," paper no. 272, CORROSION/97 (Houston, TX: NACE, 1997).
24. Y. Onel, U. Ewert, and P. Willems, "Radiographic Wall Thickness Measurement of Pipes by a New Tomographic Algorithm," paper presented at the 15th World Conference on Non-Destructive Testing, Rome, Italy, October 2000.

6 High-Temperature Sulfidic Corrosion

In this Chapter

6.1 First Glance

Carbon steel and various Cr-Mo steels are subject to high-temperature sulfidic corrosion (HTSC) in the high-temperature circuits of crude and upgrader units at metal temperatures above approximately 230 °C (450 °F). HTSC is caused by organic sulfur compounds in crudes, various crude fractions, and most product streams. With carbon steel the corrosion product is a complex iron sulfide (Fe_xS_y) scale and with low-alloy Cr-Mo steels a mixed iron chromium molybdenum ($Fe_x/Cr_y/Mo_z$) sulfide scale. These scales are tightly adhering to the corroded metal surface and reduce long-term corrosion to acceptable levels. Scale stability and the degree of protection provided by the scale grow with increasing chromium content of the steel. Therefore the only practical way of controlling HTSC is by alloy upgrading. As a general rule, 12Cr stainless steel is fully resistant to HTSC.

6.2 Cause of High-Temperature Sulfidic Corrosion (HTSC)

Sometimes erroneously called sulfur (S) or hydrogen sulfide (H_2S) corrosion, HTSC is caused by certain organic sulfur compounds in crude, collectively known as reactive sulfur (RS) compounds. Reactive sulfur compounds include aliphatic sulfides (C_n-S-C_m), aliphatic disulfides (C_n-S-S-C_m), polysulfides (C_n-S_n-S-C_m), and mercaptans (C_n-SH). Dissolved elemental sulfur (S_{xn}) and hydrogen sulfide (H_2S) sometimes, but not always contribute to HTSC if present at sufficiently high concentrations. Noncorrosive, nonreactive sulfur (NRS) compounds consist of various thiophenes (aromatic 5-membered heterocyclic compounds with one sulfur atom). The corrosivity of different reactive sulfur groups depends on the specific types of compounds present. Corrosivity generally increases with metal temperature, but not necessarily at the same rate or by the same amount. In practice, HTSC is primarily a problem in units that charge crude containing over approximately 0.6% S and/or 0.2% RS.

Crude assays normally specify total sulfur content, and most plant laboratories are only equipped to determine total sulfur content of the crude mix charged to the unit. The following general observations apply, as far as the distribution of RS, NRS, and total S contents are concerned:

- While total sulfur content generally increases with the boiling point of different crude fractions, RS content generally decreases.
- The relative amounts of RS and NRS differ from crude to crude.
- The relative amounts of RS and NRS differ from fraction to fraction of given crude.

6.3 Corrosion Mechanism

HTSC is an electrochemical process that proceeds in the liquid phase and not in the vapor phase of hot crude and its fractions at the temperatures normally encountered in crude units and upgraders. It should not be confused with sulfidation by hot gases. Corrosion is thought to proceed in two steps: Dissociation of RS compounds catalytically generates hydrogen sulfide (H_2S) that, in turn, corrodes hot metal surfaces to form iron sulfide (Fe_xS_y) scale and hydrogen (H). For example, in the case of mercaptans [C-$(CH_2)_n$-SH], hydrogen sulfide is catalytically generated, as follows:

$$C\text{-}(CH_2)_n\text{-}SH \Rightarrow C\text{-}(CH_2)_{n\text{-}3} - C_2H_3 = CH_2 + H_2S\uparrow \qquad \text{(Eq.6-1)}$$

The catalytically generated H_2S then reacts with bare metal surfaces to form ferrous ions (Fe^{+2}) at anodic areas, and sulfide ions (S^{-2}) and atomic hydrogen (H_o) at cathodic areas. The overall reaction is formation of iron sulfide (Fe_xS_y) scale and evolution of molecular hydrogen (H_2). Initially, the electrolyte is entrained and dissolved salt brine in the crude. Expressed in most simplistic terms, the partial corrosion reactions are:

$$Fe \Rightarrow Fe^{+2} + 2\,e^- \quad \text{(at anodic areas)} \qquad \text{(Eq.6-2)}$$
$$H_2S + 2e^- \Rightarrow S^{-2} + 2H_0 \quad \text{(at cathodic areas)} \qquad \text{(Eq.6-3)}$$

The overall corrosion reaction is:

$$Fe + H_2S \Rightarrow Fe_xS_y\downarrow + H_2\uparrow \qquad \text{(Eq.6-4)}$$

Once iron sulfide scale has formed, it becomes the "electrolyte." Ferrous ions now form at anodic areas of the metal/scale interface, while hydrogen forms at cathodic areas of the scale/hydrocarbon interface. Only catalytically formed hydrogen sulfide participates in the corrosion reaction: Dissolved hydrogen sulfide in crude does not affect corrosion, unless present in amounts well over 37L/m^3 (0.5ft^3/100 gal). While iron sulfide scale has been represented simply as Fe_xS_y in Equation 6-4, sulfide scale usually consists of several layers of mixed iron, chromium, and molybdenum sulfides, depending on the process conditions and metallurgy involved. As a general rule, however, all sulfide scales exhibit parabolic growth behavior, i.e., an initially high corrosion rate diminishes as protective scale forms and grows thicker with time. The corrosion rate ($\Delta x/\Delta t$) is inversely proportional to scale thickness, as shown in Figure 6-1.

6.3.1 Scale Formation in Crudes with TAN below 0.3. These crudes contain essentially no naphthenic acids and form protective sulfide scales that reduce long-term corrosion to acceptable levels. Long-term corrosion rates (months, years?) are only a fraction of the short-term corrosion rates (days, weeks?) measured with corrosion probes. The thickness, volume, and stability of sulfide scales can vary significantly, depending on the process conditions. Some scales are thin, tightly adhering to the underlying metal, and difficult to detect visually because of their metallic-black color. Other scales are voluminous, porous, and form loose deposit layers. As a general rule, the thickness of scale formed by HTSC of carbon steel is 3-4 times the thickness of metal corroded away.

FIGURE 6-1 PARABOLIC KINETICS FOR HTSC OF STEELS

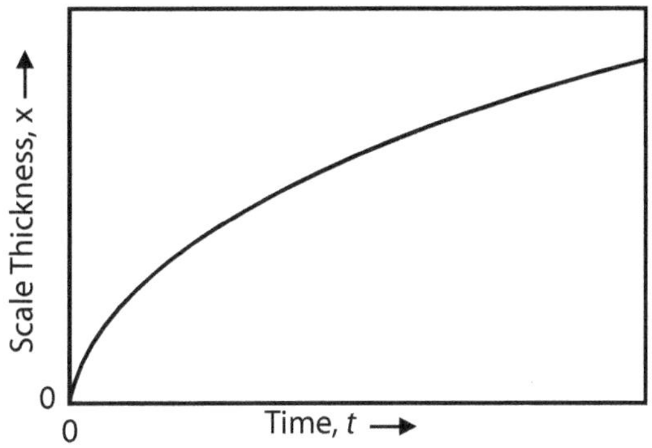

6.3.2 Scale Formation in Crudes with TAN above 0.3. These acidic crudes contain sufficient amounts of naphthenic acids to dissolve any protective sulfide scales and shift the principal corrosion mechanism to mixed HTSC/NAC and ultimately to NAC, as discussed in greater details in Chapter 7 of the *Guide.*

6.3.3 Factors Affecting Scale Stability. The stability of sulfide scales and, therefore, the degree of protection afforded against further corrosion, grows with increasing chromium (Cr) content of the steel. Carbon steel forms less protective iron sulfide scales compared to those formed by low-alloy Cr-Mo steels, such as 5Cr-0.5Mo. The latter steels form sulfide scales that are also highly resistant to turbulent flow conditions.

Flow velocity and, in the case of mixed liquid/vapor streams, the type of two-phase flow regime affect corrosion, because any protective scale is eventually loosened and removed by spalling as a result of mechanical and/or thermal stresses. Spray-type flow can be expected to do the most damage, while laminar flow the least. In most units, a minimum of 5Cr-0.5Mo steel is required for piping subjected to high flow rates and turbulence. Often, 9Cr-1Mo is required to provide good protection for transfer lines and fired heater tubes. With heater tubes, HTSC depends also on the magnitude of the heat flux through the tube wall. Corrosion is most severe on the side facing the burner flames, because it is subjected to intermittent wetting and drying by hot crude at the vapor/liquid interface.

In general, 12Cr stainless steel provides complete protection against HTSC, but not NAC. The latter requires Type 316 or Type 317 stainless steel for complete protection. Some 12Cr stainless steel chemistries can become subject to metallurgical changes at temperatures above approximately 400 °C (750 °F). Known as 475 °C (885 °F) embrittlement, the affected steels become brittle upon cooling to ambient temperatures below approximately 95 °C (200 °F) during a turnaround. While embrittlement can be reversed by subsequent annealing, many plants avoid using 12Cr stainless steel for heater tubes, return bends, or headers.

6.4 Components Subject to HTSC

The major crude unit and corresponding upgrader components subject to HTSC are listed in detail in the Chapter 4 of the *Guide.* They include primarily the following components operating at metal temperatures above approximately 230 °C (450 °F):

- The preheat section, including piping, hot preheat exchangers HPEs), fired heater tubes, outlet manifolds, transfer lines, lower columns, and lower pump-around loops;

- Some piping and components for the HAGO and LAGO sidestream columns;
- All piping for reduced crude (atmospheric resid), vacuum heater tubes, outlet manifolds, transfer lines, and lower vacuum columns and internals;
- Most piping and other components for the HVGO and LVGO pump-around loops; and
- Most piping and heat exchangers for vacuum resid (asphalt).

6.5 Susceptible Metals

Among steels, carbon steel is susceptible to HTSC at metal temperatures above approximately 230 °C (450 °F). Various low-alloy Cr-Mo steels containing less than 12% Cr become susceptible to HTSC at increasingly higher temperatures, depending on their chromium content. Copper and high-nickel alloys, such as Admiralty brass, Alloy 400, Copper Nickel 70-30, or Alloy 625 are used against overhead corrosion on many units, but should never be used to protect against HTSC. Even if their strength limit is not exceeded, these alloys should not be used, because no protective sulfide scales are formed on the copper or nickel constituent elements.

6.6 Resistant Metals

In many applications, metal selection depends primarily on past experience, relative cost, and availability on short notice. Usually, there is little opportunity or time after a failure to evaluate alternative metals. While carbon steel is still used extensively on many crude and upgrader units, high enough corrosion allowances (CA) should be specified because of its limited resistance to HTSC. Low-alloy Cr-Mo steels are the principal replacement material for carbon steel. The resistance of low-alloy Cr-Mo steels toward HTSC increases in direct proportion to their chromium content. For example, substituting 5Cr-0.5Mo steel for 2.25Cr-1Mo steel can be expected to reduce corrosion by 50% or more. Other metals used selectively to protect against HTSC depending on location include:

- 12Cr stainless steel despite the possibility of 475 °C (885 °F) embrittlement;
- Various aluminum alloys for vacuum column internals and packing, provided their strength limits are not exceeded;
- Alonized™ carbon and low-alloy Cr-Mo steels;
- Various austenitic stainless steels, including Types 316 and 317; and
- Alloy 20, Alloy C-276, and similar Ni-Cr-Mo alloys.

Silicon-killed carbon steel and 5Cr-0.5Mo piping containing over 0.15% Si present a special case, as far as resistance toward long-term HTSC is concerned. Based on the results of API and NACE surveys shown in Figure 6-2, approximately one-third of all HTSC piping failures between 1963 and 2003 (40 years) were attributed to what has become known as the "low-silicon issue." Most piping failures occurred in 25-50 year-old straight pipe segments made of ASTM A53 / API 5L steel that was not Si-killed and may or may not have had the 0.10% Si content or more needed to control long-term HTSC. As a result, some pipe segments corroded at much higher rates than other segments in the same circuit. In contrast, fittings such as ells and tees were usually made in accordance to ASTM A234 or A105 that call for 0.1% Si. Consequently, piping failures due to HTSC almost always occurred in straight segments.

FIGURE 6-2 CAUSES OF PIPE FAILURES BY HTSC BETWEEN 1963 AND 2003 (NACE AND API SURVEYS)

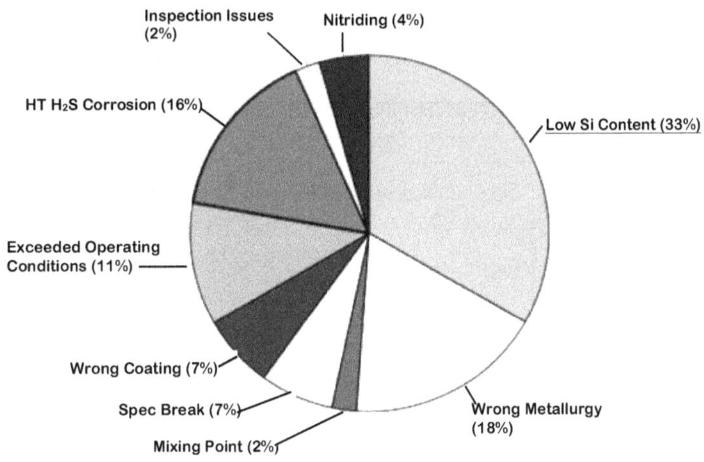

Adapted from B. Jack and C. White, "Sulfidation of Low Silicon Steels – An Inspection Challenge," presentation at API 2013 Inspection Summit (Galveston, TX: API, 2013).

As a general rule, recent construction calls for piping that is Si-killed and triple stamped to meet ASTM A53B/API 5LB/ASTM A106B and has an adequate Si content. Large diameter pipe and fittings made from plate may or may not have an adequate Si content, depending on whether fabricated from ASTM A516 plate or ASTM A285 plate. Based on reported past inspections, small pipe segments with low Si content can be expected to corrode at rates 2-10 times higher than pipe sections with high Si content. Unless repeat thickness measurements are located at an affected pipe section, thinning components are difficult to spot during repeat inspections. The photograph in Figure 6-3 clearly shows the beneficial effects of using silicon-killed steel with more than 0.15% Si.

FIGURE 6-3 HTSC OF CARBON STEEL PIPE SHOWING THE BENEFICIAL EFFECT OF Si

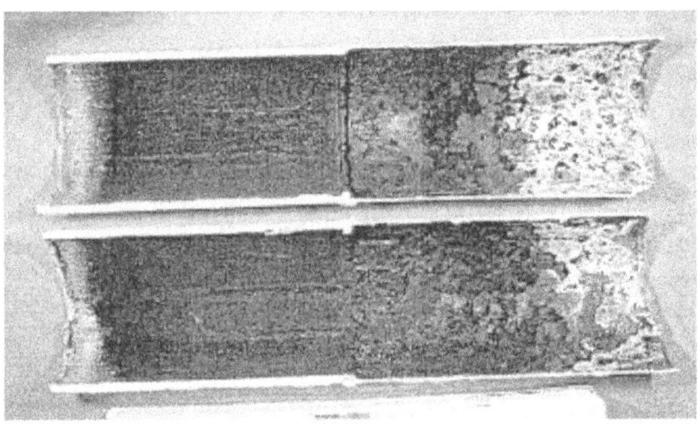

← Silicon-killed Steel →← Carbon Steel →

Figure 6-4 from API RP939-C shows the experiences reported by one refining company for A53B pipe with different Si contents operating in the same piping circuit and at approximately the same temperature. After decades of service, even minor differences in corrosion can make a big difference in the life of these components. While actual corrosion rates may be only 0.05-0.1mm/y (2-4mpy), piping is bound to spend its corrosion allowance over a 25-50-year period and likely fail by rupture.

FIGURE 6-4 EFFECT OF SI CONTENT ON HTSC OF CARBON STEEL PIPE AND FITTINGS IN FCC SLURRY PIPING AT 345-370 °C (650 -700 °F)

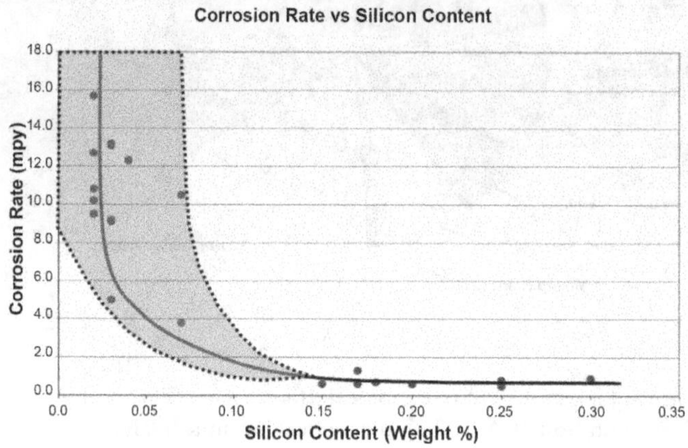

Adapted from *API Recommended Practice 939-C: Guidelines for Avoiding Sulfidation (Sulfidic) Corrosion Failures in Oil Refineries* (Washington, DC: API).

6.7 Tools for Predicting HTSC

Various curves correlating HTSC corrosion data for refinery steels at different operating temperatures have been proposed over the years. Among those, the McConomy Curves have found wide acceptance throughout the industry. While the McConomy curves, like other curves, are useful for estimating corrosion changes as a result of varying process conditions, all tend to be less reliable for predicting absolute corrosion rates.

6.7.1 McConomy Curves. The original McConomy Curves are based on data collected and correlated by H. F. McConomy of Atlantic Refining Company on behalf of the API Subcommittee on Corrosion in the early 1960s.[1] The Curves correlate field corrosion rates versus temperature for common refinery steels in crude and reduced crude streams containing 0.6% S. The McConomy Curves are also useful for establishing the maximum temperatures for various refinery steels in hydrocarbon services when reactive organic sulfur (RS) compounds are present. For example, assuming an initial 3mm (0.125in.) corrosion allowance, the maximum temperature for a 25 and 12.5-year service life, respectively, can be estimated from the appropriate curves, as listed in Table 6-1.

1. H. F. McConomy, "High Temperature Sulfidic Corrosion in Hydrogen- Free Environment," *Proc. API* 43, III (1963).

TABLE 6-1 MAXIMUM TEMPERATURE FOR STEELS IN HYDROCARBON SERVICES CONTAINING RS COMPOUNDS[2]

Steel	Maximum Metal Temperature in C (F) for 25-Year Life	12.5-Year Life
Carbon Steel	270 (520)	290 (550)
2.25Cr-1Mo	290 (550)	310 (600)
5Cr-0.5Mo	310 (600)	330 (625)
7Cr-0.5Mo	330 (625)	360 (680)
9Cr-1Mo	360 (680)	400 (750)

The original McConomy Curves were later found to be too conservative and subsequently modified by various petroleum companies in order to decrease the predicted corrosion rates by a factor of 1.5-2.5 or more. One example of the resulting modified McConomy Curves is shown in Figure 6-5 and has been found to predict HTSC by low-sulfur sweet crude reasonably well. Correction factors, shown in Figure 6-6, can then be applied for crudes with lower or higher amounts of sulfur (S). As can be seem, HTSC increases with temperature and corrosion reaches a maximum at approximately 425-450 °C (800-850 °F), depending on the steel. At higher temperatures, a large fraction of the reactive organic sulfur (RS) compounds has been converted to hydrogen sulfide (H_2S) and as a result, corrosion decreases. In addition, a more or less continuous coating or layer of tar and coke deposits on components prevents corrosion of the underlying metal, as seen in Figure 6-7.

Measured corrosion rates, including those on which the McConomy Curves are based, are usually reported in terms of process temperature instead of metal temperature, because metal temperature in most cases is not known[3]. As a result, fired heater tubes can be expected to corrode at higher rates than process piping at the same process temperature. Similarly, preheat exchanger (PHE) tubes with cold crude on the tube side and a hot product stream on the shell side can be expected to corrode at higher rates on the crude side than process piping.

FIGURE 6-5 MODIFIED MCCONOMY CURVES FOR PREDICTING HTSC OF COMMON REFINERY STEELS

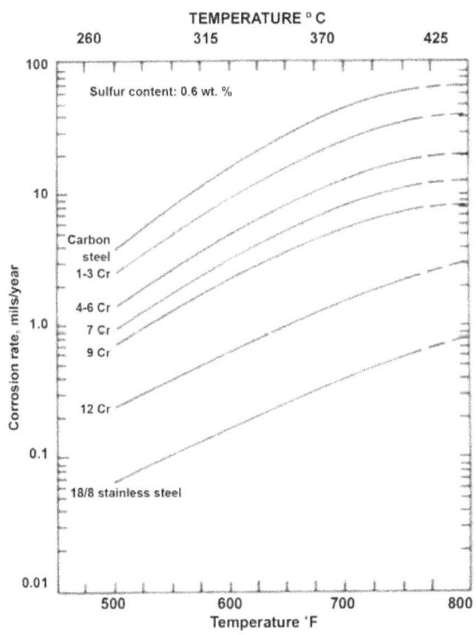

Adapted from J. Gutzeit, "High Temperature Sulfidic Corrosion of Steels," in *Process Industries Corrosion - The Theory and Practice* (Houston, TX: NACE, 1986).

2. Based on a 3mm (0.125in.) corrosion allowance.
3. As a starting point, assume metal temperatures of fired heater tubes are 85-110 °C (150-200 °F) degrees higher than the process temperature.

FIGURE 6-6 EFFECT OF SULFUR CONTENT ON HTSC BETWEEN 290 °C (550 °F) AND 400 °C (750 °F)

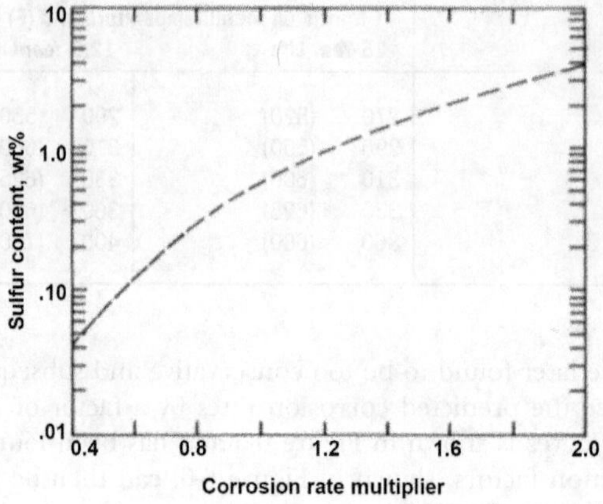

Adapted from J. Gutzeit, "High Temperature Sulfidic Corrosion of Steels," in *Process Industries Corrosion - The Theory and Practice* (Houston, TX: NACE, 1986).

FIGURE 6-7 PARTIALLY REMOVED TAR AND COKE LAYER ON BUBBLE CAP FROM BOTTOM TRAY OF VACUUM COLUMN AT 385 °C (725 °F)

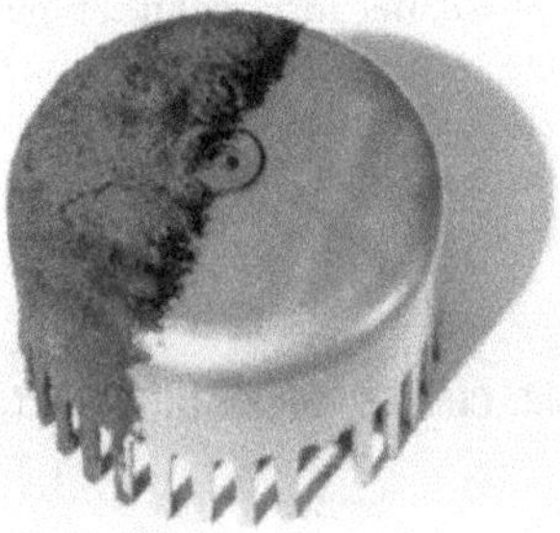

6.7.2 Reactive Sulfur Corrosion Curves. Besides the McConomy Curves, corrosion rate curves have been developed by various companies in an attempt to account for differences in corrosivity between different organic sulfur groups in crude. None of these have been published as such in the open literature, but some of the results can be inferred from personal communications, internal company reports, and various published reviews[4].

Typically, laboratory and field corrosion data are correlated, as follows:

$$CR = F\,[a \cdot C_{H2S} + b \cdot C_S + c \cdot C_{RSR} + d \cdot C_{RSH} + e \cdot C_{RSSR}] \qquad \text{(Eq.6-5)}$$

where:

CR = corrosion rate (mm/y or mpy)

F = coefficient depending on type of steel, nature of the hydrocarbon stream (vapor, liquid, or mixed flow), velocity, and degree of localized turbulence

C = concentration of different reactive sulfur groups (%).

a - e = coefficients depending on type of steel and metal temperature

The main issue with this and similar models is that the concentrations of the principal sulfur groups have to be obtained for each crude charged and for any other stream of concern. Most in-house laboratories are simply not equipped to obtain these values. Moreover, the relevant analytical methods are not necessarily standardized and there has been

4. Z. A. Foroulis, "Corrosion and Corrosion Inhibition in the Petroleum Industry," *Werkst. Korros* 33, 2 (1982).

considerable controversy over the accuracy of some of the methods used in the past. Another issue is determining the various coefficients, using both laboratory and field data. Keeping these limitations in mind, an appropriate approach is to determine only the total sulfur (S) and nonreactive sulfur (NRS) contents and then arrive at a value for the reactive sulfur (RS) content by difference. Laboratory and field corrosion rate data can then be correlated, as follows:

$$CR = a \cdot b \cdot c \cdot d \cdot C_{TRS} \quad \text{(Eq.6-6)}$$

where:

CR = corrosion rate (mm/y or mpy)
a = overall correlation coefficient
b = alloy coefficient
c = velocity and flow regime coefficient
d = temperature and concentration coefficient
C_{TRS} = total reactive sulfur concentration (in %)

While less complex than the previous model (Eq.6-5), obtaining the various coefficients from corrosion data can still present problems. An example of sulfur distribution data for four representative crudes charged at one refinery is shown below in Table 6-2. It is apparent that the reactive sulfur (RS) content represents only 30-50% of the total sulfur (S) content, indicating HTSC would be expected to be correspondingly lower than originally thought.

TABLE 6-2 SULFUR DISTRIBUTION IN FOUR REPRESENTATIVE CRUDES

Crude	Sulfur Content, wt.%		
	Total S	Reactive S	Non-Reactive S
CY Blend	2.42	0.73	1.69
SLC Blend	0.50	0.15	0.35
Landers	2.95	1.00	1.95
Trinidad	0.49	0.25	0.24

Similar data for some 17 domestic and international crudes (not specifically identified) are shown below in Figure 6-8. These data show that many high-sulfur crudes (for example, some Venezuelan crudes shown on the left side of the plot) contain only 0.2-0.3% RS, while others (for example, some West Texas crudes shown on the right side of the plot) contain as much as 1.1% RS. In practice, corrosion rates for carbon steel and the lower Cr-Mo steel grades predicted by the modified McConomy Curves tend to be significantly higher than those predicted by a reactive sulfur model. The reactive sulfur models also predict corrosion rates for as high as 540 °C (1000 °F), even though these temperatures are not encountered in crude units.

6.7.3 Predict-Crude 2.0™ Program. This software tool is based on a joint industry program (JIP) on naphthenic acid research supported by a number of designer, contractor, and petroleum companies and was subsequently expanded to include the effects of H_2S.[5] The original program is described in more detail in Chapter 7 of the *Guide*. Regarding crude oil corrosion, the program claims to be able to differentiate between HTSC, NAC, and mixed HTSC/NAC. The program is expensive to purchase and requires an annual subscription fee, and the principal users are the companies that participated financially in the original JIP. For others, various versions of the McConomy Curves and Reactive Sulfur Corrosion Curves are alternative options for predicting HTSC that have been found to be useful over the years. There appears to have been no independent evaluation of the Predict-Crude 2.0TM program by an unbiased party.

5. Many studies over the years have shown, however, that HTSC is not caused by H_2S but by organic sulfur compounds.

FIGURE 6-8 TOTAL SULFUR AND REACTIVE SULFUR CONTENT FOR VARIOUS DOMESTIC AND INTERNATIONAL CRUDES

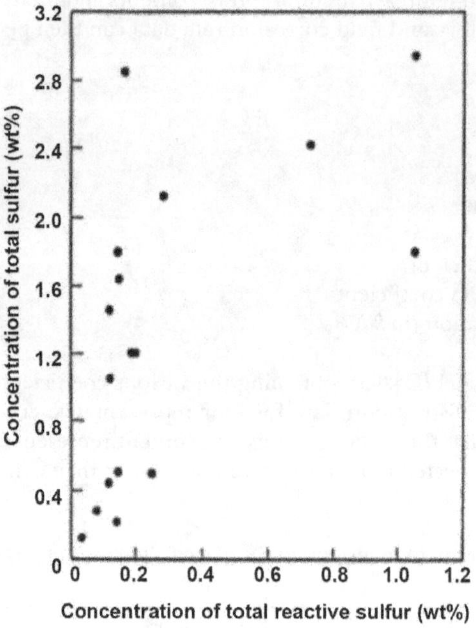

6.7.4 Other Tools for Predicting HTSC. A commercially available database ASSET™ is a prediction tool that includes information for over 50 commercial alloys and various forms of corrosion by hot corrosive gases at temperatures between 300 °C (575 °F) and 900 °C (1650 °F) and exposure times up to 6,000 hours. Aside from sulfidic corrosion, six additional corrosion mechanisms are covered: oxidation, carburization, sulfidation/oxidation, metal dusting, cyclic oxidation, and Cl_2/HCl corrosion. The ASSET™ database also includes the results of relevant joint industry programs, but clearly does not cover HTSC by organic sulfur compounds as it occurs in crudes. In practice, ASSET™ appears to be primarily a research tool and is of limited use for predicting HTSC in plant applications. The key conclusions for sulfidic corrosion by hot sulfur-containing gases are that corrosion follows a parabolic model after 1,000 hours, doubles with a 68 °C (122 °F) increase in temperature, and doubles with a 10-fold increase in H_2S content assuming all other conditions remain constant.

6.7.5 Effect of Flow Velocity. At liquid flow rates below 3-6m/s (10-20fps), HTSC of carbon steel was found to be independent of velocity. Droplet impingement[6] can occur, however, in mixed liquid/vapor streams at vapor flow rates above approximately 30m/s (100fps) and 60–65% vaporization. Depending on piping geometry, droplet impingement increases carbon steel corrosion by up to one order-of-magnitude at velocities in the 30-60m/sec (100-200fps) range and up to two orders-of-magnitude at velocities above 60m/s (200fps) in return bends, headers, heater tubes, elbows, tees, and at circumferential welds of transfer lines[7]. Corrosion of 12Cr stainless steel is not affected, but corrosion of various low-alloy Cr-Mo steels falls somewhere between these two extremes, as illustrated in Figure 6-9. By and large, the severity of corrosion in piping and heater tubes depends on the type of two-phase flow regime. Corrosion is most severe with spray and bubble-type flow patterns that cause droplet impingement at bends, elbows, tees and similar locations. In contrast, corrosion is least severe with flow patterns that cause smooth or annular flow patterns that protect the pipe walls with a liquid layer. The type of flow patterns depends on a number of complex variables, including the pipe direction (horizontal or vertical).

6. The onset and severity of damage by droplet impingement in many environments has been found to depend on the 4-5th power of velocity.
7. Z. A. Foroulis, "Method for Estimating the Sulfidation Corrosivity of Crude Oils and Their Fractions," *Standard Oil Company User's Manual*, Exxon R&E, 1976.

FIGURE 6-9 EFFECT OF VAPOR VELOCITY AND VAPOR FRACTION ON HTSC OF CRUDES IN HEATER TUBES AND TRANSFER LINES

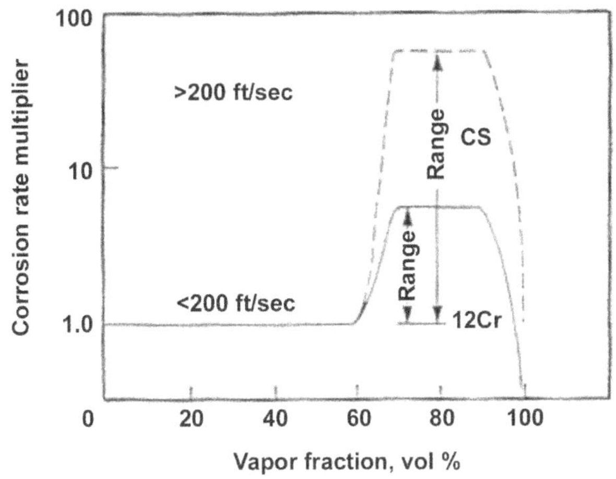

6.8 Appearance of Corroded Metal

Over time, HTSC can produce copious amounts of porous, brittle, dark-gray to black iron sulfide scale deposits, usually over a sublayer of thin, hard, black scale with a metallic sheen. The latter is often hard to distinguish from the underlying metal. Scale composition depends on process conditions and should be confirmed by chemical analysis, whenever possible. Iron sulfide scale deposits can be three to four times as thick as the metal corroded away. When loosened due to spalling from mechanical or thermal stresses, significant quantities of scale are likely to end up in downstream components and plug filters, trays, and tube bundles. Since long-term HTSC is determined, by and large, by the presence of protective sulfide scale, it stands to reason that anything that damages or destroys the scale accelerates corrosion. In this connection it should be noted that vapor flow velocities in vacuum transfer lines can reach 70-80% of sonic values and any suspended residual oil droplets at that speed can produce mechanical forces and effects similar to those produced by hydro-blasting.

Depending on the specific process conditions, metallurgy, and flow regimes, HTSC occurs in the form of uniform wall thinning, lopsided wall thinning, localized pitting, lake or island-type attack, erosion corrosion, or droplet impingement. Uniform wall thinning is the most prevalent damage mode seen on carbon and low-alloy Cr-Mo steel components, including piping and pressure vessel walls. Lopsided wall thinning is seen primarily in fired heater tubes, depending on the magnitude of the heat flux through the tube wall. In Figure 6-10, corrosion on the hot side of the radiant tube was accelerated by high turbulence at the vapor/liquid interface (the maximum heat flux on the hot side was estimated to have been over 200% of the average heat flux).

FIGURE 6-10 HTSC OF 125MM (5IN.) IPS CARBON STEEL CRUDE HEATER TUBE

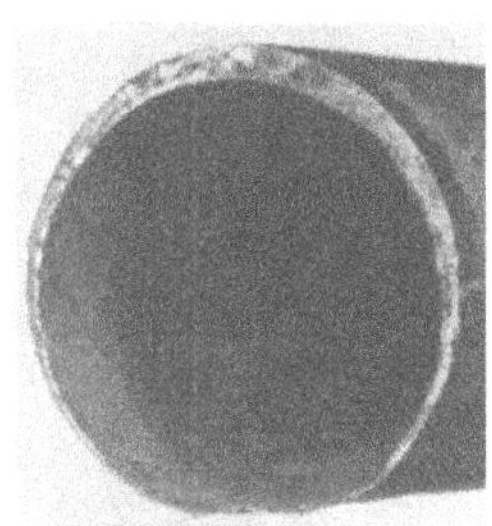

Adapted from *API Guide for Inspection of Refinery Equipment*, Chapter II, "Conditions Causing Deterioration or Failures," 2nd ed. (Washington, DC: API, 1973).

Sulfide scale and coke build-up, in combination with lopsided wall thinning, are often seen in fired heater tubes and depend on the magnitude of the heat flux through the tube wall. In Figure 6-11, excessive tube-metal temperatures due to flame impingement not only increased HTSC and sulfide scale formation, but also resulted in considerable coke build-up. The scale and coke deposits (shown in black) measured 13-32mm (0.5-1.25in.) on the fire side, but only 3mm (0.125in.) on the heater wall side. The tube eventually failed by stress-rupture at the thinnest wall section.

Erosion corrosion and droplet impingement are seen at elbows and tees of transfer lines and in fired-heater outlet tubes subjected to high flow velocities and turbulent two-phase flow regimes, as seen in Figure 6-12. The original wall thickness of the header was 12mm (0.5in.), including the external sleeve installed to build up the wall thickness. Both were eventually perforated by impingement attack at 425 °C (800 °F) and near-sonic velocities.

Lake or island-type attack is especially pronounced downstream of elbows, tees, and circumferential welds of transfer lines, as seen in Figure 6-13. This type of attack was initially blamed on naphthenic acid corrosion (NAC), but careful review of crude assays indicated that the crude mix charged contained only negligible amounts of naphthenic acids. The actual cause of the attack was a 20% increase in crude throughput to make up for gasoline shortages, which pushed vapor velocities well beyond 60m/s (200fps). General corrosion in the form of localized pitting occurs primarily on 12Cr stainless steel, such as column internals and shell cladding.

FIGURE 6-11 COKE BUILD-UP AND WALL THINNING OF 140MM (5.5IN.) IPS x14MM (0.55IN.) WALL 5CR-0.5MO CRUDE HEATER TUBE

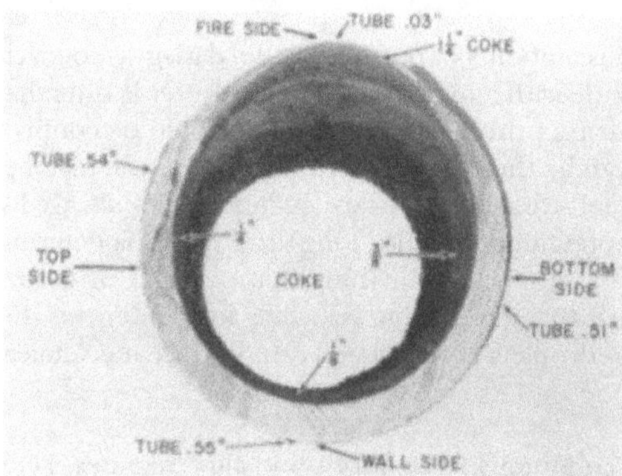

FIGURE 6-12 HTSC OF SLEEVED 500MM (20IN.) IPS 5Cr-0.5Mo VACUUM HEATER OUTLET HEADER

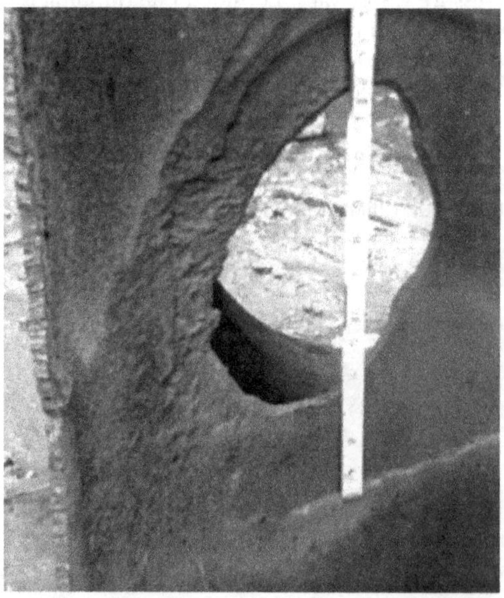

FIGURE 6-13 HTSC OF 460MM (18IN.) IPS CARBON STEEL TRANSFER LINE DOWNSTREAM OF CIRCUMFERENTIAL WELD BEAD

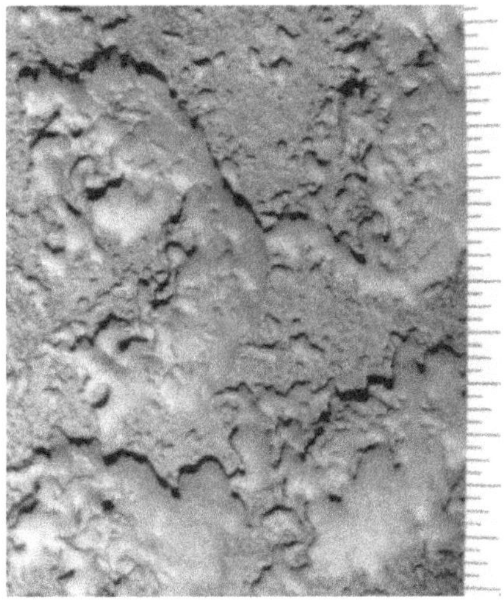

6.9 Location of Corrosion Damage

HTSC occurs on metal surfaces exposed to turbulent liquid and/or mixed liquid/vapor streams, but not to all-vapor streams. Corrosion occurs in areas where the normally protective sulfide scale was removed by high liquid/vapor velocities and/or turbulence, for example at elbows, tees, and protruding circumferential weld beads in piping and transfer lines. The outer (long) radius) of pipe elbows is especially prone to accelerated attack. Accelerated attack also occurs in fired heater tubes on the side facing the flames, in tube bends or header boxes of fired heaters, in flash zones and at downcomers of distillation columns, and at entrance and discharge nozzles of heat exchangers and condensers.

6.10 Inspection Methods

The principal inspection methods include visual inspection of components during shutdown, onstream and offstream UT thickness measurements of vessel and piping walls depending on temperature, and onstream or offstream radiography of vessel and piping walls.

6.11 Similar Corrosion Mechanisms and Damage

HTSC should not be confused with NAC that can produce similar corrosion damage, as discussed in detail in Chapter 7 of the *Guide*. The corrosion mechanism of HTSC should also not be confused with high-temperature H_2S/H_2 corrosion that occurs in various hydrotreating processes.

6.12 Repair Methods

When repairs are required, the modified McConomy curves can be used to determine the minimum chromium (Cr) content necessary for long-term protection. Unless corrosion rates are low, corroded carbon steel components should be upgraded with low-alloy Cr-Mo steel components, such as 5Cr-0.5Mo. Often, carbon steel or the wrong grade of low-alloy Cr-Mo steel is supplied, because of mix-ups during shipping, warehousing, and handling.[8] Therefore, replacement steels should be analyzed for proper composition using an x-ray fluorescent spectrograph that offers a combina-

8. During construction of a hydrotreating unit, approximately 30% of incoming materials were found to be the wrong steel or otherwise failed to meet specifications and had to be returned to the supplier for exchange.

tion of both alloy identification and elemental analysis. Large surface areas exposed to high flow turbulence such as vessel walls should be clad or weld overlaid with 12Cr or 300 series stainless steel for improved corrosion resistance. Corroded piping should be externally sleeved "in kind" or preferably with 12Cr or 300 series stainless steel, until it can be replaced during the next shutdown.

6.13 Precautions When Inspecting Corroded Components

A number of precautions apply when components are first opened for inspection or maintenance. Components containing sulfide scale evolve poisonous hydrogen sulfide (H_2S) gas when subjected to cleaning solutions. H_2S gas is heavier than air and can flow out of the bottom manholes of a column with deadly consequences when unprotected personnel try to enter for shutdown inspection.[9] Some iron sulfide scales are pyrophoric and ignite when exposed to air and allowed to dry out during shutdown. To avoid this issue, components with sulfide scale should be routinely water washed and kept wet prior to follow-up cleaning. Alternatively, sulfide scales can be oxidized with a spray of hot alkaline potassium permanganate ($KMnO_4$) solution.[10] Components made from 300 series stainless steels, such as some heat exchanger tubes or pressure vessel internals, may chloride stress corrosion crack (CSCC). Cracking occurs at temperatures above approximately 65 °C (150 °F) when components are first opened to the atmosphere and lying on the ground in the sun and waiting to be cleaned. CSCC is not a problem during normal operations, since air is excluded. Prior to, during, and after cleaning, shutdown precautions in accordance to NACE RP0170 can be used to prevent CSCC of 300 series stainless steels. As a general rule, polythionic acid stress corrosion cracking (PASCC) of austenitic stainless steels has not been a problem with crude units and upgraders.

6.14 Chapter Summary

HTSC of carbon steel and various low-alloy Cr-Mo steels occurs in the high-temperature circuits of crude units and upgraders operating above approximately 230 °C (450 °F). Depending on the steel, the corrosion products are complex iron sulfide or iron/chromium/molybdenum sulfide scales. Sometimes referred to as sulfur or hydrogen sulfide corrosion, HTSC is actually caused by organic sulfur compounds dissolved in the crude. This type of corrosion occurs primarily in units charging crudes containing more than 0.6% sulfur. HTSC should not be confused with high-temperature H_2S/H_2 corrosion found in various hydrotreating processes.

HTSC is an electrochemical process that occurs in the liquid phase and is caused by catalytically generated hydrogen sulfide (H_2S). The resulting sulfide scales become the "electrolyte" for the reaction. Initially high corrosion rates diminish over time as more protective scales form. HTSC increases with temperature and occurs primarily on metal surfaces exposed to liquid or mixed liquid/vapor streams instead of all-vapor streams. Corrosion occurs at areas where the protective sulfide scale is damaged or removed by high flow turbulence. Low-alloy Cr-Mo steels form more protective sulfide scales compared to carbon steel. The benefits of low-alloy Cr-Mo steels increase in direct proportion to their chromium (Cr) content and these steels are the principal upgrade option for carbon steel on many units. At a minimum, 5Cr-0.5Mo steel is required for piping subjected to high flow rates. The rate of HTSC of various steels can be reasonably well predicted with the Modified McConomy Curves. However, these curves are best used for estimating changes in corrosion rate as a result of changing process conditions, rather than for predicting absolute corrosion rates.

9. At one refinery, an inspector passed out upon trying to enter a bottom manhole without a self-contained breathing apparatus (SCBA). A second inspector, also without a SCBA, passed out on the access platform while attempting to pull out his coworker. Both men died despite all efforts by the refinery metallurgist who wore his SCBA, grabbed them, and hauled them off to the refinery emergency station on the back of his pickup truck.

10. J. Gutzeit, "Cleaning of Process Equipment and Piping," MTI publication no. 51 (St. Louis, MO: Materials Technology Institute of the Chemical Process Industries, 1997).

6.15 Read More

1. API Division of Refining, "High-Temperature Crude Oil Corrosivity Studies, publication 943 (Washington, DC: API, 1974).
2. H. A. Cataldi, K. J. Askevold, and A. F. Harnsberger, "Estimation of Corrosivity of Crude Oils," *Proc. API* 33, III (1953).
3. A. S. Couper and A. Dravnieks, "High Temperature Corrosion by Catalytically Formed Hydrogen Sulfide," *Corrosion* 18, 8 (1962).
4. A. S. Couper, "High Temperature Mercaptan Corrosion of Steels," *Corrosion* 19, 11 (1963).
5. Z. A. Foroulis, "High Temperature Degradation of Structural Materials in Environments Encountered in the Petroleum and Petrochemical Industries: Some Mechanistic Observations," *Anti-Corrosion* 12, 11 (1985).
6. H. F. McConomy, "High Temperature Sulfidic Corrosion in Hydrogen- Free Environment," *Proc. API* 43, III (1963).
7. J. Gutzeit, "High Temperature Sulfidic Corrosion of Steels" in *Process Industries Corrosion - The Theory and Practice*, B.J. Moniz, W.I. Pollock, eds. (Houston, TX: NACE, 1986).
8. C. H. Hudgins, Jr., "Review of Sulfide Corrosion Issues in the Petroleum Industry," *MP* 8, 1 (1969).
9. NACE Task Group 176, "Overview of Sulfidic Corrosion in Petroleum Refining," publication 34103 (Houston, TX: NACE, February 2004).
10. R. L. Piehl, "Correlation of Corrosion in a Crude Distillation Unit with Chemistry of the Crudes," *Corrosion* 16, 6 (1960).
11. G. R. Port, "Hydrogen Sulfide Corrosion in a Distilling Unit," *Proc. API* 41, III (1961).
12. J. E. Slater et al., "High-Temperature Crude Oil Corrosivity Studies," paper no. 1, CORROSION/75 (Houston, TX: NACE, 1975).
13. C. Claesen and S. A. Lordo, "Sulphidic Corrosion in Lab Evaluations and Petroleum Refinery Applications," EUROCORR 2007.
14. API Recommended Practice 939-C, "Guidelines for Avoiding Sulfidation (Sulfidic) Corrosion Failures in Oil Refineries" (Washington, DC: API, 2008).
15. B. Jack and C. White, "Sulfidation of Low Silicon Steels – An Inspection Challenge," presentation at API 2013 Inspection Summit (Galveston, TX: API, 2013).
16. D. Farrell and L. Roberts, "A Study Of High Temperature Sulfidation Under Actual Process Conditions," paper no. 10358, CORROSION/2010 (Houston, TX: NACE, 2010).
17. J. de Jong et al., "Effect of Mercaptans and Other Organic Sulfur Species on High Temperature Corrosion in Crude and Condensate Distillation Units," paper no. 07565, CORROSION/2007 (Houston, TX: NACE, 2007).
18. R. A. Mottram and J. T. Hathway, "Some Experiences in the Corrosion of a Crude Oil Distillation Unit Operating with Low Sulphur North African Crudes," paper no. 39, CORROSION/71 (Houston, TX: NACE, 1971).

7 Naphthenic Acid Corrosion

In This Chapter

7.1 First Glance

Carbon steel, low-alloy Cr-Mo steels, and 12Cr stainless steel are subject to naphthenic acid corrosion (NAC) in the high-temperature circuits of crude and upgrader units at metal temperatures above approximately 230 °C (450 °F). NAC corrosion is caused by naphthenic acids in certain crudes, particularly crudes from production wells and fields in California, South America, Norway, Russia, West Africa, China, and Indonesia. Heavy crudes from geologically young formations have the highest acid content, while paraffinic crude has the lowest acid content. Naphthenic acids are also a major contaminant of bitumen from the Athabasca oil sands in Alberta, Canada. The only practical way of controlling NAC is by crude blending and alloy upgrading even though corrosion inhibitors have been used under some specific conditions.

7.2 Cause of Naphthenic Acid Corrosion (NAC)

Despite ongoing research into all aspects of NAC during the past 40 years, no comprehensive model of NAC has been developed. What is known is that nearly all crudes contain some amount of naphthenic acids, but usually at levels below those considered to be corrosive. At sufficiently high levels, however, naphthenic acids will attack 12Cr stainless steel and even austenitic 300 series stainless steels unless the latter contain at least 3% molybdenum (Mo). In contrast to HTSC, NAC forms only soluble corrosion products and no protective scales. As a result, NAC does not decrease with time, or with increasing chromium (Cr) content in low-alloy Cr-Mo steels. Moreover, NAC can take on various forms that have resulted in false and misleading reports on the root cause of corrosion seen in many components. For example, naphthenic acids are known to have attacked:

- Heater tubes and transfer lines in the form of deep localized attack and scoop-like grooving due to velocity effects;
- 12Cr stainless steel cladding in the flash zone of atmospheric columns in the form of metal thinning, grooving, and pitting due to turbulence;
- 12Cr and 300 series stainless steel cladding of vacuum columns in the form of metal thinning, uniform etching, or pinhead type pitting due to acid condensation; and
- 12Cr and 300 series stainless steel bubble caps and the underside of bubble-cap trays in vacuum columns in the form of large, round, and deep pits due to acid condensation.

7.2.1 Types of Naphthenic Acids in Crude Receipts. The common factor in all cases of naphthenic acid corrosion is a high concentration of certain organic acids in crude receipts. The acids known as naphthenic acids are a natural component of crude, i.e., they are not a result of crude extraction, treating, or distilling operations. As shown in Figure 7-1, they are found primarily in distillate and lube oil fractions with BP temperatures below 350-450 °C (660-840 °F) of some crudes. None are found in gasoline fractions. Naphthenic acid content is usually reported in terms of a neutralization number or total acid number (TAN). All crudes contain some amount of naphthenic acids, typically corresponding to a TAN below 0.3. In contrast, problem acidic crudes are crudes with TAN above 0.5, often with TAN ranging to 6 or higher. The simplest form of naphthenic acid is cyclopentane acetic acid. The general structure of naphthenic acids is R- $(CH_2)_n$ –COOH, where:

R = cyclic nucleus of 1 to 6 cyclopentane, cyclohexane, or benzene rings.
n = 1 - 5 (or greater).

FIGURE 7-1 DISTRIBUTION OF NAPHTHENIC ACIDS IN FRACTIONS OF FOUR CRUDES

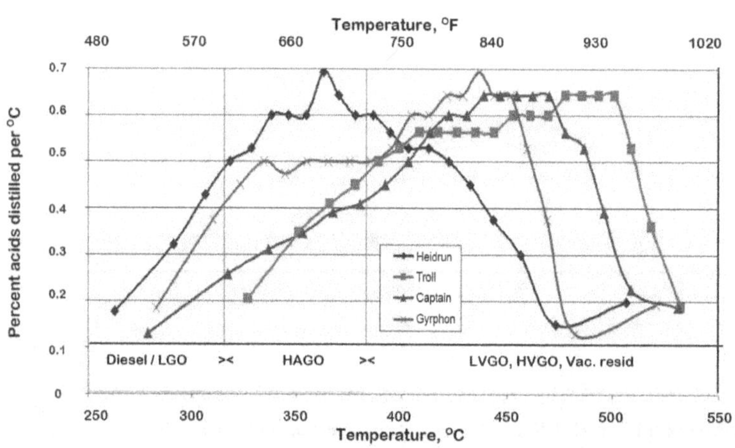

Adapted from T. Lu, "Challenges Faced in High TAN Crude Processing," Presentation at Asia Pacific International Conference 2012 (New Delhi, India: 2012).

Commercial forms of naphthenic acids have a TAN between 160 and 230 and are oily liquids or waxy solids. Examples of the individual acids identified in crudes are shown in Table 7-1 and range from hexanoic acid ($C_5H_{11}COOH$) with a BP of 205 °C (401 °F) to stearic acid ($C_{17}H_{35}CO_2H$) with BP of 361 °C (682 °F).

TABLE 7-1 EXAMPLES OF NAPHTHENIC ACIDS FOUND IN CRUDES

Name	Molecular Weight
Hexanoic acid	116.16
Benzoic acid	122.12
Cyclohexane carboxylic acid	128.17
1,2,3,4-Tetrahydro-2-naphthoic acid	148.20
3-Cyclohexanepropionic acid	156.22
Cyclohexanebutyric acid	170.25
1-Naphthoic acid	172.18
Stearic acid	284.48

7.2.2 Definition of TAN. By definition, TAN represents the number of milligrams of potassium hydroxide (KOH) required to neutralize all acids other than strong acids in a 1-gram sample of crude, crude fraction, or other hydrocarbon product stream. Strong acids, primarily hydrochloric acid (HCl) and sulfuric acid (H_2SO_4), are water soluble and are extracted from the sample prior to the analysis. Depending on the color of the sample, two analytical methods are commonly used to determine TAN: colorimetric analysis in accordance with ASTM D974 or UOP[1] Method 687 and potentiometric analysis in accordance with ASTM D664 or UOP Method 565. Assuming an average molecular weight of 200 for the acids, crude with a TAN of 1.0 contains approximately 3,500 mg/L (ppm) acids. It should be noted that TAN is not an accurate measure of the naphthenic acid content of crude. Aside from the true naphthenic acids, various other carboxylic acids, mercaptans, hydrolysable esters, phenols, lactones, acid nitrogen compounds, as well as certain additives and corrosion inhibitors used during upstream and midstream operations can affect the magnitude of TAN. As a result, there is only an approximate relationship between TAN and naphthenic acid content, as shown in Figure 7-2 below.

FIGURE 7-2 RELATIONSHIP BETWEEN TAN AND NAPHTHENIC ACID CONTENT OF CRUDES AND CRUDE FRACTIONS

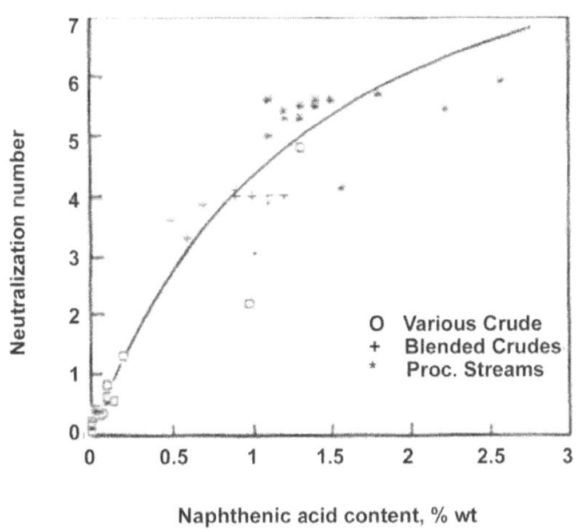

Adapted from E. Babaian-Kibala et al., "Naphthenic Acid Corrosion in Refinery Settings," *MP* 32, 4 (1993).

1. UOP LLC, formerly known as Universal Oil Products.

7.2.3 TAN, NAN, and SCAN. In an effort to improve the relationship between TAN and naphthenic acid content, at least one process additives supplier uses a naphthenic acid number (NAN) test that is more specific for active naphthenic acids compared to the TAN test. In the NAN test, the water-soluble strong acids are first extracted from the sample as is done in the TAN test, but followed by preferential solvent extraction of other organic acids and esters prior to the analysis for active naphthenic acids. The sample is then titrated with KOH to yield the NAN value, which is thought to be a more accurate assessment of the potential corrosivity of the fluid. Typically, reported NAN values are 40–70% of TAN values for API 20 whole crude and 50-80% for various unidentified product streams[2] of a small 1,000T/d (7,000 BPD) specialty refinery producing naphthenic acid based lubricating oils. Another process additives supplier uses a specific carboxylic acid number or SCAN that too is supposed to be a better predictor of naphthenic acid corrosion than TAN. In the SCAN test, only the acidity from carboxylic acids is measured: Other acids, esters, and additives in crudes are excluded. Typically, reported SCAN values are 40-50% of TAN values for whole crude and 60-80% for various gas oil streams.

While the difference between TAN, NAN, and SCAN appears to be fairly consistent for a large number of crudes, it was not necessarily so for any single crude. All in all, using TAN rather than NAN or SCAN has the advantage in that most historical data on naphthenic acid corrosion were reported in terms of TAN and can be used for comparison. Also, using the whole crude TAN instead of the TAN of individual product streams remains the best approach, because of the limited amount of data available for individual product streams.

7.2.4 Relative Corrosiveness of Naphthenic Acids. Not all naphthenic acids are equally corrosive, but the differences between weaker and stronger naphthenic acids have yet to be clearly defined. Despite claims to the contrary and various alternative definitions for naphthenic acid content as noted above, NAC does not correlate well with the naphthenic acid content of various process streams. In general, heavy cuts have a high acid content of less aggressive acids, while light cuts have a low acid content of more aggressive acids. In other words, the higher-boiling naphthenic acids in the vacuum section tend to be more corrosive than the lower-boiling naphthenic acids in the atmospheric section. For the most part, these differences can probably be attributed to differences in temperature, vapor velocities, and flow regimes. Also, the higher concentrations of reactive sulfur compounds (RS) in the atmospheric section compared to the vacuum section can be expected to produce protective iron sulfide scales that counteract NAC in the atmospheric section.

7.3 Corrosion Mechanism

Like many other corrosion processes, NAC is an electrochemical process that occurs in the liquid instead of the vapor phase of crude and its fractions at the temperatures normally encountered in crude units and upgraders. Most likely, several mechanisms apply to NAC, depending in part on the relative amounts and types of naphthenic acids and reactive sulfur (RS) compounds present. Some crudes from California and China have unusually high naphthenic acid contents (whole crude TAN up to 6), but relatively low concentrations of RS. These crudes are highly corrosive, because insufficient sulfide scale is produced to counteract NAC. In contrast, some crudes from Venezuela have intermediate naphthenic acid contents (whole crude TAN below 3), but relatively high concentrations of RS. These crudes are found to be less corrosive, because sufficient sulfide scale forms to counteract NAC.

7.3.1 Corrosion at Low Sulfur Concentrations. NAC occurs at anodic areas of the metal and hydrogen evolves at cathodic areas once any preexisting sulfide scale from prior runs is removed by high flow turbulence or other factors. The electrolyte is brine entrained and dissolved in the crude. Expressed in most simplistic terms, the partial corrosion reactions are:

$$Fe \Rightarrow Fe^{+2} + 2\,e^{-} \text{ (at anodic areas)} \qquad \text{(Eq.7-1)}$$

$$\begin{aligned} &2\,R\text{-}(CH_2)_n\text{-}COOH + 2e^{-} \Rightarrow \\ &2R\text{-}(CH_2)_n\text{-}COO^{-} + 2H_0 \Rightarrow \\ &2R\text{-}(CH_2)_n\text{-}COO^{-} + H_2\uparrow \text{ (at cathodic areas)} \end{aligned} \qquad \text{(Eq.7-2)}$$

2. T. Jackson, M. C. Winslow, and M. Wilson, "Prolonged Experience Processing High Acid Crude," Cross Oil & Refining Company, case history by GE Water & Process Technologies.

The overall corrosion reaction is:

$$Fe + 2R\text{-}(CH_2)_n\text{-}COOH \Rightarrow [R\text{-}(CH_2)_n\text{-}COO]_2\,Fe + H_2\uparrow \qquad \text{(Eq.7-3)}$$

At low sulfur concentrations, the kinetics of NAC is largely controlled by diffusion processes, such as flow velocity or turbulence. No protective sulfide scale is formed and corrosion does not decrease with time. NAC is also not reduced by increasing the chromium content of the steel, because the benefits of low-alloy Cr-Mo steels are the result of Fe/Cr/Mo sulfide scale formation.

7.3.2 Corrosion at High Sulfur Concentrations. Limited evidence suggests that a corrosion continuum exists in crude units at elevated temperatures, ranging from HTSC on one hand to NAC on the other. A mixed corrosion mechanism would be expected to apply under intermediate conditions. The ratio "total metal loss / metal loss as sulfide scale" is sometimes taken as a semiquantitative indicator of the prevailing corrosion reaction. If this mechanism is correct, an increase in naphthenic acid content of a given crude slate can be expected to remove previously formed sulfide scale and increase corrosion in the high-temperature circuits. In the case of carbon steel, the overall corrosion reaction now becomes:

$$FeS\downarrow + 2\,R\text{-}(CH_2)_n\text{-}COOH \Rightarrow [R\text{-}(CH_2)_n\text{-}COO]_2\,Fe + H_2S\uparrow \qquad \text{(Eq.7-4)}$$

In contrast, an increase in the sulfur content of given crude can be expected to produce protective sulfide scales where none existed before and thus decrease corrosion in the high-temperature circuits. In the case of carbon steel, the overall corrosion reaction now becomes the reverse of that described by Eq.7-4:

$$[R\text{-}(CH_2)_n\text{-}COO]_2\,Fe + H_2S \Rightarrow FeS\downarrow + 2R\text{-}(CH_2)_n\text{-}COOH \qquad \text{(Eq.7-5)}$$

In laboratory tests, the two opposing reactions were defined in terms of a naphthenic acid corrosion index (NAC.I.), which is simply the ratio of the overall corrosion rate (in mpy) to the amount of corrosion scale (in mg/cm^2) formed. The guideline values shown in Table 7-2 were found to apply[3]:

TABLE 7-2 PREDICTED CORROSION MECHANISM IN TERMS OF NAPHTHENIC ACID CORROSION INDEX (NAC.I.)

Naphthenic Acid Corrosion Index (NAC. I.)	Predominant Corrosion Mechanism
< 10	HTSC
10 - 100	Mixed Mechanism (NAC partially inhibited by HTSC)
> 100	NAC

In practice, experience on at least one crude unit has shown that corrosion in the high-temperature circuits was reduced when the sulfur content (in % S) reached approximately three times the TAN value of the crude slate. It is important to note, however, that increasing the sulfur content may decrease NAC, but at the same time is likely to increase HTSC of insufficiently alloyed components. For example, components fabricated from carbon steel or one of the lower-alloy Cr-Mo steels, such as 2.25Cr-1Mo, are likely to become attacked and may have to be upgraded to one of the higher-alloy Cr-Mo steels, such as 5Cr-0.5Mo or 9Cr-1Mo to prevent HTSC.

A mixed corrosion mechanism implies that both maximum and minimum allowable sulfur levels for a given naphthenic acid content and vice versa depend on the specific metallurgy involved. It also implies that some forms of NAC can be reduced by increasing the chromium (Cr) content of low-alloy Cr-Mo steels. This seems to be contrary to conventional wisdom, but has been shown to apply to corrosion in crude transfer lines and elsewhere when certain Venezuelan

3. H. L Craig, "Naphthenic Acid Corrosion in the Refinery," paper no. 333, Corrosion/95 (Houston, TX: NACE, 1995).

crudes were charged[4]. Depending on temperatures, flow conditions, and metallurgy of the various components involved, a mixed corrosion mechanism is likely to affect the severity, but not the characteristics of NAC.

7.4 Components Subject to NAC

The major crude unit and corresponding upgrader components subject to NAC are listed in detail in Chapter 4 of the *Guide.* They include primarily the following circuits and components operating at metal temperatures above approximately 230 °C (450 °F):

Atmospheric section and upgrader DRU:

- Preheat section, including piping, preheat exchangers, fired heater tubes, outlet manifolds, transfer lines, lower columns, inlet flash zone and some internals of the lower crude column, and the lower pump-around loop(s).
- Inlet flash zone and the lower column of the sidestream strippers and some piping.

Vacuum section and upgrader VDU:

- Piping for reduced crude, vacuum heater tubes, outlet manifolds, transfer lines, and intermediate shell and internals of the vacuum column. Piping and other components for the HVGO and LVGO pump-around loops.
- Demister pad, trays, and other internals in the upper vacuum column above 200 °C (390 °F).
- Bubble cap trays and packed beds with random or structured packing operating in the 290 °C (550 °F) to 290 °C (650 °F) temperature range.
- Piping and heat exchangers for vacuum resid (asphalt).

In addition to NAC, some overhead corrosion problems in both sections have been attributed to acetic acid (CH_3COOH), formic acid (H-COOH), or proprionic acid (CH_3CH_2-COOH) that were formed as a result of naphthenic acid breakdown in the fired heaters.

7.5 Susceptible Metals

Among steels, carbon steel and various low-alloy Cr-Mo steels up to and including 12Cr stainless steel are susceptible to NAC at metal temperatures above approximately 230 °C (450 °F). Copper and copper nickel alloys, such as Admiralty brass, naval brass, Alloy 400, or Copper Nickel Alloy 70-30, while widely used against overhead and cooling water corrosion on many units, should never be used to upgrade steel against NAC.

7.6 Resistant Metals

In many applications, metal selection depends primarily on past experience, relative cost, and availability on short notice. Usually, there is little opportunity or time during a turnaround or after a failure to evaluate alternative metals. While carbon steel is still used extensively on many units, high enough corrosion allowances (CA) should be specified, because of its very limited resistance to NAC. There are a few applications where the higher-chromium grades including 7Cr and 9Cr have shown reasonably good resistance to NAC. The same applies to 12Cr stainless steel. While these steels would no longer be specified for new construction or even as replacement materials, they are often left in place until they can be upgraded at the next opportune time. Other metals that have been used selectively against NAC include:

- Various aluminum alloys for tray components, structured packed beds, and demister pads in the upper part of the vacuum column;
- Alloy 20;
- Various austenitic stainless steels, including Types 316 and 317; and
- Alloy C-276 and similar Ni-Cr-Mo alloys.

In general, Type 316 stainless steel having the maximum required 3.0% Mo content of the specification range, or Type 317 stainless steel is required in severe cases of NAC in components of the vacuum section. Alonized™ steel would

4. B.E. Hopkinson and L. E. Penuela, "Naphthenic Acid Corrosion by Venezuelan Crudes," paper no. 502, CORROSION/97 (Houston, TX: NACE, 1997).

be expected to be resistant to NAC, but in practice the very thin Al-Fe layer has been found to be inadequate for that purpose.

7.7 Tools for Predicting NAC

NAC occurs in at least three distinctly different forms, each of which has its own set of characteristics and potential remedies: 1. General corrosion at relatively low rates in piping and on vessel internals; 2. Velocity-accelerated corrosion in furnace tubes and transfer lines; and 3. Corrosion by condensing acids on trays and shells of vacuum columns. As a general rule, the refining industry has predicted the corrosivity of crudes based primarily on prior experience (85%), sulfur content and TAN of crude fractions (70%), and whole crude TAN (62%).[5] Except for the Predict2™ software program, no corrosion data are available for predicting when, where, and to what extent the different forms of NAC is likely to occur. The following general guidelines list the type of crude blends and general process conditions that are known to cause NAC:

- With crude blends having a whole crude TAN between 0.5 and 1.0, NAC can be expected to occur in certain components depending on the organic sulfur content. As a general rule, NAC is not a problem in the atmospheric section and in the upgrader DRU. Severe corrosion can be expected, however, in the vacuum section and upgrader VDU where the heater outlet tubes, U-bends, outlet headers, and transfer line are attacked. The 12Cr trays in the vacuum column corrode at temperatures between 290 °C (550 °F) and 340 °C (650 °F) due to condensing naphthenic acids. The 12Cr cladding of the column shell in this temperature range is also attacked. These components may have to be replaced with Type 316 having the maximum required 3.0% Mo content of the specification range, or with Type 317 stainless steel.
- With crude blends having a whole crude TAN above 1.0, severe NAC can be expected in the crude unit vacuum section and upgrader VDU, as noted above. NAC can also be expected in a number of components of the atmospheric section and DRU, including the heater outlet tubes, U-bends, outlet headers, transfer line, crude column flash zone, and inlets to sidestream strippers operating above 230 °C (450 °F).
- NAC is unlikely to occur with crude blends having a whole crude TAN below 0.3 to 0.5.

7.7.1 Predicting NAC based on TAN. NAC at relatively low rates in piping and on vessel internals is fairly predictable, using published correlations between TAN, temperature, and metallurgy. As a general rule, NAC increase with increasing TAN of crudes. However, the correlations are mostly based on laboratory tests and agreement between correlations is not good, as shown in Figure 7-3. Under certain limited circumstances, upgrading a low-alloy Cr-Mo steel with the next or next two higher grades of Cr-Mo steel can lower general NAC to an acceptable level.

FIGURE 7-3 EFFECT OF TAN ON CORROSION OF CARBON STEEL IN VARIOUS CRUDES AT 230 °C (450 °F)

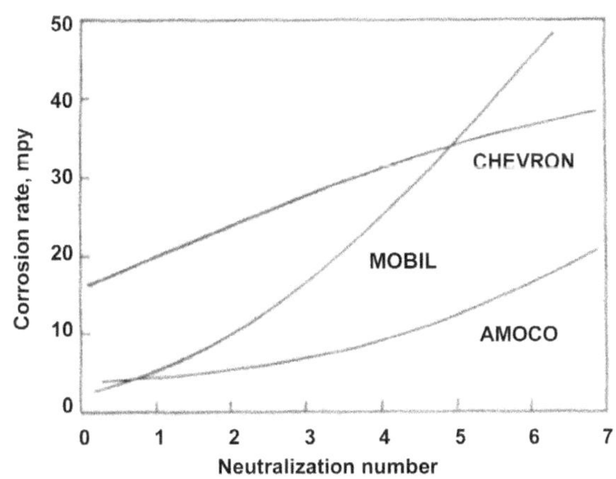

Based on data reported by members of NACE T8 Committee on Refining Industry Corrosion.

5. Industry Survey conducted in 1996.

7.7.2 Predicting NAC based on Velocity. Velocity-accelerated NAC occurs at flow velocities above 50m/s (165fps) in furnace tubes and transfer lines in the form of grooving and droplet impingement, but is entirely unpredictable. In all cases, NAC is associated with areas of high velocity or turbulence once the naphthenic acids have dissolved any HTSC scale that was present. TAN, temperature, organic sulfur content, velocity, and degree of vaporization are some of the factors that affect the severity of this type of NAC. Typically, 5Cr-0.5Mo U-bends for furnace tubes and elbows in transfer lines have holed through due to impingement attack and replacement U-bends should be at least 9Cr-1Mo steel. Other than case histories linked to specific units and components, no useful corrosion correlations exist. As a general rule, replacing low-alloy Cr-Mo steels with an austenitic 300 series stainless steel is necessary.

7.7.3 Predicting NAC based on Acid Condensation. Naphthenic acids condensing on trays and shells of vacuum columns between 290 °C (550 °F) and 340 °C (650 °F) have caused severe NAC. As a general rule, upgrading low-alloy Cr-Mo steels and 12Cr stainless steel with an austenitic 300 series stainless steel is required. However, trays and shell areas above 370 °C (700 °F) are often (but not always) protected from NAC by a physical barrier coating of tar, asphalt, and coke deposits that form at these temperatures. Trays and shell areas operating below 200 °C (400 °F) are also not subject to NAC per se, but may be subject to corrosion by short-chain organic acids, such as acetic acid (CH_3-COOH), formic acid (H-COOH), or proprionic acid (CH_3CH_2-COOH). These acids are formed by decomposition of naphthenic acids in the vacuum heater. When corrosion issues do arise, aluminum (Al) structural components and structured packing have been used in vacuum columns with some degree of success provided the temperature limits for aluminum are observed.

7.7.4 Predicting NAC based on PREN[6]. Since NAC of stainless steel and other alloys often is in the form of pitting, the use of pitting resistance equivalent numbers (PREN) has been proposed for screening alternative alloys for potential use in naphthenic acid services.[7] While PREN cannot be used to predict whether an alloy grade is suitable for any given service, it should be useful for comparing the resistance against pitting during naphthenic acid condensation. As shown in Table 7-3 and confirmed by plant experience, the higher the PREN, the more resistant an alloy can be expected to be toward pitting by naphthenic acids in vacuum columns.

TABLE 7-3 PREN FOR COMMON ALLOYS USED AGAINST NAPHTHENIC ACID CORROSION

Alloy	Chromium, %	Molybdenum, %	Nitrogen, %	PREN
304	19	0	0.11	20
316	17	2.5	0.11	27
317	19	3.5	0.11	32
904L	21	4.5	0.11	38
Alloy 20	20	2.5	0.11	30
Alloy C-276	15.5	16	0.11	70

7.7.5 Predicting NAC with Predict2™ Program. This software tool is based on a joint industry program (JIP) on naphthenic acid research supported by a number of designer, contractor, and petroleum companies and was subsequently expanded to include HTSC. The program is expensive to purchase and requires an annual subscription fee, and the principal users are the companies that participated financially in the original JIP. Because of user confidentiality agreements and restrictions, there appears to have been no independent evaluation of the program by an unbiased party. As far as the petroleum industry's acceptance is concerned, users reported the tool to be primarily useful for exploring various options in sensitivity studies when specifying materials, pipe diameters, velocity limits, and temperature parameters for unit revamps and equipment replacements.

6. PREN = %Cr + 3.3%Mo + 16%N.
7. Communication from Emilio Maluff of YPF (Argentina).

Like similar software programs, Predict2™ is short of actual and reliable corrosion data and long on assumptions and extrapolations. For example, corrosion rates were measured with static and flow-through weight-loss coupons/spools in purified Hot Oil to which various amounts of high-purity naphthenic acid were added to obtain different TAN levels. In some tests, H_2S was also added[8]. Corrosion tests typically lasted only 48 hours, which is too short to account for the effect of scale formation with time. The program takes into account the effects of TAN, H_2S content, temperature, hydrocarbon content, flow regime, and wall shear stress and then ties various crude assays to corrosion data and effective mitigation methods for piping systems, atmospheric columns, and vacuum columns. An example of the kind of raw data on which the program is based is shown in Figure 7-4.

FIGURE 7-4 IMPINGEMENT ATTACK FOR 5CR STEEL IN HOT OIL AT 650 °F AS A FUNCTION OF TAN, VELOCITY, AND H_2S CONTENT

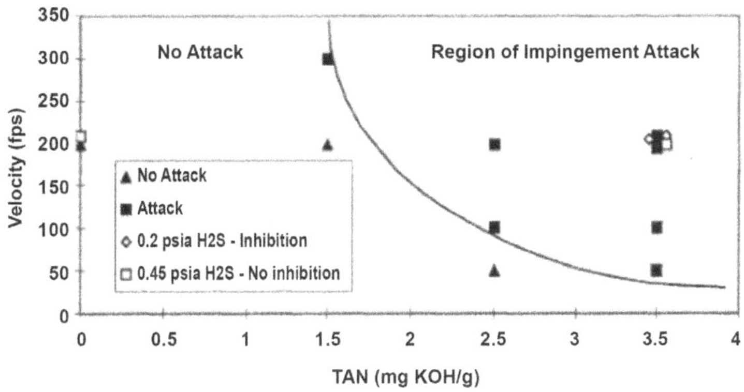

Adapted from R. D. Kane and M.S. Cayard, "A Comprehensive Study of Naphthenic Acid Corrosion," paper no. 02555, CORROSION/2002 (Houston, TX: NACE, 2002).

7.8 Appearance and Location of NAC

NAC is often highly localized, sharp edged, and deep. In all cases, NAC is defined by complete absence of visible sulfide scale. Depending on location, process stream, and metallurgy, various forms of corrosion damage have been reported:

- Striations, grooving, and "scooping" in the flow direction in crude heater outlet tubes, as in Figure 7-5.
- Pitting and lake-type attack in crude heater outlet headers and crude column inlet nozzles, as in Figure 7-6 and 7-7.
- Striations and grooving in the flow direction in vacuum heater tube elbows and return bends, as in Figure 7-8.
- Pitting and lake-type attack just downstream of thermowells and circumferential weld beads in vacuum transfer lines at velocities above 50m/s (165fps), as in Figure 7-9.
- More-or-less uniform corrosion in HVGO and vacuum resid piping at flow velocities below 6m/s (20fps).
- Deep, round, 6mm (0.25in.) diameter pits on the top of bubble caps and underneath tray decks due to naphthenic acid condensation, as in Figure 7-10 and Figure 7-11.
- Run-down condensate corrosion on vacuum column shells and internals, as in Figure 7-12 and 7-13.
- Overall pinhead pitting on Type 316 or Type 317 stainless steel overlay or cladding on vacuum column shells.
- Destruction of demister pads or structured packing beds in vacuum columns, as in Figure 7-14.

8. Extensive research over many years has shown that using H_2S, while convenient, is no substitute for organic sulfur compounds when evaluating the interaction between NAC and HTSC.

FIGURE 7-5 NAC OF 7CR-0.5MO TUBE FROM LAST PASS OF HEATER AT 370 °C (700 °F)

Adapted from J. Gutzeit, "Naphthenic Acid Corrosion – An Old Enemy of the Refining Industry," API Operating Practices Committee (OPC) Meeting (Houston, TX: 1992).

FIGURE 7-6 NAC OF 5CR-0.5MO OUTLET HEADER FROM HEATER AT 370 °C (700 °F)

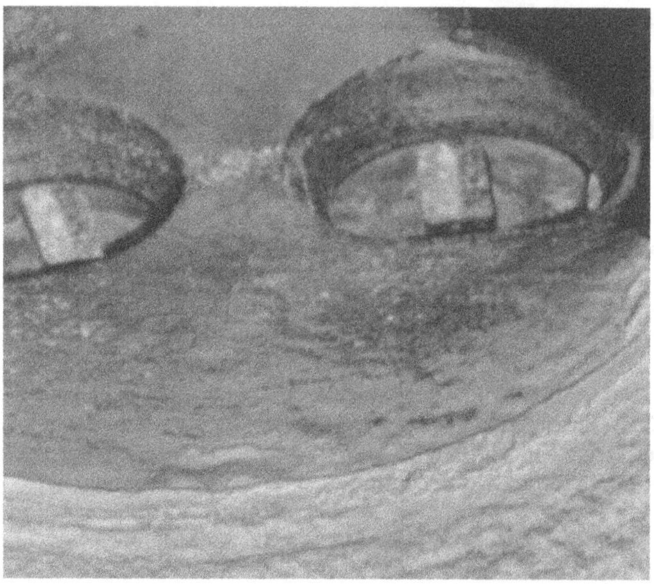

Adapted from J. Gutzeit, "Naphthenic Acid Corrosion – An Old Enemy of the Refining Industry," API Operating Practices Committee (OPC) Meeting (Houston, TX: 1992).

FIGURE 7-7 NAC OF 12CR LINER OF COLUMN INLET NOZZLE AT 370 °C (700 °F)

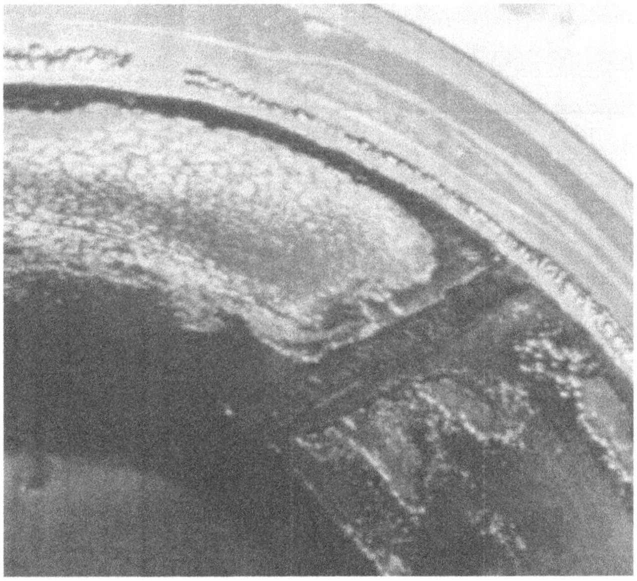

Adapted from J. Gutzeit, "Naphthenic Acid Corrosion – An Old Enemy of the Refining Industry," API Operating Practices Committee (OPC) Meeting (Houston, TX: 1992).

FIGURE 7-8 NAC OF 5CR-0.5MO RETURN BEND FOR VACUUM HEATER AT 400 °C (750 °F)

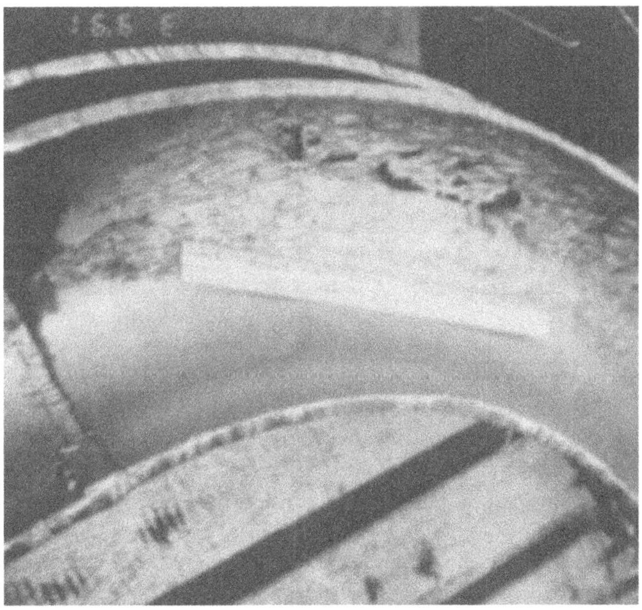

Adapted from J. Gutzeit, "Naphthenic Acid Corrosion – An Old Enemy of the Refining Industry," API Operating Practices Committee (OPC) Meeting (Houston, TX: 1992).

FIGURE 7-9 NAC IN ELBOW OF 5CR-0.5MO VACUUM HEATER TRANSFER LINE AT 370 °C (700 °F)

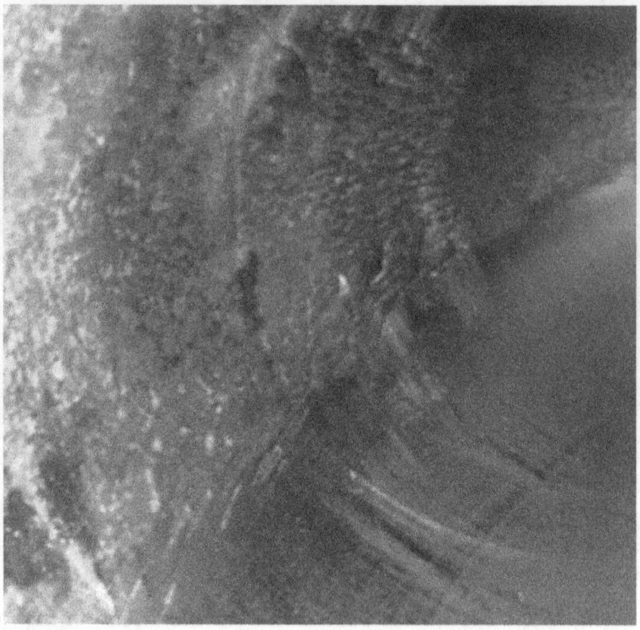

FIGURE 7-10 NAC OF TYPE 317 SS BUBBLE CAPS AND TRAY IN VACUUM COLUMN AT 360 °C (680 °F)

Adapted from J. Gutzeit, "Naphthenic Acid Corrosion in Oil Refineries," *MP* 16, 10 (1977).

FIGURE 7-11 DETAIL OF NAC OF TYPE 316 SS BUBBLE CAP IN VACUUM COLUMN AT 340 °C (640 °F)

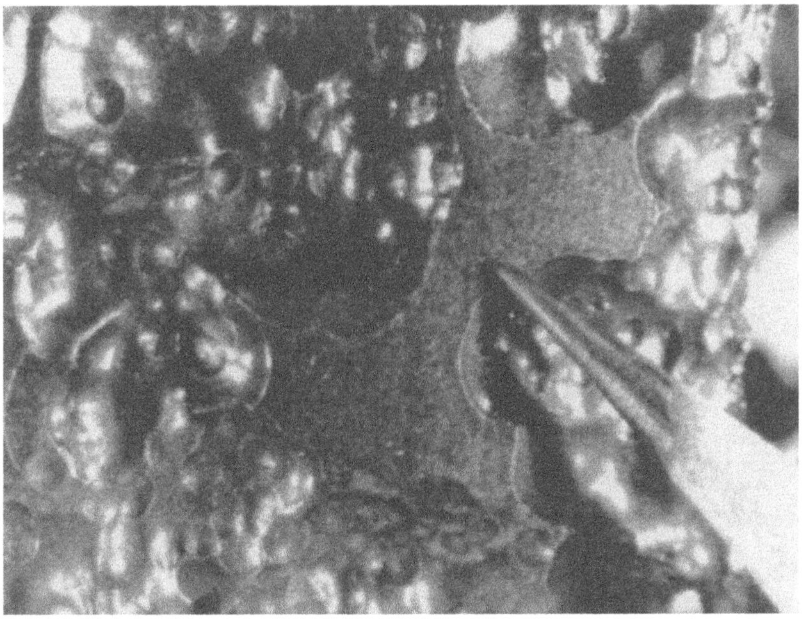

Adapted from J. Gutzeit, "Naphthenic Acid Corrosion in Oil Refineries," *MP* 16, 10 (1977).

FIGURE 7-12 NAC OF 12CR SHELL LINER IN VACUUM COLUMN AT 360 °C (680 °F)

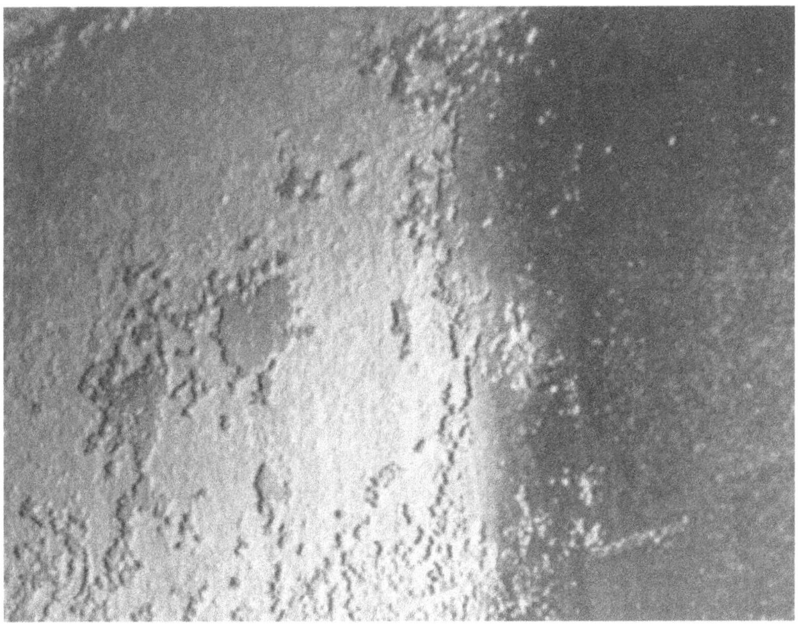

FIGURE 7-13 NAC OF 12CR SHELL LINER AND INTERNALS IN VACUUM COLUMN AT 360 °C (680 °F)

FIGURE 7-14 NAC OF TYPE 316 SS DEMISTER PAD AND SUPPORT GRID IN VACUUM COLUMN AT 290 °C (560 °F)

Adapted from J. Gutzeit, "Naphthenic Acid Corrosion in Oil Refineries," *MP* 16, 10 (1977).

7.9 NAC Inhibitors

Several types of corrosion inhibitors and organic neutralizers have been promoted over the years as a means for reducing NAC. The usual filming-amine inhibitors cannot be used for this purpose, because of the elevated temperatures of NAC. Even though some filming-amine inhibitors are stable at temperatures as high as 370 °C (700 °F), their rate of desorption exceeds their rate of adsorption at temperatures above approximately 150-180 °C (300-350 °F)[9]. As a result, process additives suppliers are offering alternative chemistries as corrosion inhibitors against NAC, but only limited field data on actual inhibitor performance are usually available. Typical commercial formulations are based on organic polysulfides, thiazolines, or alkyl phenol sulfonic acids. Some corrosion inhibitors against NAC are based on various types of phosphate and other esters that react with steel at elevated temperatures to produce a dense surface layer.

9. J. Gutzeit, C. Samans, and F.B. Bulls, "Evaluating Refinery Process Side Corrosion Inhibitors," *MP* 12, 7 (1973).

The inhibitors are not volatile and stay with resid leaving the crude and vacuum columns and do not provide protection for lighter fractions. This means additional inhibitors may have to be injected to protect the piping and equipment handling VGO and HVGO, where naphthenic acids tend to concentrate. The inhibitors also cannot protect column trays in contact with a condensing naphthenic acid vapors, nor can they protect the high-velocity areas in transfer lines that are susceptible to severe NAC.

In the case of phosphate esters, the surface layer is probably similar to that obtained with some commercial phosphatizing treatments and conversion coatings. Phosphate esters ($R\text{-}H_2PO_4$) react with entrained and dissolved brine in crude to form phosphoric acid (H_3PO_4) on steel surfaces. The acid then reacts with the steel and any existing iron sulfide (FeS) scale to form iron phosphate ($FeHPO_4$). Expressed in most simplistic terms, the stepwise reactions are:

$$R\text{-}H_2PO_4 + H_2O \Rightarrow H_3PO_4 + R\text{-}OH \qquad \text{(Eq.7-6)}$$

$$Fe + H_3PO_4 \Rightarrow FeHPO_4 \downarrow + H_2 \quad \text{(with bare steel)} \qquad \text{(Eq.7-7)}$$

$$FeS + H_3PO_4 \Rightarrow FeHPO_4 \downarrow + H_2S \quad \text{(with iron sulfide scale)} \qquad \text{(Eq.7-8)}$$

An alternative mechanism is the direct reaction of the ester with steel to form a protective iron phosphonate inhibitor film:

$$R\text{-}H_2PO_4 + 2\,Fe \Rightarrow Fe\text{-}O\text{-}PO_2\text{-}O\text{-}Fe + H_2 \qquad \text{(Eq.7-9)}$$

Because inhibitor concentrations as high as 25-100mg/L (7-30PTB) or more are usually required for a reasonable degree of protection, these corrosion inhibitors are too expensive for routine addition to crude and most other streams. The cost of the inhibitors can be as high as US$ 0.10/BBL treated fluid, resulting in an annual cost of about US$ 3-6MM for a refinery processing 27T/d (200K BPD) if the entire crude stream is treated.[10] Phosphate ester corrosion inhibitors have been blamed for fouling in the crude, vacuum, and downstream units. Consequently, they are used primarily in spot applications to temporarily extend the life of piping that has almost reached the end of its corrosion allowance. This provides sufficient time for locating or ordering replacement pipe to be installed at the next opportune shutdown.

7.10 NAC Neutralizers

As a general rule, the usual liquid organic neutralizers cannot be used, because they are too weak and expensive to neutralize the amount of acids found in many acidic crudes[11]. At least one refinery attempted to use caustic to counteract potential corrosion by acidic crude. Typically, dilute caustic is added to desalted crude by injection into the transfer line from the crude heater. Reported addition rates were as high as 1,700mg/L (500PTB) crude charged. Not surprisingly, severe corrosion and fouling of the transfer line occurred downstream of the caustic injection point. As a band-aid fix and to facilitate rapid repair, the refinery ended up stocking replacement spool pieces fabricated from Alloy C-276. As discussed in the chapter on caustic addition to control overhead corrosion, experience has shown that caustic addition must be limited to less than 15mg/L (5PTB) and preferably less than 10mg/L (3PTB) to minimize caustic attack and premature coking in downstream components.

7.11 Recovery of Naphthenic Acids

Naphthenic acids are a raw material for numerous commercial processes, including detergents, lubricants, fuel oil additives, insecticides, fungicides, and, ironically, corrosion inhibitors. The acids are usually obtained by oxidation of

10. S. D. Kapusta, A. Ooms, A. Smith, F. van den Berg, and W. Fort, "Save Processing of Acid Crudes," paper no. 04637, CORROSION/2004 (Houston, TX: NACE, 2004).

11. A number of years ago, the author was asked by a refinery to review a case history submitted by its process additives supplier in support of a new proprietary organic neutralizer to control naphthenic acid corrosion. The supplier had claimed that by adding only 3mg/L (1PTB) of the neutralizer to the crude charge, another refinery had successfully prevented naphthenic acid corrosion despite the fact that its crude had a TAN of 1.0. When the author asked the salesman to explain to him how only 3mg/L of his organic neutralizer could be equally or more effective than 1,000mg/L potassium hydroxide (KOH) in accordance with the definition of TAN, he was at a loss. So was his manager and so was the "expert" at the home office who had prepared the case history in the first place. The case history was quietly withdrawn and no more was said about the new proprietary organic neutralizer.

the naphtha fractions of acidic crudes. The acid composition varies with the crude composition and the conditions during refining and oxidation. A number of processes to remove naphthenic acids from crude and its fractions have been proposed over the years, but none have been applied to routine treatment of acidic crude, as for example, in the case of desalting or H_2S scavenging. Examples are:

- The Dresser Process[12] where naphthenate soaps from caustic treating operations are recovered by continuous acidification and centrifuging of spent caustic solutions with sulfuric acid.
- The Merichem Napfining™ process where naphthenic acids are removed from jet fuel and diesel streams by caustic washing[13].
- The Amoco Oil Potassium Hydroxide (KOH) Process where naphthenic acids are recovered by passing distillates over solid KOH, which prevents the formation of hard-to-break emulsions normally seen with caustic (NaOH) extraction.
- The Amoco Oil Agitated Alkali Process where crude or distillates are continuously mixed with KOH or NaOH granules in countercurrent flow, followed by water washing and centrifuging[14].

Other patented processes include Exxon's Thermal Destruction Process, Unipure's Lime Destruction Process, UOP's Adsorption Process, and ExxonMobil's Adsorption Process (probably originally by Pure Oil/Mobile).

7.12 Inspection Methods

Due to its unpredictable occurrence, NAC is often difficult to detect or monitor while a unit is on stream. In some cases, the corrosion products are in the form of coordination complexes $[R\text{-}(CH_2)_n\text{-}COO]_2\text{-}Fe$ or oxides $[R\text{-}(CH_2)_n\text{-}COO]_6\text{-}O\text{-}Fe_3$ that can be detected in hydrocarbon streams through analytical testing. In other cases, increased levels of Fe, Cr, and Ni in process streams are an indicator of NAC provided certain caveats are kept in mind, as discussed in Chapter 5 on corrosion monitoring. Since some amount of hydrogen sulfide (H_2S) is always present in process streams, a portion of soluble iron naphthenates is converted to suspended iron sulfide (FeS) particles that are difficult to differentiate from sloughed-off iron sulfide scale. Except for radiography, the basic detection and inspection methods are carried out offline and include visual inspection and UT measurements during shutdown.

7.13 Similar Corrosion Damage

NAC should not be confused with HTSC in furnace tubes and transfer lines, as discussed in detail in Chapter 6 on high-temperature sulfidic corrosion. When components are improperly acid cleaned, corrosion damage due to the acid cleaning solutions is sometimes blamed on NAC. General NAC at relatively low rates in piping and on vessel internals can be positively identified by installing retractable coupon holders with coupons ranging from carbon steel to 12Cr stainless steel. All coupons would be expected to corrode at similar rates when NAC is involved.

7.14 Repair Methods

Upgrading with a suitable alloy is usually the only approach available to ensure long-term protection against NAC. Unless corrosion rates are below 0.1-0.2mm/y (4-8mpy), corroded carbon steel components should be upgraded with a low-alloy Cr-Mo steel, 12Cr stainless steel, or a 300 series stainless steel. The final choice depends on the corrosion severity, damage mechanism, component involved, and steel availability. Often, carbon steel or the wrong grade of low-alloy or stainless steel is installed, because of mix-ups during shipping, warehousing, and handling. To avoid this, all replacement steels should be analyzed for proper composition using an alloy analyzer or similar x-ray fluorescent spectrograph that offers a combination of alloy identification and elemental analysis. Large shell areas of vacuum columns exposed to condensing naphthenic acid droplets should be clad or weld overlaid with Type 316 or Type 317 stainless steel. Vacuum column trays, bubble caps, or structured tray packing should also be made from these stainless steels. Corroded piping should be sleeved in kind or preferably with a 300 series stainless steel. In all cases, Type 316

12. D.B. Todd and F.C. Rac, "Recover Naphthenic Acids Continuously," *Hydroc. Proc.* 46, 8 (1967).
13. N. Kittrell, "Removing Acid from Crude Oil," Presentation at Crude Oil Quality Group Meeting (New Orleans, LA: 2006).
14. Unpublished information made available to the author.

stainless steel containing the maximum required 3.0% Mo content, or Type 317 stainless steel should be used. Type 304 stainless steel is not resistant to NAC and should be avoided.

7.15 Precautions When Inspecting Corroded Components

A number of precautions apply when high-temperature components are opened for inspection or maintenance, as follows:

- Components containing sulfide (FeS) scale will evolve poisonous hydrogen sulfide (H_2S) gas when subjected to cleaning solutions.
- Hydrogen sulfide (H_2S) gas is heavier than air and can flow out of the bottom manholes of a column with deadly consequences when unprotected personnel try to enter.
- Some iron sulfide scales are pyrophoric and ignite when exposed to air and allowed to dry during shutdowns. To avoid this problem, components with sulfide scale should be water washed and keep wet prior to cleaning. Alternatively, sulfide scales should be oxidized by spraying with a hot alkaline potassium permanganate ($KMnO_4$) solution.
- Components made of 300 series stainless steels, including piping, furnace tubes, heat exchanger tubes, or pressure vessel internals, may chloride stress corrosion crack (CSCC). Cracking occurs at temperatures above approximately 65 °C (150 °F) when these components are first opened to the atmosphere while hot or lying on the ground in the sun waiting to be cleaned. CSCC is not a problem during normal operations since air is excluded. Prior to, during, and after cleaning operations, shutdown procedures in accordance to NACE RP0170 should be used to prevent CSCC. As a rule, polythionic acid stress corrosion cracking (PASCC) is not a problem in crude units and the corresponding upgrader units.

7.16 Chapter Summary

NAC of carbon steel, various low-alloy Cr-Mo steels, and 12Cr stainless steel occurs in the high-temperature circuits of crude and corresponding upgrader units operating above 230 °C (450 °F). The corrosion products are oil-soluble Fe/Mo/Cr naphthenates; no protective scales are formed. NAC is caused by dissolved naphthenic acids in some types of crude having a TAN above 0.5. NAC should not be confused with HTSC, which, under some conditions, produces similar corrosion damage. NAC should also not be confused with corrosion damage caused by improper acid cleaning of components.

Like many other corrosion processes, NAC is an electrochemical process. Corrosion occurs at anodic areas of the metal, while hydrogen evolves at cathodic areas. The electrolyte is entrained and dissolved brine in the crude. Not all naphthenic acids are equally corrosive. There appear to be weaker and stronger naphthenic acids, but the differences between these have yet to be clearly defined. As a general rule, however, the higher-boiling naphthenic acids are more corrosive than the lower-boiling naphthenic acids.

NAC increases with TAN and temperature and occurs primarily in liquid or mixed liquid/vapor streams instead of all-vapor streams. NAC occurs in at least three distinctly different forms: general corrosion at relatively low rates in piping and on vessel internals; velocity-accelerated impingement corrosion in heater tubes and transfer lines; and corrosion by naphthenic acid-rich condensate in vacuum columns.

7.17 Read More

1. E. Babaian-Kibala and M. J. Nugent, "Naphthenic Acid Corrosion Literature Survey," paper no. 378, CORROSION/99 (Houston, TX: NACE, 1999).
2. E. Babaian-Kibala et al., "Naphthenic Acid Corrosion in Refinery Settings," *MP* 32, 4 (1993).
3. H. J. de Bruyn, "Naphthenic Acid Corrosion in Synthetic Fuels Production," paper no. 576, CORROSION/1998 (Houston, TX: NACE, 1998).
4. H. L. Craig, Jr., "Naphthenic Acid Corrosion in the Refinery," paper no. 333, CORROSION/95 (Houston, TX: NACE, 1995).

5. H. L. Craig, Jr., "Temperature and Velocity Effects in Naphthenic Acid Corrosion," paper no. 603, CORROSION/96 (Houston, TX: NACE, 1996).
6. B. Danilov, "The Control of Corrosion in Refinery Vacuum Plants," *Anti-Corrosion* 22, 8 (1975).
7. B. Danilov, "Examples of Corrosion Control, Part 2 - Vacuum Crude Distillation," *Hydroc. Proc.* 60, 3 (1981).
8. W. A. Derungs, "Naphthenic Acid Corrosion - An Old Enemy of the Petroleum Industry," *Corrosion* 2, 12 (1956).
9. J. Gutzeit, "Naphthenic Acid Corrosion in Oil Refineries," *MP* 16, 10 (1977).
10. J. Gutzeit, "Naphthenic Acid Corrosion - An Old Enemy of the Refining Industry," paper presented on behalf of the API Subcommittee on Corrosion and Materials, API Operating Practices Committee (OPC) Meeting, Sept. 28, 1992, Houston, TX.
11. J. Gutzeit, "Studies Shed Light on Naphthenic Acid Corrosion," *Oil & Gas J.* 74, April 5 (1976).
12. J. L. Hau et al., "Classifying Crude Oils According to Corrosivity Using the Fe Powder Test," paper no. 00699, CORROSION/2000 (Houston, TX: NACE, 2000).
13. J. J. Heller, "Corrosion of Refinery Equipment by Naphthenic Acid," *MP* 2, 9 (1963).
14. D. Johnson, G. McAteer, and H. Zuk, "The Safe Processing of High Naphthenic Acid Content Crude Oils - Refinery Experience and Mitigation Studies," paper no. 03645, CORROSION/2003 (Houston, TX: NACE, 2003).
15. R. D. Kane and M.S. Cayard, "A Comprehensive Study on Naphthenic Acid Corrosion," paper no. 02555, CORROSION/2002 (Houston, TX: NACE, 2002).
16. S. D. Kapusta et al., "Safe Processing of Acid Crudes," paper no. 04637, CORROSION/2004 (Houston, TX: NACE, 2004).
17. B. Messer, B. Tarleton, and M. Beaton, "New Theory for Naphthenic Acid Corrosivity of Athabasca Oil sands Crudes," paper no. 04634, CORROSION/2004 (Houston, TX: NACE, 2004).
18. Naphthenic Acid Corrosion, *Refin·Cor 7.0*, Transcript of 48 Years of Shared Solutions to Refinery Corrosion Problems, Minutes of Committee T-8 (STG34) (Houston, TX: NACE, 2005).
19. M. J. Nugent, "Experience with Naphthenic Acid Corrosion in Low TAN Crudes," paper no. 577, CORROSION/1998 (Houston, TX: NACE, 1998).
20. R. L. Piehl, "Naphthenic Acid Corrosion in Crude Distillation Units," *MP* 27, 1 (1988).
21. E. Slavcheva, B. Shone, and A. Turnbull, "Factors Controlling Naphthenic Acid Corrosion," paper no. 579, CORROSION/1998 (Houston, TX: NACE, 1998).
22. S. Tebbal, "Critical Review of Naphthenic Acid Corrosion," paper no. 380, CORROSION/99 (Houston, TX: NACE, 1999).
23. M. Z. Zetlmeisl, "Naphthenic Acid Corrosion and its Control," paper no. 218, CORROSION/96 (Houston, TX: NACE, 1996).
24. T. Jackson, M. C. Winslow, and M. Wilson, "Prolonged Experience Processing High Acid Crude, Cross Oil & Refining Company," case history by GE Water & Process Technologies.
25. B. E. Hopkinson and L. E. Penuela, "Naphthenic Acid Corrosion by Venezuelan Crudes," paper no. 502, CORROSION/97 (Houston: TX: NACE, 1997).
26. J. Gutzeit, C. Samans, and F.B. Bulls, "Evaluating Refinery Process Side Corrosion Inhibitors," *MP* 12, 7 (1973).
27. D. B. Todd and F.C. Rac, "Recover Naphthenic Acids Continuously," *Hydroc. Proc.* 46, 8 (1967).
28. N. Kittrell, "Removing Acid from Crude Oil," presentation at Crude Oil Quality Group Meeting (New Orleans, LA: 2006).

8 Tank Settling of Crude

In this Chapter

8.1 First Glance

Crude is repeatedly gravity settled in tankage during upstream and midstream operations in order to lower shipping costs and add value by removing as much water and salt as practical. Further gravity settling of crude in tankage at the refinery or upgrader, while not considered an actual part of the desalter operations, plays an important role in removing residual water, salt crystals, contaminants, and suspended solids prior to final blending and thereby minimizing desalter upsets.

8.2 Tank Settling of Crude Receipts

Free water and entrained water in crude receipts are present in the form of saturated salt brine. Therefore, gravity settling in tankage helps lower the salt content of crude receipts even though desalter water is added later on. Adequate gravity settling is particularly important when heavy crudes with API gravities below 20 are charged because their high density and viscosity interfere with brine droplet coalescing and gravity settling in the downstream desalter. The main problem with gravity settling in tankage is the fact that it can take considerable time (days and weeks), depending on the API gravity and water content of crude receipts. This means that crude inventory is temporarily tied up in tankage and cannot be immediately blended and sent to the desalter. Like cars sitting unsold on a dealer's lot, tied-up crude inventory costs money. While an emulsion breaker can be and is sometimes added to speed up gravity settling, it is often counter-productive because the emulsion breaker is likely to interfere later on with emulsion formation in the mix valve for the desalter.

8.3 BS&W Content of Crude

The overall goal of gravity settling is to reduce the bottoms sediment and water (BS&W) content of the crude blend going to the desalter to less than 1.0% in order to minimize desalter upsets due to periodic slugs of entrained brine and suspended salt crystals. BS&W consists of brine-in-oil emulsions that can contain entrapped brine, salt crystals, sand, silt, drilling mud, crude, sediment, wax particles, asphaltenes, corrosion products, hard water scale, and other particulates from upstream and midstream production, gathering, and transportation operations. Depending on location, BS&W can also contain entrapped rain water and melted snow that entered the holding tanks through leaking tank roof seals or defective tank roof drains. Particulate matter (the sediment portion of BS&W) typically ranges from 20-200µ (0.8-8mil) in size. High levels of particulate matter eventually settle out and form a sludge layer in crude tanks that causes under-deposit corrosion of tank bottoms, as shown in Figure 8-1. If unattended, the bottom of brand-new crude receiving tanks can and have corroded, to the point where the first leaks forms after only 8-10 years. Most desalter upsets occur when this sludge layer is disturbed during normal tank activity and picked up intermittently by the tank's crude suction pipe. On the way to the desalter, a portion of the entrained sludge is deposited in the crude distribution lines and headers and causes under-deposit corrosion in those components.

FIGURE 8-1 TANK BOTTOM CORROSION CAUSED BY SLUDGE BUILD-UP IN CRUDE RECEIVING TANK

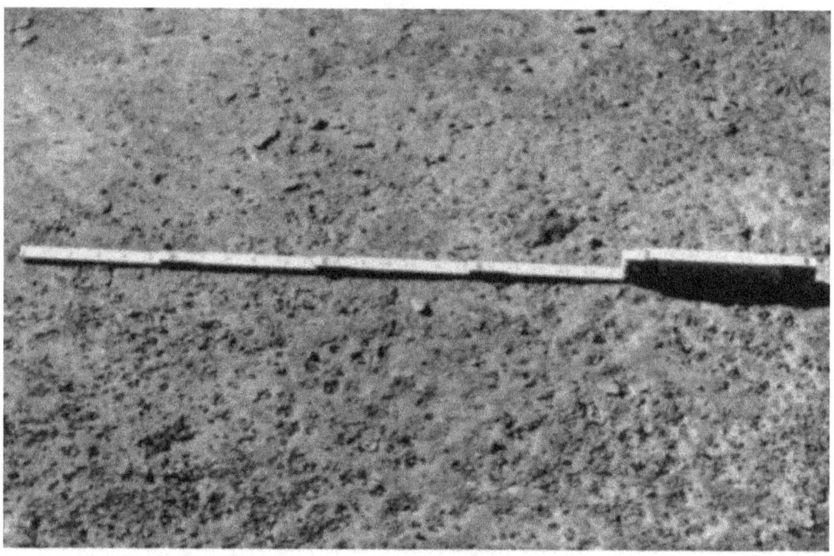

8.4 Controlling Sludge Build-up in Crude Tanks

All crudes contain a certain amount of sludge that settles to the bottom of crude tanks. Sludge consists of organic and inorganic materials, including sand, silt, drilling mud, wax particles, asphaltenes and corrosion products that originate in upstream and midstream operations. Since sludge is heavier than crude, it will settle and form a solid layer on the bottom of a tank. As a result, crude becomes difficult to pump out of a tank and periodic slugs of sludge interfere with downstream desalter operations. Various methods are available to keep sludge more or less uniformly suspended in crude tanks including swivel-angle propeller mixers and rotating submerged jet nozzles to delay sludge build-up.

8.4.1 Propeller Mixers. Propeller mixers are mounted horizontally near the bottom of the tank and produce a spiraling circulation pattern. The propellers swivel as they operate to cover a larger surface area of the tank bottom. Multiple units are often used in a single tank to overcome the limited power capacity of the mixers. The number of units depends on the diameter of the tank. As a general rule, one mixer can handle tanks measuring up to 30m (100ft) in diameter. Each 15m (50ft) diameter increase requires adding an additional mixer. To assure proper mixing, all mixers should have sufficient power to meet the mixer power/crude volume guideline of approximately $2KW/Mm^3$ (0.4HP/MBBL). The energy required to prevent sludge formation in medium and heavy crudes is twice that much or approximately $4KW/Mm^3$ or 0.8HP/MBBL.

8.4.2 Submerged Jet Nozzles. Submerged jet nozzles direct a constant force on the tank bottom to dislodge any settled sludge. With large crude storage tanks, the jet nozzles need to rotate to apply this force to the entire tank bottom. While single jets can be used, double jets are preferred because the forces on the mixer are better balanced. The Butterworth P43™ machine shown in Figure 8-2 operates by pumping crude oil though a jet nozzle. The nozzle is designed to provide sufficient velocities at the furthest point in the tank. The machine can be mounted in the center of a tank or on the sides. The center location allows only one jet mixer to be used, while multiple mixers are often used mounted on the sides. The rotating jet mixers are expected to prevent settling in the first place and if necessary resuspend any settled sludge. The sludge is mainly composed of long-chain hydrocarbons that behave as thixotropic fluids. For a jet to resuspend deposited sludge particles, the kinetic energy of the jet must overcome the intermolecular forces. The critical velocity for keeping sludge particles in suspension is approximately 0.6 to 1.2 m/sec (2-4FPS) for most crudes.

FIGURE 8-2 ROTATING JET NOZZLES FOR USE IN CRUDE RECEIVING TANKS TO PREVENT SLUDGE BUILD-UP

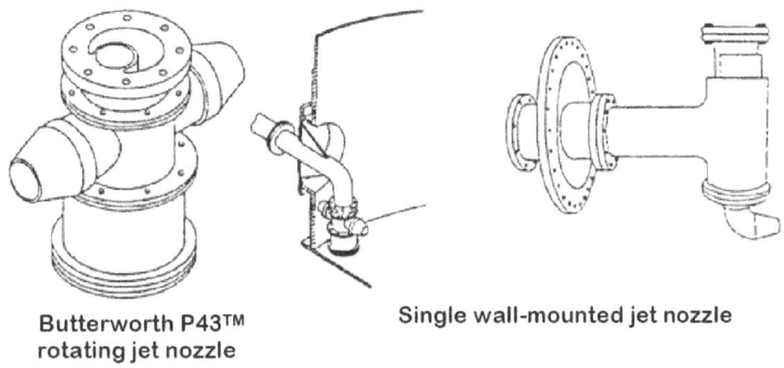

Adapted from R. R. Hemrajani, "Fluid Mixing Technology in the Petroleum Industry," chapter in *Handbook of Industrial Mixing*, E. L. Paul et al., eds. (Hoboken, NJ: Wiley, 2004).

8.4.3 Bottoms Draining. Even with tank agitation, all tank bottoms should be drained at regular intervals to keep the height of the water-and-sludge level below 150-300mm (6-12in.). Periodic removal of water-and-sludge bottoms used to be normal practice in the past, but nowadays tends to be avoided because of stringent environmental regulations on waste disposal. Some tank bottoms are never drained, resulting in predictable consequences. A major problem with draining tank bottoms manually is the likelihood that some amount of crude ends up in the water-and-sludge bottoms going to the API separator. However, by installing one or two AGAR probes™, the location of the oil/water interface can be accurately determined and controlled from outside the tank, as shown in Figure 8-3.

FIGURE 8-3 EQUIPMENT FOR CONTROLLING THE OIL/WATER INTERFACE LEVEL IN CRUDE TANKS WITH AGAR PROBES™

Courtesy of AGAR Corporation, Houston, TX.

AGAR probes™ use energy absorption technology to detect water-in-oil or oil-in-water phases and have been found to be highly suitable and reliable in this application. Once the probe detects free water, the tank-bottoms dump valve is opened. As water and sludge are drained off, the oil/water interface begins to drop. When the probe detects the oil/ water interface, it signals the dump valve to close again and prevent oil from being discharged. In some installations, an upper probe is used for control, while a lower probe provides an alarm and insures positive closure of the dump valve. A typical probe setup is shown in Figure 8-4.

FIGURE 8-4 PROBE SETUP FOR CONTROLLING OIL/WATER INTERFACE LEVEL IN CRUDE TANKS WITH AGAR PROBES™

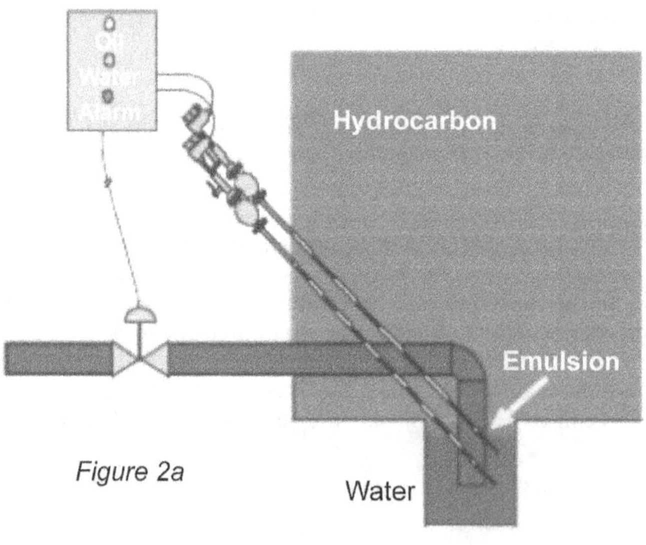

Courtesy of AGAR Corporation, Houston, TX.

8.5 Periodic Cleaning of Crude Tanks

Depending on the type and quality of crude receipts, a sludge layer that measures as much as 1.5m (5ft) or more in thickness can build up on the tank bottom. Thorough cleaning of crude tanks will eventually become necessary despite regular water and sludge removal[1]. Cleaning of large crude tanks is a major, costly, and time-consuming operation that is usually done semimanually by a combination of dredging, shoveling, hydro jetting, and vacuum truck operation. The preferred method is to circulate a volume of heated crude containing suitable dispersants over the bottom of the nearly empty tank to help redissolve settled sludge. The resulting sludge-enriched crude is then treated as slop oil and added to the main crude charge by trickle feeding, as discussed later in the Chapter 10 on slop oil and desalter upsets in the *Guide.*

8.6 Monitoring BS&W Content of Crude

Most refineries rely on shippers and pipeline companies for routine BS&W analyses of crude shipments. These may or may not be augmented with spot BS&W analyses by the refinery laboratory before a crude tank is brought on line. In either case, BS&W content of a crude sample is measured by separating the water/solids phase from the oil phase by centrifuging. A better approach is to monitor the BS&W content of incoming crude continuously, using an AGAR probe™ mounted in the crude line upstream of the desalter. AGAR probes™ have been found to be highly reliable in this application. Typically, the AGAR probe™ is located as far upstream of the desalter as possible (generally near the tank farm). A properly mounted probe can warn the plant operators nearly 30 minutes before the electrical current (in amps) of the lower desalter grid rises due to wet crude emulsion entering the desalter as a result of slop oil addition or tank switches. A typical probe readout is shown in Figure 8-5.

FIGURE 8-5 AGAR PROBE™ READOUT OBTAINED DURING BS&W MONITORING OF INCOMING CRUDES

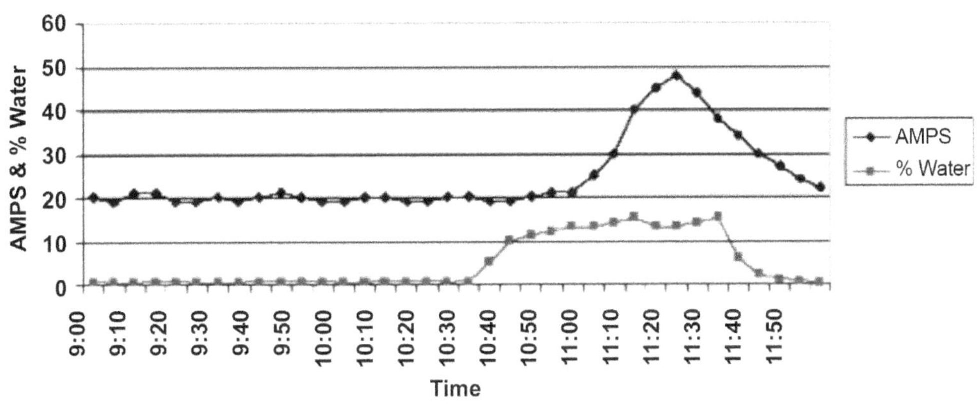

Adapted from G. Fransen, "New Control System Detects Desalter Problems Before Upsets Occur," AICHE 2004 Spring National Meeting, April 2004.

8.7 Adjusting BS&W Content of Crude

Shipping specifications for light and intermediate crude usually call for a BS&W content of 1.0%, or less. In contrast, shipping specifications for heavy crude may call for BS&W contents up to 3.0%. To ensure that all crude receipts going to the desalter have the correct BS&W contents, some refineries have detailed tank settling procedures. These usually

1. For a detailed description of tank cleaning procedures, refer to Section VIII in the manual *Cleaning of Process Equipment and Piping* by J. Gutzeit, MTI publ. no. 51 (St. Louis, MO: Materials Technology Institute of the Chemical Process Industries,1997).

call for each receiving tank to be sampled and analyzed repeatedly for BS&W content prior to crude blending. If the BS&W content is low enough, the crude is signed off and accepted by a chief operator on the unit and the tank is cut into the crude charge. If the BS&W is too high, the crude is rejected and stays in the receiving tank for further gravity settling. When incoming crudes have excessive BS&W values, several measures can be taken to avoid desalter upsets and the resultant overhead corrosion problems: 1. Crude with high BS&W content is diverted and mixed with crude with low BS&W content to achieve an acceptable crude charge; 2. Variations in BS&W content of incoming crude are minimized by continuous tank agitation to keep sediment and water more-or-less uniformly suspended throughout the crude most of the time; and 3. The time used for tank switches is extended by several hours or days to minimize sudden changes in BS&W content and the resulting desalter upsets.

8.8 Chapter Summary

Gravity settling of crude in tankage at the refinery plays an important role prior to final crude blending, because it can remove as much residual water, salt crystals, and suspended solids as possible. The overall goal of tank settling is to reduce the BS&W content of light and intermediate crudes going to the desalter to 1.0% and heavy crudes to 3.0%. In time, a sludge layer builds up in all crude tanks. Most desalter upsets occur when this sludge layer is disturbed during normal tank activity and picked up intermittently by the tank's suction pipe. Depending on the type and quality of crude receipts, periodic cleaning of crude tanks becomes necessary despite regular water and sludge removal. BS&W content is normally measured by analyzing spot samples from crude tanks. A better approach is to monitor BS&W content of incoming crude continuously using an AGAR probe™ mounted in the crude charge line well upstream of the desalter.

8.9 Read More

1. Marginal Well Commission, *The Lease Pumper's Handbook* (Oklahoma City, OK: State of Oklahoma).
2. Petrolite Equipment and Instruments Group, *Oil Field Dehydration and Desalting of Crude Oil,* 16 pgs. (Houston, TX: Petrolite Corp., 1986).
3. G. Fransen, "New Control System Detects Desalter Problems Before Upsets Occur," AICHE 2004 Spring National Meeting, April 2004.
4. J. Gutzeit, *Cleaning of Process Equipment and Piping*, MTI publ. no. 51 (St. Louis, MO: Materials Technology Institute of the Chemical Process Industries, 1997).
5. G. M. Heath et al., "Paraffinic Sludge Reduction in Crude Oil Storage Tanks through the Use of Shearing and Resuspension," *Acta Montanistica Slovaca* 9, 3 (2004).
6. R. R. Hemrajani, "Fluid Mixing Technology in the Petroleum Industry," chapter in *Handbook of Industrial Mixing*, E. L. Paul et al., eds. (Hoboken, NJ: Wiley, 2004).

9 Desalting of Crude

In this Chapter

9.1 First Glance

The root cause of corrosion and fouling in overhead systems is various inorganic salts in the incoming crude charge. These salts are partially hydrolyzed in the hot preheat exchangers (HPEs) and fired heater and form hydrogen chloride

(HCl) vapor that dissolves in aqueous overhead condensate to become hydrochloric (HCl) acid. HCl vapor also reacts with ammonia (NH_3) from organic nitrogen compounds in crudes, or added as a neutralizer to control acid corrosion. Desalting decreases the amount of HCl formed by decreasing both the salt and BS&W content of incoming crudes. However, desalting does not reduce the amount of organic chlorides which are insoluble in water and are not extracted in the desalter. Besides corrosion, desalting also reduces salt deposition in the HPEs and in the flash zones of the crude column and DR column; chloride carryover in the upper sidestreams of the columns; and the amount of sodium chloride (NaCl) in atmospheric resid or bitumen going to the vacuum column and VDU, respectively.

After tank settling, crude desalting is the second and probably most important step towards controlling overhead corrosion. Without proper desalting, all subsequent steps including caustic injection, overhead pH control, corrosion inhibitor injection, or water washing are largely compromised. Untimely failures of overhead lines, condenser bundles, and other components cannot be prevented without proper desalting. Moreover, product quality is also compromised when chloride salts, corrosion products, and cooling water enter product streams due to condenser tube leaks and carryover from receiver drums. At the same time, however, crude desalting is one of the least appreciated process operations on a crude unit. It is often taken for granted, unless major overhead corrosion and fouling issues affect refinery margins. The situation is made more difficult by the fact that many crude units have to do with outdated, undersized, and poorly performing desalters. Most likely, the desalter came with the original plant and a larger, updated desalter is rarely a high-priority item when a plant is revamped.

9.2 Desalter Operation

All crudes contain various and varying amounts of inorganic salts and brine that are sent to the plant after upstream and midstream treating and gravity settling to remove the bulk of production water, salt particles, and suspended solids. Typically, crude receipts still contain 100-400mg/L (30-120PTB) inorganic chloride salts, but some contain considerably more than that. Crude receipts can also contain 1-5% BS&W. Since overhead corrosion is usually caused by hydrogen chloride (HCl) which evolves in the fired heater by hydrolysis of inorganic chloride salts, the amount of salt and brine in crude must be lowered as much as possible by further desalting at the refinery. After initial tank settling and blending, the crude slate is preheated to 125-150 °C (260-300 °F) in a series of cold preheat exchangers (CPEs) prior to desalting. The inorganic chloride salts are then removed from the preheated crude by continuous contacting and extracting with water before the crude is sent on to the hot preheat exchangers (HPEs) and fired crude heater. The overall goal of desalting is to decrease the concentration of inorganic chloride salts in crude to 4-10mg/L (1-3PTB), and the amount of BS&W to less than 0.3%.

9.3 Desalter Performance

In general, crude desalting is a relatively simple and straight-forward operation, as far as most processes are concerned. For this reason, and probably because desalting chemicals are usually purchased from suppliers of other process additives, most plants just as soon have the supplier look after the desalter operation. Still, plant operators often ask what kind of performance goals they should reasonably expect from their desalter and how and what desalting parameters should be monitored. As a general rule, desalters are designed and purpose-built to meet certain overall performance specifications, depending primarily on source, composition, API gravity, viscosity, salt content, and BS&W content of the crude normally charged to a unit. A properly operating, adjusted, and maintained desalter easily meets these specifications, and some desalters perform a lot better than that. The overall performance goals listed in Table 9-1 are typical of what to expect from a desalter. They assume that the incoming raw crude slate was first gravity settled in tankage and has acceptable salt and BS&W contents.

While the plant operators and laboratory staff are qualified to adjust desalter operations as needed on a "per shift" or daily basis and carry out routine analytical tests, major changes are often based on prior testing and recommendations by the supplier. It makes some sense to have one supplier held responsible for the entire corrosion control program on a unit including desalting, but there are also potential issues to be considered:

- Suppliers on all but the largest plants are not always available, and certainly not around the clock.
- Midsize and small plants, especially those away from major plant locations, are visited by the supplier only once or twice a week – and sometimes even less than that.

- Even daily visits (actually meaning 5days/wk) are normally insufficient to identify and address desalter upsets and other issues in a timely fashion.
- Despite all good intentions, suppliers are sometimes tempted to cut corners or recommend measures that may or may not be the best choice for solving a given problem.

There are definite justifications for installing instrumentation that allows the ongoing monitoring and automatic control of at least a number of critical desalter variables by the plant operators. Inadequate monitoring of the desalter operation is probably the major cause of desalter upsets on many units. Without ongoing monitoring, insufficient data are accumulated to allow pinpointing the underlying causes of subsequent corrosion and fouling issues on the unit.

TABLE 9-1 OVERALL PERFORMANCE GOALS FOR SINGLE AND 2-STAGE DESALTERS

Variable	Light Crude (API Gravity>30)		Intermed. Crude (API Gravity 20-30)		Heavy Crude (API Gravity<20)	
	1-Stage	2-Stage	1-Stage	2-Stage	1-Stage	2-Stage
Salt Removal, %	90-95	98-99	85-90	97-98	85-90	90-95
Dewatering, %	>95	>95	>90	>90	>85	>85
Solids Removal, %	60-90	60-90	50-80	50-80	50-80	50-80
Water in Desalt. Crude, %	<0.3	<0.3	<0.3	<0.3	<0.3	<0.3
BS&W in Desalt. Crude, %	<0.3	<0.3	<0.3	<0.3	<0.3	<0.3

Some variables are already monitored routinely on the control board of most crude units including crude throughput, desalter water rate, oil-in-brine content, and various stream temperatures. Other variables are or can be checked on a daily basis or as needed. Ideally, the variables listed below in Table 9-2 should probably be included in a regular monitoring program.

TABLE 9-2 MONITORING DESALTER PERFORMANCE

Variable	Frequency
Salt and BS&W (raw crude)	Daily check
Mix valve ΔP	Daily check
Water in oil phase	Daily check
Interface level and thickness	Daily check
Demulsifier rate/dosage	Daily check
Electrical parameters	Daily check
Salt and BS&W (desalted crude)	Daily check
Desalting efficiency, %	Daily calculation
Water removal efficiency, %	Daily calculation
Solids removal efficiency, %	Daily calculation

9.4 Properties of Raw Crude

For the purpose of desalting, crudes can be divided into three main categories: light crude with API gravity above 30, intermediate crude with API gravity between 20 and 30, and heavy crude with API gravity below 20. In practice, however, most crudes with API gravity below 30 present some desalting issues regardless of what they are called. Light crudes are relatively easy to desalt, because their water-in-oil emulsions separate much faster than those of any other type of crude. The major problem with light crude is, of course, that they are difficult to find these days. Available crude supplies have not only become scarcer and costlier, but also heavier, sourer, dirtier, and more difficult to desalt. This trend is likely to continue and will probably get worse and take its toll of desalter operations. Intermediate and heavy crude are an entirely different matter. While there are other problem crudes including so-called opportunity crudes, most desalter and consequently many overhead corrosion issues are caused by crudes with API gravity below 30, and especially API gravity below 20. Heavy crudes also contain 20-50% residuum and are almost as heavy as water. They are nearly impossible to separate from water by straight-forward gravity settling. Heavy crudes also contain impurities that tend to stabilize emulsions, increase oil carryunder in the brine discharge, and water carryover in the desalted crude. More specifically, intermediate and heavy crudes present desalting issues whenever the salt content reaches 100–400mg/L (30-120PTB) and the filterable solids content 30–200mg/L (10-60PTB). Other potential problems are:

- High BS&W content, free-water, vanadium (V), and other metals contents.
- Poor heat transfer in combination with excessive fouling of the cold preheat exchangers (CPEs) that decreases desalter inlet temperatures to as low as 105 °C (220 °F).
- Tight emulsion/brine interface layers ("cuff" or "rag") that are difficult to break.

Certain measures can be taken to minimize these issues:

- Increasing the difference in the specific gravity of the brine and crude emulsion (ΔĐ) by blending heavy crude with light crude, light oil distillates, or even naphtha.
- Increasing the desalting chemical dosage to 12-24mg/L or twice the usual rate for other crudes.
- Using desalting chemicals containing special wetting agents.
- Reducing the crude viscosity by blending, raising the desalter temperature, or increasing the desalter water rate provided the caveats listed below are kept in mind.
- Keeping slop oil out of the desalter feed and adding it at a controlled trickle rate downstream of the desalter.

Crude blending can have negative side effects by creating emulsions that are difficult to break. Moreover, indiscriminate blending with naphthenic acid crudes can destabilize asphaltenes and encapsulate salt crystals that become difficult to remove. The same compounds that interfere with gravity settling can also cause electrical conductivity problems. With most crudes, the electrical field strength is proportional to conductivity, but with highly conductive crudes, electrical field strength changes exponentially with conductivity. As a result, the voltage drop across the electrodes decreases precipitously. To avoid overloading and damaging the electrical equipment, some desalters have adjustable electrodes that permit changing the voltage drop across the electrodes. When problem crudes are run, electrode spacing is increased by anywhere from 25-100mm (1-4 in.) to increase voltage drop.

9.4.1 Salt Content of Raw Crude. The average salt content of raw crude going to the desalter should normally not exceed 100mg/L (30PTB), while the maximum salt content should normally not exceed 250mg/L (80PTB). Salt content is determined preferably by potentiometric titration. Conductometric methods are unreliable at low salt contents, because the inherent crude conductivity instead of salt content is measured. Sudden increases in raw crude salt content result in desalter upsets, unless the desalter parameters can be immediately adjusted which is rarely the case. Excessive salt contents usually reflect insufficient treating during crude production, gathering, transportation, or intermediate tankage operations. The reasons should be investigated and corrected, if possible.

9.4.2 BS&W Content of Raw Crude. The BS&W content of raw crude regardless of origin or API gravity should normally not exceed 1.0%, but heavy crude may have BS&W contents up to 3.0%. Often, BS&W is reported separately as bottoms sediment (BS) and water. In that case, BS is the difference between the results of two separate analyses: a standard BS&W analysis by centrifuging a diluted crude sample, and a standard water analysis by distillation. As noted in the chapter on reducing salt in raw crude by tank settling, an AGAR probe™ can be used to continuously monitor BS&W of incoming raw crude. AGAR probes™ have been found quite suitable for this purpose and provide a continuous 4-20mA output signal that can be made available on the plant's control board.

9.4.3 Filterable Solids Content of Raw Crude. Filterable solids content of raw crude should not exceed 60mg/L. Filterable solids range from 1-20μm (0.04-0.8mil) in size and include mineral matter and corrosion products, such as iron sulfide (FeS) or iron oxide (Fe_2O_3) present in crude. Filterable solids are determined with a standard filtration test at 85 °C (180 °F), using a 1μm (0.04mil) filter and a hot toluene and hot water wash, respectively.

9.5 Properties of Desalted Crude

To minimize overhead corrosion, the salt content, BS&W content, and filterable solids content of desalted crude must be properly controlled. In practice, the average salt content of desalted crude should not exceed 3.5mg/L (1PTB), while the maximum salt content should not exceed 10mg/L (3PTB). The BS&W content of desalted crude should not exceed 0.3% while the filterable solids content should not exceed 60mg/L.

9.5.1 Salt Content of Desalted Crude. Given the salt removal specifications listed earlier and assuming the desalter is working properly, the salt content of desalted crude is directly proportional to the salt content of raw crude, as shown in Tables 9-3 through 9-5. As a general rule, two-stage desalters are needed to achieve the recommended average salt levels of 1PTB or maximum salt levels of 1PTB in desalted crudes, except when light crudes containing less than 70mg/L (20PTB) salt are charged. As an alternative to two-stage desalting, 3–5PTB caustic (NaOH) can be added to desalted crude downstream of a single-stage desalter. Caustic addition lowers the amount of hydrogen chloride (HCl) going overhead and is equivalent to raising the salt removal efficiency to 99%, or approximately that of two-stage desalting. The use of caustic is discussed in detail in Chapter 12 of the *Guide*.

9.5.2 BS&W Content of Desalted Crude. With proper desalting, the BS&W content of desalted crude should not exceed 0.3% corresponding to at least 70% removal from raw crude. When BS&W of desalted crude is reported separately as bottoms sediment (BS) and water, the results of each should be less than 0.15%. To achieve these low BS&W values, the brine droplet settling rate must exceed the upward flow rate of desalted crude in the desalter vessel to minimize water carryover with desalted crude.

9.5.3 Filterable Solids Content of Desalted Crude. As with raw crude, the filterable solids content of desalted crude should not exceed 60mg/L. Filterable solids range from 1-20μm (0.04–0.8mil) in size and include mineral matter and corrosion products, such as iron sulfide (FeS) or iron oxide (Fe_2O_3) in crude. Filterable solids are determined with a standard filtration test at 85 °C (180 °F), using a 1μm (0.04mil) filter and hot toluene and water washes.

TABLE 9-3 EXPECTED SALT CONTENT OF DESALTED LIGHT CRUDE

Salt Content of Raw Crude, PTB	Salt Content of Desalted Light Crude, PTB		
	1-Stage Desalter	1-Stage Desalter + Caustic (NaOH)	2-Stage Desalter
20	1 - 2	<1	<1
30	1.5 - 3	<1	<1
40	2 - 4	<1	<1
60	3 - 6	<1	<1 - 1.5
80	4 - 8	1	1 - 2

TABLE 9-4 EXPECTED SALT CONTENT OF DESALTED INTERMEDIATE CRUDE

Salt Content of Raw Crude, PTB	Salt Content of Desalted Intermediate Crude, PTB		
	1-Stage Desalter	1-Stage Desalter + Caustic (NaOH)	2-Stage Desalter
20	2 - 3	<1	<1
30	3 - 4.5	<1	<1
40	4 - 6	<1 - 1	<1 - 1
60	6 - 9	1	1 - 1.8
80	8-12	1.5	1.5 - 2.5

TABLE 9-5 EXPECTED SALT CONTENT OF DESALTED HEAVY CRUDE

Salt Content of Raw Crude, PTB	Salt Content of Desalted Heavy Crude, PTB		
	1-Stage Desalter	1-Stage Desalter + Caustic (NaOH)	2-Stage Desalter
20	2 - 3	<1	<1
30	3 - 4.5	<1	<1
40	4 - 6	<1 - 1	<1 - 1
60	6 - 9	1	1 - 3
80	8 - 12	1.5	1.5 - 4

9.6 Desalting Objectives

Desalting is based on the principles that water is heavier than crude; water and crude are immiscible and form stable water-in-crude emulsions; and the water-in-crude emulsions can be subsequently separated (broken) with an electrical field and/or suitable desalting chemicals. This is accomplished by a mix valve for forming the water-in-crude emulsion, by the desalter vessel proper for separating the water from crude, and by equipment for creating an electrical field to aid in the separation of water and crude. The process consists of several interconnected steps:

- Adding desalter water upstream of the special mix valve that forms the water-in-oil emulsion.
- Injecting a desalting chemical (usually a demulsifier) to aid in the desalting operations.
- Extracting the salt crystals and brine pockets from the crude into the water droplets of the emulsion.
- Combining the water droplets by applying an electrical field.
- Gravity settling the water droplets to form a separate salt brine layer.
- Sending the accumulated brine layer to a brine settling tank for further treatment or disposal.
- Sending the desalted crude to the hot preheat exchangers (HPEs) for further heating.

Depending on throughput rate, API gravity, salt content of the incoming crude, and desired salt and water content of the desalted crude, several desalter vessels may have to be configured in parallel (multitrain desalters), in series (multistage desalters), or both to achieve these objectives. On most units, however, either single-stage desalters are used as shown in Figure 9-1, or two-stage desalters shown in Figure 9-2 are used.

FIGURE 9-1 SCHEMATIC OF TYPICAL SINGLE-STAGE CRUDE DESALTER

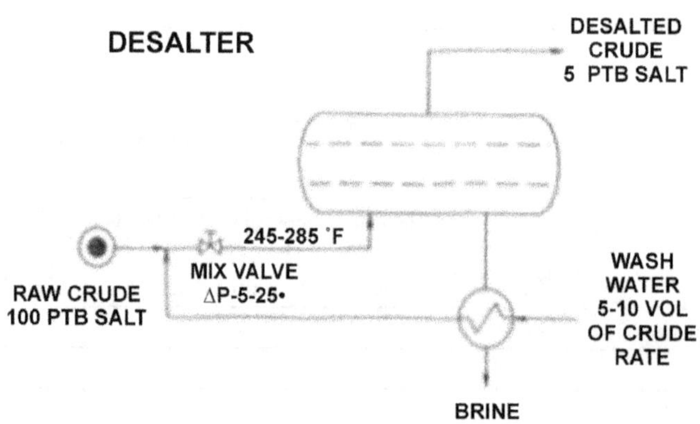

FIGURE 9-2 SCHEMATIC OF TYPICAL TWO-STAGE CRUDE DESALTER

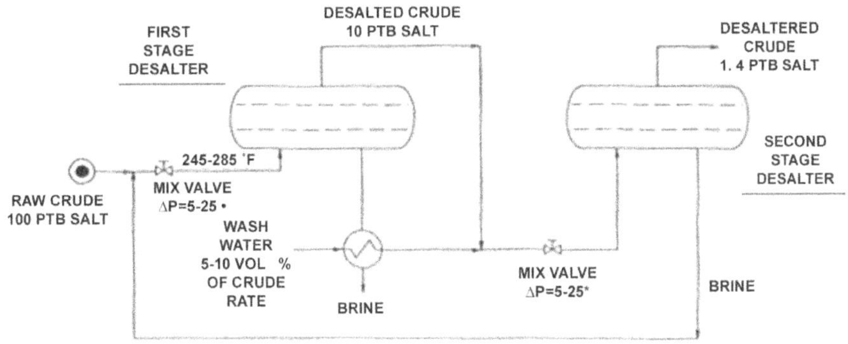

9.7 Mix Valves

The first and most vital step in desalting is the even dispersion of desalter water throughout the incoming crude to form a water-in-crude emulsion. This is accomplished with the mix valve that is a specially designed globe valve located just upstream of the desalter vessel. A slight rotation of the valve stem produces a relatively long and narrow circular opening. The width of the opening determines the pressure drop across the valve that, in turn, determines the degree of viscous shear. Ideally, the crude and water phases are stretched out in the form of alternating "sheets" of liquid. The thickness of the oil sheets ultimately represents the contacting efficiency. Adjusting the mix valve setting is an important aspect of desalter operation and is usually the responsibility of the supplier of desalting chemicals on the unit.

9.7.1 Mix Valve Settings. The mix valve produces, over a brief contact time, a very high shearing action between crude and desalter water and forms the very small water droplets needed for desalting. With some desalters, another valve or a static mixing element is placed ahead of the mix valve for better blending of crude and water. If another valve is used for this purpose, it should be designed for blending rather than shearing. A static mixing element can also be placed after the mix valve to prolong the contact time between crude and desalter water, or to blend a desalting chemical into the emulsion. Variable-speed inline dynamic mixers downstream of the mix valve and upstream of the desalter vessel can be used to improve mixing, but dynamic mixers are costly. The goal of proper mix-valve settings is to maximize salt removal with minimum brine carryover in desalted crude, and minimize oil carryunder in brine effluent. This is achieved with water droplets averaging 0.5μ (0.02mil) in size and approximately two droplet-diameters apart while obtaining the best possible mixing without creating an emulsion that cannot be separated (broken) in the desalter.

9.7.2 Mix Valve Pressure Drop. The optimum mix valve pressure drop (ΔP) is 0.7-2bar (10-30psi) for light crude, 0.5-1.5bar (7-22psi) for intermediate crude, and 0.2-1bar (3-15psi) for heavy crude. To establish optimum mix-valve settings for different crudes, the usual procedure is to change the mix valve setting incrementally at 30-minute intervals and obtain a ΔP versus salt versus BS&W profile for the desalter, as shown by the "bathtub" curves in Figures 9-3 and 9-4. For proper desalting, the mix-valve setting is then decreased approximately 0.14bar (2psi) beyond the point where the BS&W starts to rise. Mix valves are available to allow manual and remote adjustment. Mix valve should never be used to balance crude flow into parallel desalter trains. Only specially designed Globe valves should be used as mix valves. For greater accuracy, a differential pressure gage instead of two separate gages should be used to determine mix-valve Δp. Intermediate and heavy crude are more sensitive to over-mixing than light crude and the normal mix-valve Δp must be reduced. Low mix-valve pressure drop cause undermixing and increase the salt content of desalted crude due to insufficient salt extraction, but have no significant effect on the BS&W content of desalted crude. High mix-valve pressure drop causes overmixing and increases the salt and BS&W contents of desalted crude, because the emulsion is difficult to break, but increases the demulsifier requirements and hence the desalting cost.

FIGURE 9-3 EFFECT OF MIX VALVE PRESSURE DROP ON SALT CONTENT OF DESALTED CRUDE

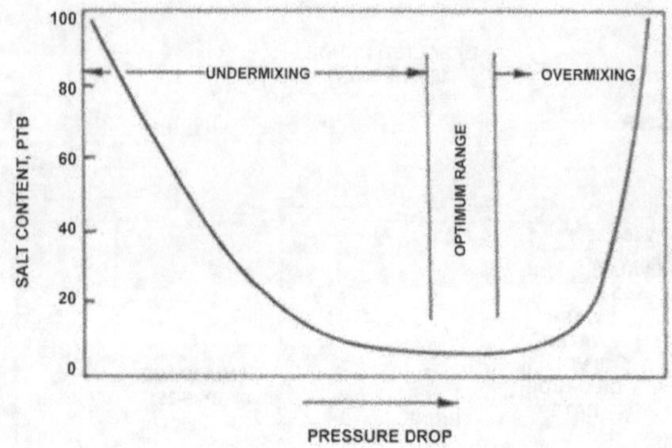

FIGURE 9-4 EFFECT OF MIX VALVE PRESSURE DROP ON BS&W CONTENT OF DESALTED CRUDE

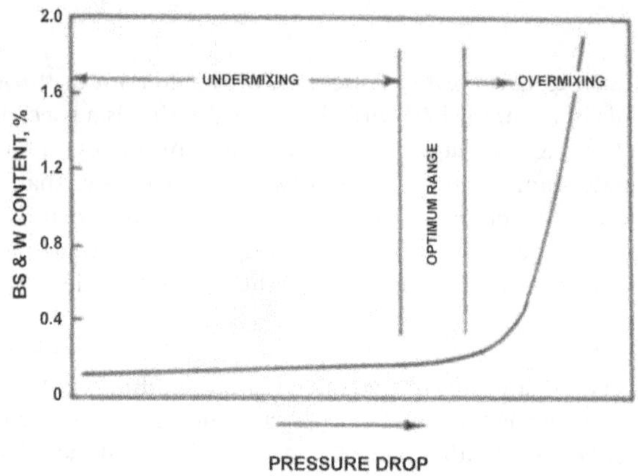

9.8 Desalter Vessels

The desalter vessel proper is designed to carry out three basic tasks: coalescing the suspended water droplets in the water-in-crude emulsion, combining the coalesced water droplets to form a salt brine layer, and separating the brine layer from the desalted crude by gravity settling. These operations normally require lengthy vessel residence times unless the process is speeded up by applying an electrical field, adding a desalting chemical (usually a demulsifier), or both. Because of the large volumes of crude involved, crude units require very large desalter vessels with special design features. They are easily recognized among other plant equipment, as shown by the example in Figure 9-5.

FIGURE 9-5 APPEARANCE OF SINGLE-STAGE CRUDE DESALTER

Courtesy of NATCO Corporation.

9.8.1 Desalter Vessel Design. Desalter vessels are constructed of carbon steel. This means that they are likely to corrode, unless the desalter water pH is maintained at 7 to 7.5. Accumulations of silt, dirt, scale, corrosion products, and other deposits must be removed at regular intervals by mud washing to avoid underdeposit corrosion and maintain vessel capacity. Numerous types, designs, and variations of desalter vessels have been used over the years. Examples are listed below in order of increasing effectiveness:

- Conventional low-velocity AC desalters with 2 horizontal electrode grids (many suppliers).
- Conventional AC desalters with 3 horizontal electrode grids (TriGrid™, TriGridmax™, TriVolt™ by NATCO).
- Conventional AC desalters with 4 horizontal electrode grids (TriVoltmax™ by NATCO).
- Cylectric™ high-velocity AC desalters with 2 horizontal electrode grids (by PETRECO).
- Bilectric™ AC desalters with 3 horizontal electrode grids (by PETRECO).
- Dual-polarity AC/DC desalters with vertical electrode plates (by NATCO).
- Electro-dynamic desalters with electrode grids of varying conductivity or voltage resulting in two stages in a single vessel (by NATCO).

9.8.2 Desalter Vessel Size. The recommended desalter vessel size depends on crude throughput, crude viscosity, specific gravity of the brine and emulsion, and desalter temperature. To check if the size of an existing desalter vessel is still satisfactory despite repeated changes in any of the above parameters, the following relationship can be used:

$$A = R / (46 \log \Delta\rho/\mu + 65) \qquad \text{(Eq.9-1)}$$

where:
A = horizontally-projected vessel area over the heads (HPAH), ft^2
R = crude throughput, BPD
$\Delta\rho$ = difference in the specific gravity of brine and emulsion, g/cm^3 at desalter temperature
μ = emulsion viscosity, poise (g/cm/sec) at desalter temperature

For example, a 10,000TPD (75,000BPD) crude unit charging API 25 crude at 150 °C (300 °F) requires a horizontally projected vessel area over the heads (HPAH) of $72m^2$ ($773ft^2$). If the vessel diameter is 3m (10ft), the overall length of the vessel should be approximately 23m (77ft), including the lengths of the heads.

9.8.3 Desalter Residence Time. Assuming semispherical heads are used, the above vessel has a capacity of 163,000L (43,000gal), not counting internals. The calculation assumes a $\Delta\rho = 0.1$ and $\mu = 0.02$ for API 25 crude at 150 °C (300 °F). The following relationship is used to determine crude residence time:

$$T = C / 1.75\, R \qquad \text{(Eq.9-2)}$$

where:
T = residence time, hr
C = vessel volume, gal
R = crude throughput, BPD

The vessel in the above example provides a residence time of approximately 20min. for crude which is somewhat lower than the 30min. sometimes recommended. Residence times as long as 60min. have been used for heavy crudes. From Equation 9-1 and 9-2 it is apparent that any significant increase in crude throughput due to unit revamp may well require an increase in desalter vessel size. The same applies to a significant decrease in API gravity of the crude charged. If the vessel size is not increased, the crude residence time is decreased and salt and water carryover into the desalted crude is increased. The brine residence time is approximately 20 times longer than the crude residence time, because the water rate is usually 5% of the crude rate. Long water residence times improve brine gravity settling and ensure that crude under-carry with effluent brine is kept as low as possible.

9.8.4 Flow through Desalter Vessels. The construction details for a typical desalter with three horizontal bar grids are shown in Figure 9-6. In most desalters, the incoming crude emulsion enters just above the oil/water interface layer and flows upward through the electrode grids to an outlet collector. As the emulsion passes through the electrical field, water droplets begin to coalesce and settle out to form the brine layer in the lower half of the vessel. Desalted crude exits the vessel at the top, while water (brine) exits at the bottom. The brine discharge pump controls the level of the oil/water interface layer.

FIGURE 9-6 DETAILS OF AC DESALTER WITH 3 HORIZONTAL BAR GRIDS

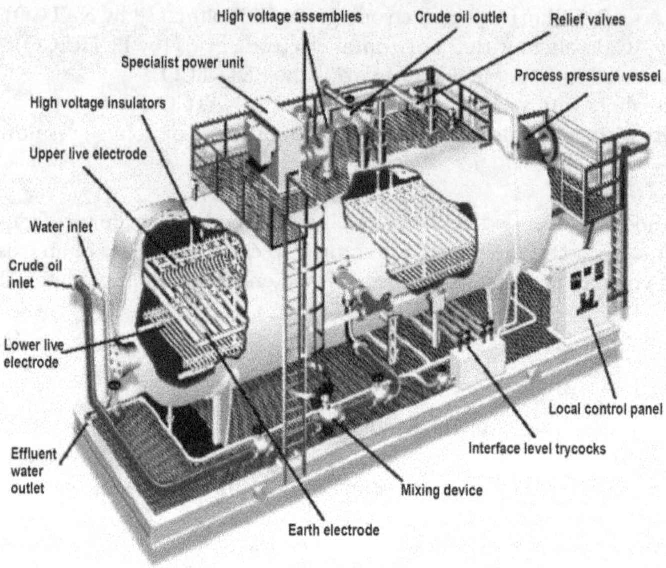

Courtesy of NATCO Corporation.

Desalter inlet and outlet nozzles can be located and sized to produce a uniform flow pattern through the charged desalter grids. It is more efficient, however, to establish uniform flow by controlling the velocity, direction, and residence time of the emulsion. For this reason, properly designed inlet distributor headers and outlet collectors are used. Some inlet distributors have adjustable gaps. Uniform placement of the emulsion between the electrical grids increases desalting efficiency. Therefore, some method is usually available for adjusting the emulsion layer within the electrical field. On some desalters, this is accomplished by moving an adjustable electrical grid up or down in response to power consumption or feedback signals from a level probe. As the emulsion nears the edge of the electric field, droplet coalescence slows down due to the increased distance between droplets. This can be partly offset by adding water in the desalted crude zone above the electrode grids. The water is usually taken from the brine layer and reinjected as relatively coarse droplets to prevent droplet entrainment and carryover. When the coarse droplets enter the electric field, they break up into stable droplets that become uniformly distributed throughout. These droplets sink through the emulsion and sweep up any entrained small brine droplets as they sink.

9.8.5 Effect of Gravity on Droplet Settling. Besides being an electrical coalescing vessel, a desalter can also be a gravity settler, albeit a rather inefficient one. Therefore, operating a desalter with inoperative electrical equipment is ill advised.[1] As the coalesced water droplets descend through the bulk emulsion phase, they combine to form a brine phase while the emulsion turns into desalted crude. Normally, the emulsion phase occupies the upper two-thirds of the vessel, while the brine phase occupies the lower one-third. The two phases are separated by an interface layer. In accordance with Stokes' Law, the rate at which droplets descend through the emulsion phase depends on droplet diameter, difference in the specific gravity of brine and emulsion, and the emulsion viscosity. The velocity of descending water droplets is proportional to droplet size. Moreover, for a given droplet size, the velocity is highest when the difference in specific gravity of brine and emulsion (ΔĐ) is high and the emulsion viscosity (μ) is low. In practice, this confirms that light oils are much easier to desalt than heavy oils. It also means the desalter water rate and temperature must be

1. The author encountered a desalter in an overseas refinery that had been inoperative for years, because of government orders not to interrupt the crude flow in order to repair the electrical equipment. As a result, the refinery had to bypass various overhead coolers and condensers at two-week intervals to replace leaking tubes. It never occurred to anyone that they might as well bypass the desalter entirely for all the good it did and then repair the electricals.

high enough for the viscosity to be sufficiently low. Finally, it means the desalter vessel must be large enough to provide a long-enough crude residence time.

9.8.6 Desalter Vessel Maintenance. Many desalter vessels are in need of regular mud washing and/or periodic cleaning. Accumulated silt, dirt, corrosion scale, and other solids not only reduce desalter capacity, but also lead to poor desalting. They also cause other problems, such as corrosion of the desalter vessel, water and solids carryover with desalted crude into downstream components, and oil carryunder with effluent brine into the brine settling tanks. During normal operation, a sludge (mud) layer builds up on the bottom of desalter vessels. Desalter mud contains the same sand, silt, drilling mud, and sediments of various types; wax particles and asphaltenes; corrosion products; and hard water scale particles that enter the desalter with raw crude. Calcium naphthenate emulsions and deposits are a recent problem seen during crude production. Calcium naphthenate formation can be suppressed by maintaining desalter pH between 5 and 6 and/or adding suitable process additives.

Regular mud washing can help keep the inside of the desalter vessel relatively clean by agitating and removing settled mud from the bottom of the desalter vessel. Regular mud washing reduces the formation of a compacted mud layer in the desalter vessel and increases the vessel volume available for the brine phase, the net desalter-water residence time, and the time between scheduled desalter cleaning. It also decreases periodic peaks of oil carryunder in effluent brine (black water), the solids content of effluent brine, and the chemicals costs at the brine settling tank.

9.9 Mud Washing Procedures

Several methods are used for mud washing desalter vessels:

- Applying repeated "shots" of steam, lasting from 15-30sec each through one or more special steam headers. However, long or continuous steam injection causes excessive stirring, interferes with droplet coalescing and settling, and wastes steam.
- Flushing repeatedly with desalter water added through downward-pointed nozzles in several long distributor pipes. Like steam injection, long or continuous water flushing causes excessive stirring and interferes with droplet coalescence and settling.
- Using various types of mechanical scraping devices to prevent, or periodically loosen and remove, any mud build-up. Eventually, these devices tend to suffer from corrosion and mechanical issues and, as a rule, are not too reliable.

To avoid desalter upsets, a desalter should never be bypassed for mud washing during normal operations, even if done only for a short period of time. Mud washing procedures should be optimized for each desalter of a two-stage desalter. While the first stage accumulates most of the mud build-up, both stages should be monitored regularly. If the electrodes start to arc during mud washing with desalter water, the interface level may have to be lowered temporarily. This should be done in steps to avoid increasing oil carryunder. With regular mud washing, 3-5yrs of uninterrupted desalter operation can be expected before a thorough cleaning at a scheduled turnaround is required. Some operators believe that having black desalter water is a sign that solids are removed with brine instead of going to the preheat exchangers with desalted crude. More likely, however, black water is caused by a growing layer of mud in the vessel that interferes with proper desalting. Desalter vessel cleaning should be carried out by the same cleaning contractor that also cleans other components on the unit during a turnaround.[2] At the same time, desalter vessel cleaning presents an excellent opportunity for repairing or upgrading the desalter electrical equipment.

9.10 Desalter Electrode Configurations

Effective desalting requires an electrical field of homogeneous field strength. Typically, the electrical field forms between two or more electrodes across the area where the water droplets are expected to coalesce and the emulsion separates. Small water droplets coalesce in an electrical field when each droplet develops an opposite electrical charge (+ or -) at its two poles. In the proximity of another charged droplet, opposite poles are drawn together and the droplets combine to form larger droplets. Crystalline salts in crude are not affected by electrical fields. Depending on desalter

2. J. Gutzeit, *Cleaning of Process Equipment and Piping – A Complete How-To Manual*, MTI publ. no. 51 (St. Louis, MO: Materials Technology Institute of the Chemical Process Industries, Inc., 1997).

type, several electrode configurations are used. Each has its own characteristics, advantages, and disadvantages. In general, any planar grid layout creates a homogeneous electrical field that promotes uniform droplet movement through the emulsion toward the brine layer. A nonhomogeneous field is less desirable, because it tends to move the droplets in the direction of the higher field strength. Typical crude desalters either use horizontal electrode grids with AC fields, or vertical perforated electrode plates with AC/DC fields. An example of each type is described below:

9.10.1 AC Desalters with Horizontal Electrode Grids. Conventional low-velocity AC desalters use an arrangement of parallel horizontal bar grids to establish an electrical field. The basic 3-grid layout consists of upper and lower charged (live) grids and an intermediate grounded grid.

Grids are spaced 150-200mm (6in.-8in.) apart. The desalter vessel itself and the brine layer are also grounded, as shown in Figure 9-7. AC fields are established between the charged grids and the grounded grid and water layer.

FIGURE 9-7 AC DESALTER WITH THREE HORIZONTAL ELECTRODE GRIDS

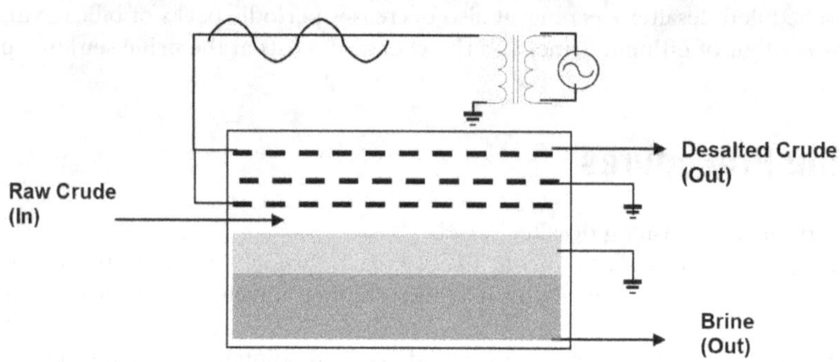

9.10.2 AC/DC Desalters with Vertical Electrode Plates. AC/DC desalters establish a pulsed DC electrical field within an array of suspended vertical plates. Two sets of perforated plates are alternately charged +/0 or 0/-, using suitable connections to rectifiers and transformers. Each set of plates is connected to the same end of the transformer winding, while the other end is connected to ground, as shown in Figure 9-8. This type of setup produces a plate-to-plate DC field and a plate-to-ground AC field. In this way, the advantages of uniform droplet charge and movement in a DC field are combined with the film-rupturing capability of an AC field, as discussed below. In addition, AC fields are less likely to cause desalter vessel corrosion compared to DC fields.

FIGURE 9-8 AC/PULSED DC DESALTER WITH VERTICAL ELECTRODE PLATES

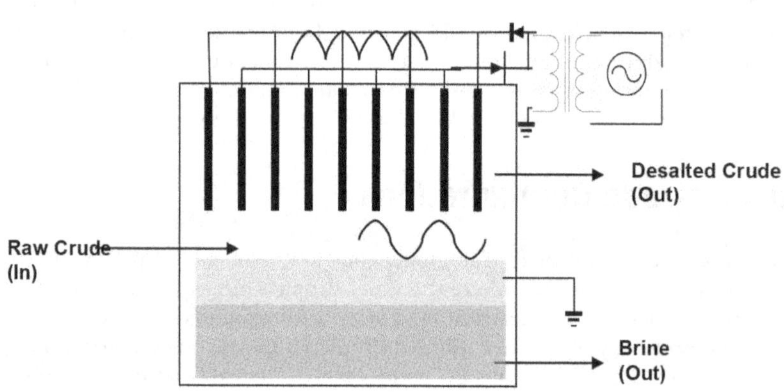

9.11 Electrical Field Strength

When the desalter electrodes become outdated and/or damaged, the desalter can no longer produce a homogenous electrical field and turns into a simple, but inefficient settling drum. Therefore, it should not be surprising to find that overhead corrosion issues often continue despite of what appears to be, at least on the surface, adequate crude desalting. The field strength must be strong enough for small droplets to coalesce without breaking larger drops apart. For any given field strength there is a range of stable drop sizes that depends on the ability of the field to transport the drops without breaking them apart. Field strength is affected by a number of factors that need to be considered when troubleshooting issues with the electrical equipment, including electrical conductivity, current hysteresis, applied voltage and frequency, and droplet size.

9.11.1 Electrical Conductivity. With most crudes, field strength is directly proportional to conductivity, while with other crudes it changes exponentially with conductivity. As a result, small increases in electrical current produce large increases in field strength and raise power consumption to the point where the electrical system of the desalter quickly exceeds its design capacity and breaks down.

9.11.2 Current Hysteresis. An alternating AC field produces oscillation and flow not only of water droplets, but also of polar and certain other molecules in the emulsion. The resulting dielectric losses cause energy dissipation due to hysteresis of the potential and current between the electrodes.

9.11.3 Applied Voltage. The applied voltage in desalters normally ranges from 16-20KV. As would be expected, voltages near the high end of the range improve desalter performance when low-conductivity crudes are charged.

9.11.4 Applied Frequency. The great majority of desalters operate at 50 or 60Hz, but frequencies up to 200Hz seem to improve desalter performance. Ideally, the electrical field has a high-intensity zone to coalesce small droplets, but not high enough to break up large droplets. This high-intensity zone is followed by a zone of gradually decreasing field strength to shift the equilibrium toward larger droplet sizes.

9.11.5 Droplet Size. Field strength should not reach the level where large water droplets break apart. This level depends on the size of the droplets and occurs with the largest droplets when field strengths exceed approximately 5KV/cm (13KV/in.).

9.12 Electrical Entrance Bushings

Regardless of desalter type, special entrance bushings are required to transfer the 16-20KV high-voltage current through the wall of a desalter vessel filled with hot crude. Entrance bushings are critical maintenance items and leakage or failure usually requires bypassing the desalter or shutting the unit down. Therefore, careful materials selection, design, assembly, and regular inspection are essential. A typical design for the piping, oil seal, and entrance connection of desalter bushings is shown in Figure 9-9. Bushing materials must be able to withstand hot crude and brine and maintain their insulating value at temperatures as high as 160 °C (320 °F) over extended periods of time. They must also have a high dielectric constant in order to reduce current leakage.

FIGURE 9-9 DESALTER HIGH-VOLTAGE ENTRANCE BUSHING

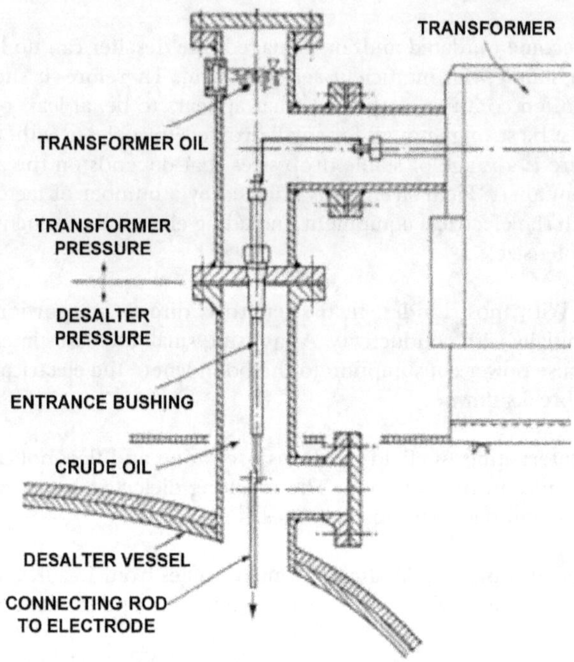

Typical bushing materials include ceramics, glass-like materials, and polytetrafluoroethylene (PTFE). All meet most desalter requirements, but glass-filled PTFE is less likely to attract surface fouling which can lead to short-circuiting. Sintered PTFE has a relatively high permeability and is likely to absorb water that can lead to short-circuiting. Pure PTFE (Teflon) has been found to be highly satisfactory when carefully machined to the correct shape and dimensions.

9.13 Oil/Water Interface Control

Special instrumentation is required for locating the emulsion/brine interface band ("cuff" or "rag") in a desalter vessel so that the level can be adjusted and maintained by adjusting the brine discharge valve. Four types of level detection devices are usually used, but two have different issues and should be avoided. Those include internal and external floats that are weighted to float at the emulsion/water interface layer and can be used for crudes over an API gravity range of 10 degrees. They are inaccurate especially with heavy crudes, because of the relatively small density differences between emulsion and brine. Moreover, the internal floats can be difficult to repair.

This leaves capacitance and AGAR probes™. Capacitance probes function by sensing the difference in capacitance between emulsion and brine and can be used with all crudes regardless of gravity. AGAR probes™ use energy absorption technology to detect water-in-oil or oil-in-water phases. The latter probes have been found to be quite reliable for continuously monitoring and controlling the desalter interface and are not affected by crude gravity. Three AGAR probes™ can be used to provide continuous 4-20mA output signals that can also control the demulsifier injection rate and brine discharge valve settings. Probe readouts can be made available on the control board.

A typical setup is shown in Figure 9-10. Here one probe controls the emulsion/brine interface level by operating the brine dump valve, one probe controls demulsifier addition rate, and one probe monitors the brine level (low-level alarm). An additional probe is often installed upstream of the desalter to monitor the BS&W of incoming raw crude, as discussed in the Chapter on reducing salt in crude by tank settling.

FIGURE 9-10 CONTROL OF EMULSION/BRINE INTERFACE, DEMULSIFIER INJECTION, AND BRINE DISCHARGE WITH AGAR PROBES™

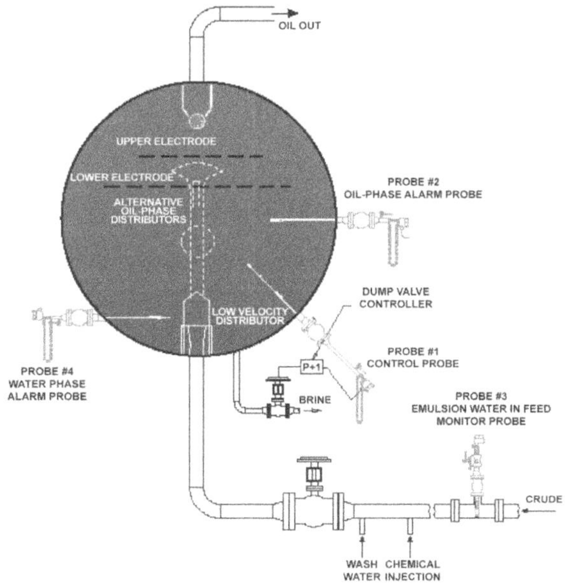

Adapted from G. Fransen, "New Control System Detects Desalter Problems Before Upsets Occur," AICHE 2004 Spring National Meeting, April 2004.

9.14 Height of Oil/Water Interface

The optimum interface height from the desalter bottom is 0.75-1.5m (2.5-5ft) with a desalter vessel having a 3.5-4.2m (12-14ft) diameter and assumes that clear water is below the 0.75m (2.5ft) level. The interface position depends on desalter type, design, and BS&W content of the crude charge, and the Operations Manual that came with the desalter should be consulted for specifics. A thick emulsion band layer at the interface interferes with interface adjustment and may shorten out the electrodes. If necessary, the standard capacitance probes need to be replaced with AGAR probes™ that can continuously monitor and control the desalter interface. The height of the interface determines the voltage across the electrodes and the oil and water residence times. Adequate water separation requires at least a 20min. oil residence time and 200min. water residence time with most crudes. Low interface levels decrease the brine residence time, increase oil carryunder with brine effluent ("black water"), cause separation issues at the API separator, and increase the solids content of the brine effluent due to mud pickup from the bottom of the desalter vessel. High interface levels decrease the oil residence time, increase salt and water carryover with desalted crude, increase the desalter current because the electrical path to ground is decreased, and cause arcing and water redispersion.

9.15 Thickness of Interface Emulsion Band

The optimum emulsion band thickness is 150mm (6in.), or less. Moderately wider emulsion band thicknesses measuring 150-300mm (6-12in.) are caused by semistable emulsions and can be corrected by increasing the demulsifier rate and/or substituting a different demulsifier. Wide emulsion bands measuring 300-600mm (1-2ft) are caused by highly stable emulsions, high solids content of raw crude, heavy crude, or insufficient demulsifier rate and can be corrected by increasing the demulsifier rate, substituting a different demulsifier, or adding a polymer or wetting agent. Some desalters have a piped-up "cuff header" for extracting a wide emulsion/brine interface band, when other measures to control thickness are ineffective. This avoids having to draw down the desalter contents to the clean-oil level to get rid of the interface band or dump the emulsion, if the above measures are ineffective.

9.16 Oil Carryunder in Effluent Brine

The maximum oil content in brine effluent should be 100mg/L with light crudes, 500mg/L with intermediate crudes, and 1,000 mg/L with heavy crude. High oil levels in brine effluent (oil carryunder) overload the brine settling tank and the API separator of the plant. The usual causes are heavy or other problem crudes, excessive or wrong type of demulsifier, and high desalter temperatures that cause increased oil solubility in the brine. To reduce oil carryunder, heavy crudes should be blended with light crudes, gas oil, or naphtha; demulsifier dosages should be decreased; different demulsifiers should be substituted until the brine effluent clears; and desalter temperature should be decreased.

9.17 Emulsion Dumping

The following general procedure is recommended when other attempts at clearing up issues with wide emulsion bands have been ineffective:

- Lower the brine level and drop the top of the emulsion band to the center sample tap on the desalter vessel.
- Increase the demulsifier rate to 3 times the normal rate for approximately 1hour.
- Reduce the demulsifier rate to twice the normal rate for 2-12 hours and observe the effect on the thickness of the emulsion band.
- If the emulsion band thickness has not decreased after 12 hours, dump the vessel content down to the clear-oil level and reestablish a new emulsion/water interface level.
- Leave the demulsifier rate at twice the normal rate as long as the problem crude is charged.

9.18 Desalter Water

Depending on location, availability, and costs, various types of water are used for desalting. Even sea water has been used effectively for crude desalting when no other water was available. As a general rule, however, desalter water with a low salt content and minimum of impurities is preferred to maximize salt extraction, avoid electrical issues with the desalter electrode grids, and minimize corrosion and fouling of the desalter vessel.

9.18.1 Water Sources. For most plants, the following water sources can be considered and are listed below in approximate order of decreasing desirability because of increasing costs:

- Overhead water condensate.
- Vacuum column ejector condensate.
- Stripped sour water.
- Steam condensate.
- Treated boiler feed water (BFW).
- Fresh water.

The following water sources should not be considered, unless there are no other options:

- Surface waters including river or pond water, because of their high levels of oxygen, bio-matter, solids, or debris.
- Brackish waters, because of their high levels of salt, oxygen, bio-matter, solids, or debris.
- Seawater, because of the high levels of salt, oxygen, or bio-matter.

The following water sources should not be used under any circumstances:

- Cooling tower blowdown, because of high levels of oxygen, solids, calcium, magnesium, and silica.
- Water condensate from cat light ends and hydrotreating units, because of high levels of ammonia (NH_3), hydrogen sulfide (H_2S), and cyanide (HCN).

9.18.2 Water Addition Points. Using alternative water addition points allows optimum flexibility when desalting different crude blends. Typically, desalter water is added to the incoming crude upstream of the crude charge pumps. This provides maximum contact between oil and water, but can result in poor control of the shearing action and create an emulsion that is difficult to break in the desalter. An alternative location is downstream of the crude charge pumps, but upstream of the cold preheat exchangers (CPEs). This avoids the above problem and has the additional benefit of providing some degree of water washing to keep these heat exchangers clean. Finally, water can be added just upstream of the mix valve to provide close control of the shearing action with heavy crudes, but may produce insufficient contact time with other crudes. In practice, desalter water for single-stage desalters is often added upstream of the crude charge pumps, with an alternative connection point provided downstream of the pumps for use as necessary. Desalter water for two-stage desalters is often added upstream of the mix valve for the second-stage desalter. The effluent brine from the second-stage desalter is then added upstream of the mix valve for the first-stage desalter. Alternatively, fresh desalter water instead of effluent brine from the second-stage desalter is added upstream of the mix valve for the first-stage desalter.

9.18.3 Water Addition Rate. The optimum desalter water rate is 4-6% for light crude, 5-7% for intermediate crude, and 6-8% for heavy crude, based on the raw crude rate. When slop oil is added upstream instead of downstream of the desalter, the calculated desalter water demand should take into account the amount of water entering with the slop oil, including dissolved water, entrained water, and periodic slugs of water. The amount of desalter water has a direct effect on the stability of the emulsion formed, as shown in Figure 9-11. Less than 4% desalter water provides insufficient dilution and increases the salt content of water droplets and the distance between dispersed water droplets. At the same time, it decreases the rate of droplet coalescence and solids removal. More than 8% desalter water promotes emulsification and increases the density of water droplets in emulsion, the conductivity of the emulsion, water carryover with desalted crude, and oil carryunder with desalter effluent brine ("black water"). At the same time, it decreases crude preheat, because the extra water has to be heated by the crude in the desalter.

9.19 Desalter Temperature and Pressure

The optimum desalter temperature is 95-120 °C (200-250 °F) for light crude, 110-125 °C (230-260 °F) for intermediate crude, and 120-150 °C (250-300 °F) for heavy crude. The optimum desalter pressure is 1.4-1.7bar (20–25psig) above the vapor pressure of crude at the desalter temperature. Insufficient pressure allows the lighter fractions of the crude to boil in the desalter vessel which interferes with gravity settling. While desalter temperature requirements for various crudes depend primarily on the API gravity, other factors, such as crude viscosity or asphaltene content, can also affect desalter temperature. Somewhat lower temperatures can be used with dual polarity desalters compared to conventional desalters. Since desalter temperature depends on the heat exchanged in the upstream cod preheat exchangers (CPEs), it is important to configure the preheat system to provide enough heat even when tube fouling occurs. To err on the safe side, desalters are often operated at temperatures higher than necessary to ensure good performance. In most cases, however, the limiting temperature is 150 °C (300 °F) that is the maximum tolerated by most electrode entrance bushings over prolonged periods of time.

9.19.1 Elevated Desalter Temperatures. The principal advantage of using higher-than-normal desalter temperatures is decreased crude viscosity. This helps compensate for the relatively small difference in specific gravity between heavy crude and water and increases the water settling rate with heavy crude. However, high desalter temperatures increase the salt content of desalted crude as more brine droplets evaporate and leave salt crystals behind. They increase the crude conductivity and electrical power consumption, which can cause the transactors to trip off. Since they also decrease the desalter voltage, droplet coalescence is decreased.

FIGURE 9-11 EFFECT OF DESALTER WATER RATE ON SALT CONTENT OF DESALTED CRUDE

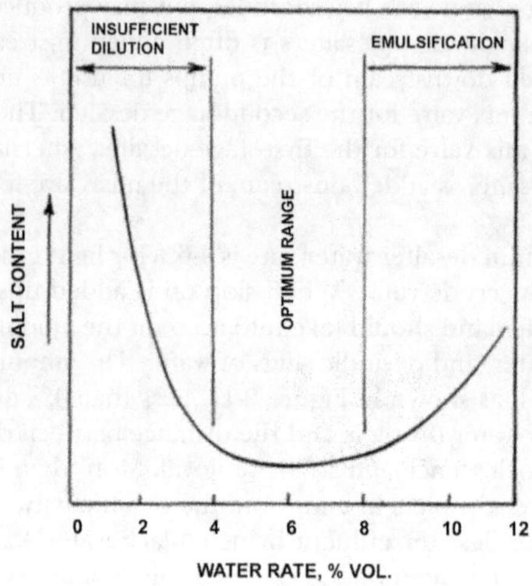

9.19.1 Depressed Desalter Temperatures. The principal advantage of using lower-than-normal desalter temperatures is the lower amount of dissolved water remaining in desalted crude, because water solubility decreases with decreasing temperature. Low temperatures also decrease desalter-vessel corrosion and increase the life of electrical components, especially entrance bushings. However, low desalter temperatures increase oil viscosity which decreases droplet coalescence and increases water carryover with desalted crudes, especially with heavy crudes. They also increase the salt content of desalted crude, because low desalter temperatures increase the amount of entrained water remaining in desalted crude.

9.20 Desalter pH

The optimum pH range for desalter water is 6-8 with most crudes and 5-6 with naphthenic acid containing crudes. Desalter pH affects the amount of acidic (H_2S, phenol, and organic acids) and basic (NH_3) species that partition into the oil phase and brine phase, respectively.

9.20.1 Elevated Desalter pH. High desalter pH affects the pH of desalted crude by decreasing H_2S, phenol, and naphthenic acid carryover, while increasing NH_3 carryover. High desalter pH also affects the effluent brine by decreasing ammonia carryunder while increasing H_2S, phenol, and naphthenic acid carryunder. Elevated pH can also increase calcium (Ca) deposition in the desalter vessel. Some acidic crudes, such as Norwegian Heidrun with a TAN of 5-6, contain approximately 200mg/L calcium that can cause major deposit issues in midstream and downstream separators, desalters, and associated piping. As the pressure is reduced during upstream recovery operations, carbon dioxide (CO_2) is released and the pH of entrained brine is increased. This causes the naphthenic acids to dissociate and react with calcium ions (Ca^{+2}) to form calcium naphthenate [$R\text{-}(CH_2)_n\text{-}COO_2Ca$] deposits. Calcium naphthenate formation in the downstream desalter can be suppressed by lowering the pH of desalter water to 5-6 with hydrochloric acid (HCl) addition. At least one process additives supplier offers a water-soluble complexing agent that can be added to desalter water to both decrease the pH and increase the Ca solubility. This product also includes a corrosion inhibitor to help control desalter corrosion at the low pH values. In one application, 95% Ca removal efficiency was claimed. If corrosion becomes excessive, however, the desalter vessel and various internals may have to be selectively protected with an organic or alloy spray coating or alloyed by weld-overlaying.

9.20.2 Depressed Desalter pH. Low desalter water pH affects the pH of desalted crude by increasing H_2S, phenol, and naphthenic acid carryover, while decreasing NH_3 carryover. Low desalter pH also affects the effluent brine by increasing ammonia carryunder while decreasing H_2S, phenol, and naphthenic acid carryunder. Low desalter water pH is often the result of improper purging of one or more wells that were acidified to improve production and recovery in upstream operations. Subsequent crude shipments are likely to retain high levels of hydrochloric (HCl) or sulfuric (H_2SO_4) acid used for acidification that depress the desalter pH and increase desalter vessel corrosion. If necessary, the pH can be raised again by adding dilute caustic (NaOH) solution to the desalter water. However, caustic addition to the desalter water upstream of the desalter for pH control should not be confused with caustic addition to desalted crude downstream of the desalter for overhead pH control, as discussed in more detail in Chapter 12 of the *Guide.* When overhead water condensate or stripped sour water is used as desalter water, dissolved ammonia (NH_3) and ammonium ($NH_3 \bullet HCl$ and $R\text{-}NH_2 \bullet HCl$) salts also raise the desalter pH.

9.21 Desalting Chemicals (Demulsifiers)

The purpose of desalting chemicals is to improve overall desalting efficiency, reduce water and solids carryover in desalted crude, and reduce oil carryunder in the effluent brine. Most desalting chemicals are demulsifiers that help break the tight emulsion formed by the mix valve and produce separate phases of desalted crude and brine, respectively. Demulsifiers function at the emulsion/brine interface by destabilizing the interface film around each water droplet so that adjacent water droplet can readily coalesce. At the same time, they decrease the thickness of the stable emulsion/brine interface layer and reduce desalter power consumption.

Many commercial demulsifiers for crude desalting are undisclosed proprietary formulations. Others are patented and their composition and method of manufacture are known in some detail. Demulsifiers are usually purchased from the same process additives suppliers that also supply antifoulants, filming-amine corrosion inhibitors, liquid organic neutralizers, and similar products for controlling overhead corrosion and fouling at crude units and elsewhere in the plant. Demulsifiers can be custom formulated to achieve a high rate of water removal from the crude, but at the cost of poor solids wetting and likely oil carryunder in the brine. Other formulations increase the rate of oil removal from the brine, but at the cost of increased water carryover with the desalted crude. Solids wetting is improved, but at the cost of water carryover with the crude. Examples of chemical groups that make up various demulsifier formulations for crude desalting include the following, listed in alphabetical order:

- Alkanolamine condensates for water removal.
- Ethoxylated resins for water, oil, and solids removal.
- Ethoxylated phenol /formaldehyde resins for water and solids removal.
- Oxylated phenols for solids removal.
- Polyglycols for general purpose products.
- Polyglycol esters for oil removal.
- Polyhydric alcohols for general purpose products.
- Sulfonated castor oil for general purpose products.
- Sulfonic acid salts for water and solids removal.

9.21.1 Treatment Rate. The optimum demulsifier rate for most crudes is 6-12mg/L (0.25-0.5GPM/BBL). Demulsifiers are blends of several types of chemicals and contain approximately 25% active ingredients. As a general rule, treatment rates below 3mg/L (0.12GPM/BBL) or 50% of minimum are ineffective, while treatment rates above 24mg/L (1GPM/BBL) or 200% of maximum not only lead to emulsification issues, but also waste money. Heavy crudes require higher treatment levels than are normally used: The optimum demulsifier rate for heavy crudes is 12-24mg/L (0.5-1GPM/BBL). A secondary product may also be required in some applications for optimum results.

9.21.2 Addition Methods. Oil-soluble or oil-based products are added directly into the crude charge, while water-soluble or water-based products are added into the incoming desalter water. In either case, a simple pipe nipple connection with shutoff and ball-check valves can be used: no injection quills or spray nozzles are needed. With heavy crudes, special demulsifiers are sometimes added to the crude receiving tanks 2-3 hrs before the crude is sent to the desalter.

9.22.3 Evaluation Methods. Demulsifiers are usually application-specific: Bench testing of competitive products may be required to find a product that produces the best results. Most process additives suppliers carry small portable test desalters in their vehicles for on-the-spot evaluation of proper demulsifier formulations and dosages. However, field evaluation is usually also required to confirm the results for a given crude mix and desalter.

9.22 Supplier Services

The process additives supplier is usually responsible for the overall performance of the desalter operation, including trouble-shooting, mud washing, and cleaning. As a rule, the supplier is expected to stock sufficient chemicals for uninterrupted desalter operation, supply and maintain the metering pumps and feed control systems for the demulsifier or any other chemical (polymers) used, and ensure that the equipment is capable of being controlled via a distributed control system (DCS). The plant is usually expected to supply and maintain storage and mixing tanks for the chemicals, and supply and maintain the equipment for continuously monitoring the oil-in-brine content and interface level with capacitance or AGAR™ probes.

In many ways, crude desalting is more "Art than Science": No two desalter operations are alike or entirely predictable. Successful desalting requires accurate record keeping and an ability to analyze past issues in order to establish new parameters for future operations. To be able to do so, certain routine activities such as daily status reports, annual summary reports, future performance and monitoring goals, and potential improvement projects are usually expected from the process additives supplier and should be included in the chemicals cost.

9.23 Chapter Summary

The principal purpose of crude desalting is to reduce overhead corrosion by reducing the amount of hydrogen chloride (HCl) evolved in the hot preheat exchangers (HPEs) and fired heater. This is carried out by adding water to the incoming raw crude and using a mix valve to form a water-in-oil emulsion. The water droplets in the emulsion extract salt crystals and brine from the crude. Since water is heavier than and immiscible with crude, the emulsion can be separated into a desalted crude phase and salt brine phase with the aid of an electrical field and suitable demulsifier chemicals.

Desalter upsets are unexpected salt spikes in desalted crude. The root cause of desalter upsets is raw crude with periodic spikes in salt and BS&W content. Another cause of desalter upsets is periodic slop oil addition to raw crude upstream instead of downstream of the desalter. In either case, unit operators are unable to adjust the desalter operation in time to prevent an increase in the chloride levels of overhead water condensate.

Desalters are designed to meet certain overall performance specifications, depending primarily on source, composition, API gravity, viscosity, salt content, and BS&W content of the crudes charged to the unit. As a rule, a properly operating, adjusted, and maintained desalter easily meets specifications. Light crudes are relatively easy to desalt compared to intermediate and heavy crudes. Heavy crudes present additional desalting issues because their low API gravity and high viscosity impede gravity settling. The tendency of heavy crudes to form tight emulsion/brine interface layers make it difficult to separate the desalted crude from brine. Target values/ranges and guidelines are provided for establishing and monitoring important desalter parameters.

9.24 Read More

1. J. W. Barnett, "Desalters Can Remove More Than Salts and Sediment," *Oil & Gas J.*, April 11 (1988).
2. D. W. Choi, "How to Buy and Operate Desalters," *Hydroc. Proc.* March (2005).
3. L. E. Fisher, G. C. Hall, and R. W. Stenzel, "Crude Oil Desalting to Reduce Refinery Corrosion Problems," *MP* 1, 5 (1962).
4. G. Fransen, "New Control System Detects Desalter Issues before Upsets Occur," paper presented at AICHE 2004 Spring National Meeting, April 2004.
5. K. Green, "Improve Processing of Opportunity Crudes," *Hydroc. Proc.* June (2004).
6. D. L. Kronenberger and D. B. Pattison, "Troubleshooting the Refinery Desalter Operation," *MP* 25, 7 (1986).
7. D. L. Kronenberger, "Corrosion Issues Associated with the Desalting Difficulties of Maya and Other Heavy Crudes," paper no. 128, CORROSION/84 (Houston, TX: NACE, 1984).
8. J. E. Lack, "An In-Depth Look at Amine Behavior in Crude Units Using Electrolyte-Based Simulation," paper no. 05570, CORROSION/2005 (Houston, TX: NACE, 2005).

9. P. M. Lindemuth, R. B. Lessard, and M. Lozynski, "Improve Desalter Operations," *Hydroc. Proc.* September (2001).
10. K. K. Mandal, "Improve Desalter Control," *Hydroc. Proc.* April (2005).
11. NPRA Q&A, "Optimizing Desalter Operations Can Help Refiners Cope with Heavy Crudes," *Oil & Gas J.*, October 27 (1997).
12. NPRA Q&A, "Poor Crudes Create Refining Problems," *Oil & Gas J.*, May 3 (1982).
13. G. J. Samuelson, "Hydrogen-Chloride Evolution from Crude Oils as a Function of Salt Concentration," *Proc. API* 34 (III) (1954).
14. S. Tebbal et al., "Analysis and Corrosivity Testing of Eight Crude Oils," paper no. 04636, CORROSION/2004 (Houston, TX: NACE, 2004).
15. L. C. Waterman, "Crude Desalting: Why and How," *Hydroc. Proc.* 44, 2 (1965).
16. S. Bieber and L. Kremer, "Rethink Desalting Strategies When Handling Heavy Feedstocks," revised presentation from the NPRA Annual Meeting, March 9–11, 2008, San Diego, CA.
17. J. Gutzeit, "Controlling Crude Unit Overhead Corrosion – Rules of Thumb for Better Crude Desalting," paper no. 07567, CORROSION/2007 (Houston, TX: NACE, 2007).
18. J. Gutzeit, "Controlling Crude Unit Overhead Corrosion by Improved Desalting," *Hydroc. Proc.* February (2008).
19. K. W. Warren and J. Armstrong, *Desalting Heavy Crude Oils – The Venezuelan Experience*, Company Bulletin (Houston, TX: NATCO Group).
20. Company Bulletin, *Oil Field Dehydration and Desalting of Crude Oil* (Houston, TX: Petrolite Corporation, 1983).
21. S. Lordo, J. M. Garcia, and S. Garcia-Swofford, "Desalter Acidification Additives and Their Potential Impacts on Crude Units," paper no. 08556, CORROSION/2008 (Houston, TX: NACE, 2008).

10 Slop Oil Addition

In This Chapter

10.1 First Glance

Indiscriminate addition of slop oil to the crude charge causes desalter upsets that can result in unexpected overhead corrosion on many units. The key to avoiding desalter upsets is to control what ends up in slop oil, to add slop oil downstream of the desalter, and to add slop oil at a constant trickle rate. Slop oil originates with various plant operations and ends up in waste water due to spills, leaks, unforeseen discharges, and plain carelessness. In addition to steam condensate and various water effluents, waste water contains oils and grease (OAG) that have to be completely recovered before the water can be discharged into surface waters (retention ponds, rivers, lakes), deep wells, or public sewer systems. In addition to OAG from waste water, a variety of other liquid wastes often find their way into the crude tanks for want of a better disposal method. These include, for example, various offspec products, oil skimmings, drum residues, marketing slops, tanker bilge oils, dock slops, solvents, and wastes from in-house shop operations.

10.2 API Separator Operation

OAG is normally recovered from waste water in gravity settlers which are based on design parameters developed originally by API. A simplified sketch of what has since then become known as an "API separator" is shown in Figure 10-1. Because of the large volume and nature of waste water generated, refineries and upgraders require large API separators with special design features, as shown in Figure 10-2. API separators usually consist of two consecutive stages enclosed in the same vessel or structure: In the first stage, a four-shaft, chain-and-flight scraper continuously collects and moves accumulated mud and other solid waste for removal through a bottom discharge nozzle that is piped to a waste slurry tank. At the same time, recovered OAG floats to the top and is picked up by an adjustable, rotating, skimmer nozzle that is piped to a dedicated slop-oil tank. In the second stage, waste water is further clarified by gravity settling. Any remaining OAG in the waste water floats to the top, while clear waste water is discharged through a bottom nozzle which is piped to a holding tank. API separators are covered with removable roof panels for safety and containment of flammable, noxious, and toxic off gases in the waste water. As a rule, they can only handle waste water with pH values between 4.5 and 9. Lower pH values cause severe corrosion of the steel separator vessel, its machinery, and roof panels while higher pH values cause foaming and emulsification problems.

FIGURE 10-1 SCHEMATIC OF TWO-STAGE API SEPARATOR FOR REMOVING OAG FROM WASTEWATER

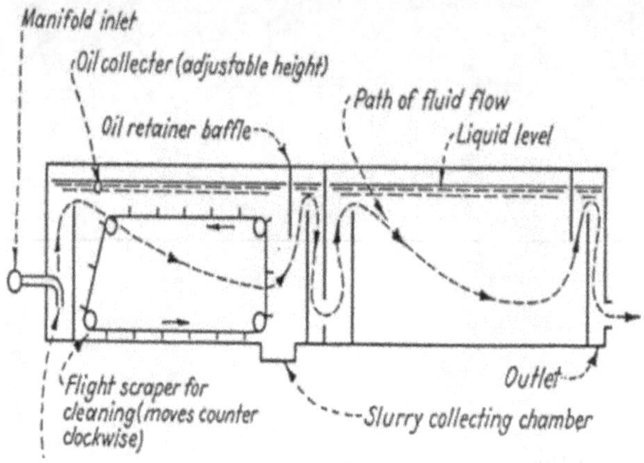

Adapted from W.L. Nelson, *Petroleum Refinery Engineering*, 4th ed. (New York, Toronto, London: McGraw-Hill, 1958).

FIGURE 10-2 TYPICAL REFINERY API SEPARATOR WITH ROOF PANELS REMOVED

Adapted from US Filter Corp. Bulletin, "API Separators: The Workhorse of Refinery Wastewater Treatment System," *Solutions for the Petroleum Industry* 6, 1 (2004).

Most API separators perform reliably and consistently well. For example, even though the OAG content of incoming waste water can vary from 100-1,100mg/L and higher, the OAG content of waste water effluent from a well-designed API separator varies only from 50-100mg/L, as shown in Figure 10-3. This means that, with a more-or-less constant flow rate, the amount of slop oil sent to the crude charge is directly proportional to the OAG content of waste water sent to the API separator. As a general rule, API separators are unable to resolve oil-in-water emulsions ("white water") without special treatment or additives, but have no issues with water-in-oil emulsions ("black water"). They can only remove oil droplets greater than 150μ (6mil) in size and some type of polishing filter is required to remove smaller oil droplets. Dissolved organic compounds are removed by biological treatment.

FIGURE 10-3 OAG REMOVAL FROM PLANT WASTEWATER BY PROPERLY DESIGNED API SEPARATOR

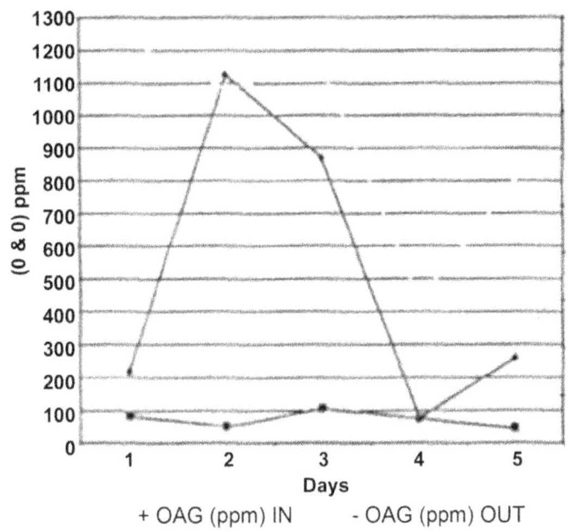

Adapted from US Filter Corp. Bulletin, "API Separators: The Workhorse of Refinery Wastewater Treatment System," *Solutions for the Petroleum Industry* 6, 1 (2004).

10.3 Slop Oil Composition

Slop oil consists primarily of OAG skimmed from various waste water streams in the API separator. Ideally, this "wet slop" should be sent to a dedicated slop oil tank for further water settling before being added to the crude charge. Unfortunately, at least from a corrosion view point, wet slop is often sent directly and randomly to one or more of the crude receiving tanks. In essence, the crude tanks and desalter become the proverbial "garbage pail" of the plant. On average, refineries produce OAG and various slop oils in quantities ranging from 0.1-1.5% of the crude intake. Higher levels occur during desalter upsets (due to increased oil carryunder in the brine) and process upsets (due to increased offspec product-make). The quantities of OAG and combined slop oils are so large that slop reprocessing at the refinery and upgrader is absolutely necessary. For example, OAG and slop oils of a typical 13,000TPD (100,000BPD) crude unit can amount to 13-195TPD (100-1,500BPD) and have a considerable cash value when added to the normal crude charge. As noted above, besides OAG just about any and all hydrocarbon waste is likely to end up in the crude tanks. In addition to the usual oil skimmings, drum residues, marketing slops, tanker bilge oils, and dock slops, hydrocarbon wastes are likely to include:

- Crude carryunder in desalter brine effluent periodically skimmed from the top of the brine settling tank.
- Other hydrocarbons periodically skimmed from the top of brine settling, caustic, amine, sour water, cooling water tanks, and the like.
- Hydrocarbons periodically skimmed from vacuum overhead hot wells.
- Vacuum truck contents.

- Off-spec product diverted from product streams (for example, rerun or "sour" naphtha).
- Spent catalysts (aluminum chloride and organic chloride solutions).
- Chemical byproducts and waste streams.
- Spent cleaning solvents and OAG waste produced in-house by maintenance and shop operations.
- Used hydraulic and engine oils returned to the plant for disposal.

Many waste streams originate within the plant or a nearby petrochemical facility[1]. Other wastes arrive with crude via pipelines due to careless or intentional dumping at a lease automatic custody transfer (LACT) unit, or via tank truck as a result of a voluntary or legislated recycling program.

10.4 Effect of Slop Oil Addition on Desalter Operations

Desalter upsets are often caused by indiscriminate and excessive slop oil addition to the crude charge. Since the amount and composition of slop oil can vary significantly on an hourly or daily basis, any effect on desalter operations is difficult, if not impossible to predict without ongoing monitoring. Slop oil has a number of components that can interfere with normal desalter operations, including:

- Dissolved and entrained slugs of water that cause desalter upsets when the slop oil is added upstream of the desalter.
- Mechanical or hydraulic issues when slop oil is added downstream of the desalter.
- Various organic and inorganic acids that change the desalter pH.
- Dissolved and entrained amines, such as monoethanolamine (MEA), that change the desalter pH and causes corrosion and fouling issues in the upper atmospheric column when present in the desalted crude charge.
- Heavy asphaltic components mixed in with a range of polymeric waste products that cause desalter upsets when slop oil is added upstream of the desalter, and fouling of the hot preheat exchangers (HPEs) when added downstream of the desalter.
- Surfactants in emulsified oil-in-water contaminants entrained in slop oil that cause desalter upsets when slop oil is added upstream of the desalter.

When slop oil contains significant amounts of water, the total amount of water (i.e., water in raw crude + water in slop oil + desalter water) going to the desalter cannot be readily established without suitable instrumentation. As a result, the desalter water rate is usually not adjusted to account for the changing water content of slop oil. Desalter operation and upsets are discussed in more detail in Chapter 9 of the *Guide*.

10.5 Slop Oil Addition to the Crude Charge

Various methods for adding slop oil to crude can be used, depending on local conditions and origin, composition, and amount of slop oil produced. When slop oil is added upstream of the desalter, it can be either added to the normal crude charge in the crude receiving tanks or injected into the crude charge line. When slop oil is added downstream of the desalter, it is either injected into the desalted or preflashed crude line, or into the flash zone of one of the columns. While there are advantages and disadvantages to either approach, downstream addition is generally preferred for a number of reasons noted below. Regardless of which location is used, however, slop oil should be added to the crude charge in small amounts (less than 1%) and at a constant rate to avoid overhead corrosion and fouling problems.

10.5.1 Adding Slop Oil Continuously Upstream of the Desalter. Slop oil added upstream of the desalter is usually mixed with crude in the crude receiving tanks, or injected through a simple nozzle connection at the crude charge line upstream of the charge pumps. In either case, slop oil is desalted along with the normal crude charge. However, desalting slop oil is normally unnecessary because slop oil comprises only a minor portion of the crude charge. For example, when 5% slop oil containing as much as 35mg/L or 10PTB salt is added to desalted crude, the salt content of the resultant crude mix is increased by only 0.5 percent. If only the maximum recommended volume of 1% slop

1. At one domestic refinery, ongoing chloride levels in excess of 200mg/L in overhead water condensate of an atmospheric column caused leaks in the carbon steel overhead vapor line just about every year. Since the desalters were operating properly, the refinery was unable to pinpoint what caused the leaks. After a thorough investigation it was finally discovered that an adjacent petrochemical plant regularly dumped its spent chloride-based catalyst streams (red oil) into the refinery's API separator for disposal. These streams then returned to the crude unit with other slop oils and were added to the crude charge. Once this practice was stopped, overhead chloride levels dropped to below 25mg/L (ppm) and the failures stopped.

oil is added, the salt content increases by only 0.1 percent. Therefore, the main benefit of adding slop oil ahead of the desalter is to use the desalter to remove water-soluble components (for example, organic and inorganic acids, amines, and slugs of water), but at the risk of causing desalter upsets.

10.5.2 Adding Slop Oil Continuously Downstream of the Desalter. When slop oil is added downstream of the desalter, water-soluble components are not removed. Again, a simple nozzle connection with the desalted crude line upstream of the preflash column (or preflash drum) will do. Connection piping approximately 2m (7ft) upstream of the injection nozzle should be alloyed to match the crude line, because of the elevated temperatures involved. Slop oil can also be injected just upstream of the fired heater. This has the advantage that no slugs of water cause pressure surges in the preflash column, where used. If caustic addition is used on the unit, slop oil can be used as a diluent for the caustic solution. However, a segment of the connection piping requires suitable alloying against hot caustic attack. Caustic addition for chloride control is discussed in greater detail in Chapter 12 of the *Guide*. Alternatively, slop oil can be injected into the flash zone of the preflash column or atmospheric column, but at the risk of pressure surges from slugs of water and volatile components in the slop oil. Such pressure surges can upset and damage trays and other column internals. Some companies have injected wet slop oil into the flash zone of atmospheric columns for many years without serious problems. The key to success here is to install devices which reduce or divert the flow, if and when slop oil contains large amounts of water. Such safeguarding can be controlled with conductivity probes or AGAR probes™. Backup probes and controls should be in place in case of instrument failure.

10.5.3 Adding Slop Oil Intermittently to the Crude Charge. Slop oil addition may have to be temporarily discontinued, if the injection system is oversized and pump constraints do not permit adjusting the injection rate to the recommended 1% or less of raw crude. Sometimes, the prescribed crude slate calls for only a small volume of slop oil. To avoid desalter upsets due to varying slop-oil addition rates, it is often best to discontinue slop oil addition entirely, put the slop oil into a dedicated holding tank, and add it to the crude charge at suitable intervals throughout the year.

10.5.4 Rerun Addition to Slop Oil. Product streams that, for one reason or another, do not meet specifications are known as "rerun." As a rule, the exact composition of rerun is not known and no samples are analyzed prior to sending rerun to the slop oil tank. Rerun is added to slop oil in quantities ranging from 0.5-3% on crude intake, but not necessarily at a controlled and uniform rate. Again, if the prescribed crude slate calls for a reduced volume of rerun, addition should be temporarily discontinued to avoid desalter upsets.

10.6 Keeping Organic Chlorides out of Slop Oil

When chloride levels in overhead condensate water reach 20–100mg/L (ppm) for short periods of time, desalter issues are the most likely cause and should be addressed according to the guidelines listed in Chapter 9 of the *Guide*. However, when chloride levels exceed 100mg/L for any length of time, organic chlorides of one type or another are probably involved. Severe overhead corrosion can be expected to occur, unless corrective action is immediately taken. If organic chlorides are suspected to be present in crude receipts, samples of all incoming crude should be immediately analyzed to help identify the source. Any crude containing more than 3mg/L (ppm) organic chlorides should be temporarily diverted to a dedicated holding tank, until the crude can either be diluted at a controlled rate by trickle feeding or rejected outright. If organic chlorides are suspected to originate within the refinery, slop oil samples should be analyzed and the slop oil temporarily diverted to the holding tank, until the source can be identified and corrective action taken. Issues with organic chlorides are discussed in some detail in Chapter 11 of the *Guide*.

10.7 Chapter Summary

Refinery waste water consists of process waters and steam condensate that contain various amounts of residual oil and grease (OAG) and other impurities. EPA and public safety require that these are completely recovered before the waste water can be discharged into surface waters, deep wells, or public sewer systems. OAG is normally recovered from refinery waste water in large gravity settlers (API separators). The resulting "wet slop oil" is usually sent to a crude receiving or dedicated slop oil tank for more water settling prior to being mixed with the crude charge. Unfortunately from a corrosion viewpoint, crude and slop oil tanks, and by extension desalters, tend to become the proverbial "garbage pail" of the plant. Besides OAG from the API separator, just about any and all process byproduct and waste streams are likely to end up as slop oil. As a result, most slop oil has a number of components that can interfere with normal desalter operations.

Several methods for continuously adding slop oil to crude are used, depending on local conditions and the origin, composition, and amount of slop oil produced. Indiscriminate and excessive slop oil addition to the crude charge upstream of the desalter often causes desalter upsets. Desalting slop oil is usually unnecessary, because slop oil comprises only a minor fraction of the crude charge. Thus the only benefit of adding slop oil ahead of the desalter is to use the desalter for removing water-soluble components and excess water, but at the risk of causing desalter upsets. Downstream addition is generally preferred. Regardless of what location is used, slop oil should be only added to the crude charge in small amounts (trickle charge) and at a constant rate to avoid desalter upsets.

10.8 Read More

1. W. L. Nelson, *Petroleum Refinery Engineering,* 4th ed. (New York, Toronto, London: McGraw-Hill, 1958).
2. US Filter Corp., "API Separators: The Workhorse of Refinery Wastewater Treatment System," *Solutions for the Petroleum Industry* 6, 1, (2004) .

11 Organic Chlorides in Crudes

In This Chapter

11.1 First Glance

In contrast to inorganic chloride salts, organic chlorides are rarely found in produced crudes as such, but instead were added intentionally or inadvertently during upstream and midstream operations. Some reports refer to organic chlorides as nonextractable chlorides, nondesaltable chlorides, phantom chlorides, and even rogue halogens to emphasize the difficulty of identifying the composition, source, and properties of organic chlorides found in crude receipts. They are insoluble in water, cannot be extracted by desalting, and stay with the desalted crude going to the crude heater. In contrast to inorganic chloride salts, most organic chlorides are not hydrolyzed to hydrogen chloride (HCl) at the relatively low temperatures of preheat exchangers (PHEs), but are partially hydrolyzed in the fired crude heater. Hydrogen chloride (HCl) from organic chlorides along with HCl normally evolved from inorganic chloride salts in desalted crude rises in columns and causes various overhead corrosion issues. The nonhydrolysable portion of organic chlorides is simply vaporized in the fired heater and ends up as dissolved organic chlorides in one of the upper side stream cuts or in light naphtha. When light naphtha is charged to a naphtha catalytic reformer, organic chlorides can cause run-away coking and shorten the life of some types of reformer catalysts. After conversion to HCl in the reformer reactor, they can also cause chloride stress corrosion cracking (CSCC) of reactor effluent coolers (REACs) made from 300 series stainless steel.

11.2 Sources of Organic Chlorides

The major source of organic chlorides in crudes is intermittent disposal of various chlorinated solvents and specialty chemicals during upstream and midstream operations, including the production, recovery, gathering, treating, and shipping of crudes. For want of other ready means of legal disposal, these chemicals are often dumped surreptitiously into crude at lease automatic custody transfer (LACT) units of pipeline stations, or into slop oil at the API separators of the refinery and upgrader. Organic chlorides identified in raw and desalted crude are insoluble in water and have a number of common characteristics. As a general rule, BP temperatures range from 66 °C (150 °F) to 110 °C (230 °F); decomposition temperatures range from 370 °C (700 °F) to 400 °C (750 °F); chlorine (Cl) content is 80 to 90% compared with 60% in inorganic chloride salts; and hydrolysable chlorine content is 40 to 45% compared with 10 to 12% in inorganic chloride salts.

Known sources of organic chlorides in raw and desalted crude include:

- Organic chloride solvents used to remove paraffin, wax, or tar deposits from crude gathering lines, tanks, treating facilities, or similar equipment. The waste solvents are then added to the crude for disposal. Examples of organic solvents that have been used for these purposes in the past include carbon tetrachloride (1,1,1-trichloro ethylene), various chlorofluorocarbons (CFC), and perchloroethylene (PERC).
- Organic chloride solvents used as fracturing fluids for well stimulation. These solvents stay with the crude throughout subsequent treating, shipping, and processing operations. The solvents hydrolyze to HCl at temperatures above 160 °C (320 °F) downstream of the desalter. Residual organic chlorides end up in the light naphtha cuts going to downstream naphtha catalytic reformers.
- Halogenated biocides injected into oil-bearing formations, such as 2,2-dibromo-3-nitrilopropionamide (DBNPA), chlorinated cyanurates, and chlorine dioxide used as downhole biocides that are or form organic chlorides upon heating.
- Marine tankers that have loaded slop oil from on-shore API separators as tanker ballast water or crude washings at certain ports-of-call. The slop oil often contains organic chlorides that contaminate subsequent crude shipments.

Known sources of organic chlorides in the refinery and upgrader include:

- Organic chlorides used to regenerate reformer catalysts.
- Spent chemical catalysts and waste oils ("black oil" or "red oil").
- Various chlorinated chemicals that end up in waste water going to the API separator.
- Various chlorinated chemicals that end up in slop oil added to raw crude for removal and disposal in the desalter. However, since organic chlorides are insoluble in water, they stay with the desalted crude and evolve HCl in the fired crude heater.
- Additives in transformer and lube oils that can form organic chlorides when heated in service or for recycling.
- Used halogenated transformer oils that have been disposed in crude despite the fact that this practice is illegal in most countries.
- Certain process additives, including emulsion breakers, flocculation aids, biocides, and tank-cleaning chemicals.
- Miscellaneous laboratory and shop waste chemicals.

11.3 Problems Caused by Organic Chlorides

The main problem with organic chlorides is the relatively large amount of hydrogen chloride (HCl) evolved in the fired heater compared to incoming inorganic salts. Even minor amounts of 2–3mg/L organic chlorides in desalted crude can increase overhead corrosion enough to upset corrosion control programs and shorten the service life of overhead exchanger and condenser tubes. Major amounts of 100mg/L and more have caused hole-through in overhead vapor lines and destroyed condenser tube bundles in a matter of days or weeks. To avoid duplication, a detailed review of all aspects of crude unit overhead corrosion is deferred to Chapter 14 in the *Guide*, as well as subsequent chapters dealing with neutralizers, filming-amine corrosion inhibitors, and water washing. However, the following should be considered at this point:

11.3.1 HCl Evolution in Fired Crude Heaters. In the absence of organic chloride contamination, approximately 0.11kg (lb) HCl per kg (lb) inorganic salt evolves in the crude heater at approximately 316 °C (600 °F) by hydrolysis. Of this amount, approximately 0.1kg (lb) HCl is formed by hydrolysis of magnesium chloride ($MgCl_2$), while 0.003kg (lb) HCl is formed by hydrolysis of calcium chloride ($CaCl_2$). In comparison, when crude is contaminated with a widely used organic chloride solvent such as carbon tetrachloride (CCl_4), approximately 4 times as much HCl or 0.46 kg (lb) HCl per kg (lb) carbon tetrachloride evolves in the crude heater. As shown in Figure 11-1, this amount is in addition to the amount of HCl evolved by hydrolysis of the inorganic salts in crude and assumes that only half of the CCl_4 is hydrolyzed in the fired heater while the remainder is vaporized and dissolved in light naphtha.

FIGURE 11-1 ESTIMATED DISTRIBUTION OF CCl_4 ORGANIC CHLORIDE IN CRUDE COLUMN OVERHEAD SYSTEM

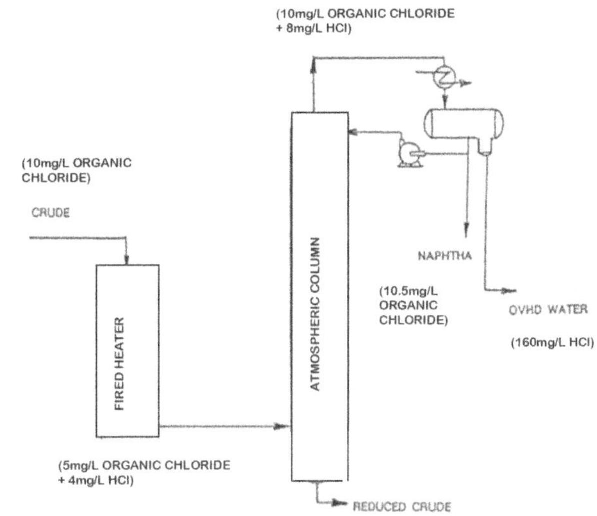

11.3.2 Chloride Level of Overhead Water Condensate. As discussed in Chapter 9 in the *Guide*, the chloride content of overhead water condensate is approximately proportional to the salt content of desalted crude. For example, with 3.5mg/L (1PTB) salt in desalted crude, the expected chloride level is approximately 16mg/L. This figure assumes that similar amounts of light naphtha are obtained from many types of crude (approximately 45%) and similar amounts of steam condensate (approximately 5%) are collected. These amounts can be easily adjusted up or down to fit particular crudes, plants, and naphtha and condensate rates. At the level of 3.5mg/L (1PTB) salt in desalted crude, the chloride level increases by approximately 16mg/L for each mg/L organic chloride present in desalted crude as CCl_4. As a result, the chloride level now is 32mg/L instead of 3.5mg/L, as shown in Table 11-1. Any time chloride levels increase above approximately 20mg/L, overhead corrosion increases proportionally.

11.3.3 Elevated Chloride Levels. Elevated chloride levels are caused by either desalter or organic chloride issues. If desalter issues can be definitely ruled out, it can be assumed that organic chlorides are entering with raw crude or finding their way into the slop oil. In most cases, unusually low pH values are the first indications that organic chloride contamination of one type or another has occurred. Organic chloride contamination of incoming crude usually shows up as sudden chloride spikes in overhead condensate water instead of consistently high chloride levels over some period of time. Chloride spikes can last from 1 to 3 days or more, depending on the volume of contaminated crude that passes through the unit. Unless the pH levels are constantly monitored, these spikes are easily missed. They will be missed for sure, if pH readings are obtained only every other day or so. In contrast, ongoing addition of small volumes of organic chloride-containing waste at the API separator usually accounts for higher-than-normal chloride levels over an extended period of time (weeks or months) as long as the contaminated slop oil is added to the crude charge or until the source has finally been discovered and shut off.

TABLE 11-1 HCl CONTENT OF OVERHEAD WATER DUE TO CCl_4 CONTAMINATION OF DESALTED CRUDE

Salt Content of Desalted Crude, mg/L	CCl4 Content of Desalted Crude, mg/L	HCl Content of Overhead Water, mg/L
3.5	0	16
3.5	1	32
3.5	2	48
3.5	5	96
3.5	10	176
3.5	100	1,616

11.4 Effect of Contaminated Crude on Plant Operations

The following example shows what can happen when contaminated crude suddenly enters the crude mix at a rate of only 20%. The problem was first discovered, when the pH of overhead water condensate from the atmospheric column dropped below 5 despite repeated increases in the overhead neutralizer (ammonia) rate. During the next several days, the pH continued to drop. The ammonia injection lines were then enlarged in order to be able to add more ammonia to maintain pH control. This too proved to be ineffective and the pH eventually plunged to 0.5 and lower. Based on information obtained from another refinery that had a similar problem, the suspect crude in the crude mix was first cut from 20% to 3%, and several days later was cut out altogether. Extensive analytical testing identified the contaminated crude. The contaminant chemical was a mixture of chlorinated hydrocarbons with a boiling point (BP) in the light through heavy naphtha range. Because of the relatively short exposure time, major corrosion damage at the overhead system was avoided. However, the high organic chloride levels in the naphtha feed to a catalytic reformer resulted in run-away coking of the reformer catalyst and the reformer had to be shut down. Some of the relevant data obtained prior to, during, and after this incident are summarized in Tables 11-2 through 11-4 below.

TABLE 11-2 MEASURED CHLORIDE LEVELS IN CONTAMINATED CRUDE

Raw Crude	Inorganic Chlorides in Raw Crude, mg/L	Organic Chlorides in Raw Crude, mg/L
Contaminated LMS Crude (Pipeline/Tank Samples)	175 - 1,050	400 - 1,400

TABLE 11-3 MEASURED CHLORIDE LEVELS IN CRUDE AND OVHD WATER DURING NORMAL AND UPSET CONDITIONS

Mixed Feed	Chlorides in Desalted Crude		In Overhead Water HCl, mg/L
	Inorganic, mg/L	Organic, mg/L	
20% uncontaminated LMS crude in mix	2.8 - 28	0	20-30
20% contaminated LMS crude in mix	3.5 - 7	150	1,500-2,000
3% contaminated LMS crude in mix	8.7 - 12	N.A.	30-40

TABLE 11-4 MEASURED CHLORIDE LEVELS IN NAPHTHA PRODUCTSDURING NORMAL AND UPSET CONDITIONS

Mixed Feed	Organic Chlorides, mg/L		
	Light Naphtha	Interm. Naphtha	Heavy Naphtha
20% LMS Crude	0 - 2	0	0
20% Contaminated LMS Crude in Mix	50 - 55	45-90	20-25
3% Contaminated LMS Crude in Mix	N.A.	10-25.	4-15

11.5 Controlling Corrosion by Organic Chlorides

Organic chloride contamination of crude should never be taken lightly. While it cannot always be avoided, certain measures can be taken at low cost to guard against future problems. For example, before crude enters a pipeline it usually passes through a LACT[1] unit as shown in Figure 11-2. LACT units are normally kept locked, but drivers of crude delivery trucks are issued a key to access the line as needed. A LACT unit not only measures the volume of crude delivered, but also samples all crude passing through the unit at regular intervals. Typically, a small sample is automatically collected in a sampling pot at hourly, daily, biweekly, or even monthly intervals, depending on the expected volume passing through the unit. Samples from various LACT units can then be retained for later analysis, as needed.

FIGURE 11-2 FLOW METER AND SAMPLING STATION OF SKID-MOUNTED CAMLACT™ LACT UNIT

Courtesy: Cameron International, Houston, TX.

1. A Lease Automatic Custody Transfer (LACT) unit is designed to facilitate the automatic transfer of ownership of crude or condensate between the buyer and seller.

If contaminated crude appears to be entering a pipeline, all crude suppliers should be required to analyze and certify further crude deliveries for organic chloride content. Many pipeline operators now require this routinely to avoid litigation with refineries. In addition, random spot samples from various crude shipments, tanks, and LACT units at pipeline stations should be analyzed periodically by the refinery for organic chlorides. At a minimum, crude samples should be collected at regular intervals and retained for possible future analyses. Contaminated crude can then be traced back to a specific crude supplier once evidence of organic chloride contamination has been established. Finally, crude suppliers who continue to provide contaminated crude should be locked out of LACT units indefinitely.

If pipeline samples are found to be free of organic chlorides, possible entry points for organic chlorides within the refinery should be searched out and eliminated. Likely problem streams are spent solvents from refinery shop and other degreasing operations. These should be picked up by a qualified waste hauler for disposal instead of being sent to the API separator. Other sources are used engine or machinery oils returned to the refinery for disposal or reclaiming. These too should be turned over to a qualified waste hauler instead of being sent to the API separator. Spent catalyst solutions and reaction byproduct streams containing organic chlorides should be returned to the original suppliers for catalyst regeneration or disposal.

11.6 Using Pipeline Routings to Identify Source of Contaminated Crude

In connection with litigation to recover corrosion-related damages, it may become necessary to clearly establish the time of arrival of contaminated crude at the refinery. Therefore, certain pertinent data and records should be obtained from the pipeline operator, including crude volumes; routings; distances between pipeline stations and junctions; size of initial and intermediate tankage; and pipeline sizes, capacities, pumping rates, and flow velocities. Depending on location, refinery crude supply routes can range from 160 to 320km (100 to 200miles) and longer; and flow velocities from 3 to 8km/hr (2 to 5mph). Considering the time spent in holding tanks at pipeline stations and junctions, 5 to 10 days may well be required for contaminated crude to arrive at a plant after entering a pipeline system at a LACT unit. Note that the organic chloride levels decrease proportionally each time contaminated crude is mixed with uncontaminated crude in holding tanks at pipeline stations and junctions.

11.7 Organic Chloride Limits for Crude Receipts

When organic chloride contamination of crude became a problem in the past, many pipeline operators and plants established stringent limits on organic chloride content of crude shipments and, by extension, crude charged to a crude unit. Some companies called for a limit of 5.0mg/L, while others allowed no organic chlorides whatsoever in the crude mix. Most companies called for 1.0 to 3.0mg/L. From a corrosion point of view, an upper limit of 3.0mg/L is considered to be too high: a more realistic limit is 1.0mg/L. Even with only that amount in the crude charge, the amount of HCl evolved in the crude heater is doubled if the crude charge is normally desalted to 3.5mg/L (1PTB) inorganic salts. Some pipeline operators and refineries have only allowed higher limits on a case-by-case basis, and then only when they received sufficient advance notice. Others flatly reject any and all crude containing organic chlorides in any amount. However, with sufficient advance notice, refineries usually have enough time to isolate a shipment of contaminated crude in a separate holding tank to be added to the normal crude charge by trickle-feeding or dilution by blending.

11.8 Chapter Summary

The major cause of organic chloride contamination in crude is intermittent disposal of chlorinated solvents and specialty chemicals from crude production, recovery, gathering, treating, shipping, and refining operations. For want of other ready means for legal disposal, waste chemicals are often dumped surreptitiously into crude at pipeline stations, or into slop oil at the refinery. Organic chlorides are insoluble in water and cannot be extracted by desalting. They are partially hydrolyzed to HCl in the fired heater that travels up downstream columns and causes increased overhead corrosion problems. The nonhydrolyzed portion is vaporized and usually ends up in various product sidestreams and overhead light naphtha. Examples of organic chloride contaminants include solvents for wax and tar deposits; downhole fracturing fluids; downhole biocides; tanker ballast slops; and refinery slop oils. Known sources of organic chlorides in refinery slop oils include emulsion breakers and other desalting chemicals, flocculation aids, biocides, organic chlorides for regeneration of reformer catalysts, chemical waste oils ("black oil" and "red oil"), lube unit wastes, tank cleaning chemicals, laboratory and shop wastes, and recycled transformer and lube oils.

Even minor amounts (2-3mg/L) of organic chlorides in desalted crude can increase overhead corrosion sufficiently to upset existing corrosion control programs and shorten the life of overhead exchanger tubes. Amounts in excess of 100mg/L have caused hole-through in overhead vapor lines and destroyed tube bundles in a matter of days or weeks. Random spot samples from various crude shipments, tanks, and LACT units at pipeline stations should be analyzed periodically by the refinery for organic chlorides. In the past, pipeline operators and refining companies have established stringent limits on organic chloride content of crude. Most companies call for a maximum of 1.0-3.0mg/L. From a corrosion point of view, an upper limit of 3.0mg/L is too high: a more realistic limit is 1.0mg/L.

11.9 Read More

1. K. Brooks, "Organic Chloride Contamination Rears Its Ugly Head Again," *Oil & Gas J.* 60, November 26 (1962).
2. J. Gutzeit, "Effect of Organic Chloride Contamination of Crude on Refinery Corrosion," paper no. 0694, CORROSION/2000 (Houston, TX: NACE, 2000).
3. NACE Task Group 274, "Effect of Nonextractable Chlorides on Refinery Corrosion and Fouling," publ. 34105 (Houston, TX: NACE, 2005).
4. D. H. Stormont, "Chlorides in Crude Oil Plague Refiners," *Oil & Gas J.* 67, April 14 (1969).
5. ASTM D4929-15 (latest revision) *Standard Test Methods for Determination of Organic Chloride Content in Crude Oil* (West Conshohocken, PA: ASTM).
6. Cameron Process Systems Bulletin, Design, Operation & Maintenance of L.A.C.T. Units, Rev. 2015 (Houston, TX: 2015).

12 Caustic Addition for Chloride Control

In this Chapter

12.1 First Glance

Caustic (NaOH) addition to desalted crude offers a convenient and relatively low-cost alternative to two-stage desalting. It should be considered when overhead corrosion cannot be controlled with single-stage desalting or when the desalter capacity has become too small for the volume or type of crude now being charged. Caustic addition controls overhead corrosion by reducing the amount of hydrogen chloride going overhead. Caustic addition also helps control chloride swings due to desalter upsets, reduces the amount of ammonia or liquid organic neutralizers needed for overhead pH control, and reduces fouling and under-deposit corrosion by neutralizer hydrochloride salts in the top of columns and in the dry sections of overhead systems.

12.2 Types of Caustic

The term caustic[1] is commonly used by the industry for sodium hydroxide (NaOH). Compared to most other process additives, caustic is cheap and readily purchased on the open market in solid form or as a concentrated solution with

1. The term caustic soda is sometimes applied to water solutions of caustic.

water. As a general rule, commercial-grade granules or bricks should be purchased if the caustic solution is to be prepared on site; commercial-grade caustic solutions should be purchased if caustic solution is to be obtained in bulk. Commercial-grade caustic contains up to 0.5% chlorides, primarily as sodium chloride (NaCl). For use in crude units, caustic does not have to be free of chlorides. The term caustic potash is commonly used for potassium hydroxide (KOH). It is no substitute for caustic because it does not tie up hydrogen chloride (HCl) as sodium chloride (NaCl) and minimize overhead corrosion. Sodium carbonate (Na_2CO_3) is also no substitute for caustic because it can evolve carbon dioxide (CO_2) that can contribute to overhead corrosion.

Spent or waste caustic solutions recovered from other processes should never be used to avoid introducing unknown chemical contaminants of varying compositions. Asides from potentially creating unexpected corrosion and fouling problems, most are extremely toxic and odorous. Examples of contaminants are high levels of:

- Hydrogen sulfide (H_2S) that is more toxic than hydrogen cyanide (HCN) to humans; death can occur when exposed to as little as 100-500mg/L in the atmosphere.
- Naturally occurring mercaptans or thiols.
- Phenols together with cresylic and naphthenic acids.
- Cresylates or phenolics.
- Naphthenes (cyclic alkanes) and cyclic nonaromatic hydrocarbons (cycloalkanes) including cyclopentane and cyclohexane.

12.3 Properties of Caustic Solutions

Caustic solutions can be prepared or diluted to various strengths depending on specific needs, as shown in Table 12-1. Solutions containing 20 or 25% commercial-grade NaOH in water are readily available in most parts of the country and are shown in bold. Their relatively high caustic content minimizes shipping, handling, and storing costs, yet does not require alloys to prevent corrosion at near-ambient temperatures. Their relatively low freezing point avoids the need for special storage and handling facilities, such as heated tanks and heat-traced lines at plants located in the northern part of the United States, but not necessarily in Canada. Solutions containing 20% NaOH in water have a freezing point of -26 °C (-15 °F), while solutions containing 25% NaOH in water have a somewhat higher freezing point of -23 °C (-10 °F). Since freezing point rapidly increases with concentrations below 20% NaOH and above 25% NaOH, those solutions are only practical for refineries located in the southern part of the United States.

TABLE 12-1 PROPERTIES OF DILUTE CAUSTIC SOLUTIONS

OBaumé	NaOH, %	Spec. Gravity	Freezing Point		Boiling Point	
			OC	OF	OC	OF
1.4	1.0	1.01	-1	30	100	212
2.9	2.0	1.02	-2	28	100	212
4.5	3.0	1.03	-3	27	100	212
7.4	5.0	1.05	-5	23	101	214
10.2	7.0	1.08	-3	27	103	217
14.2	10.0	1.11	-10	14	104	219
20.4	15.0	1.16	-17	1	107	225
26.1	**20.0**	**1.22**	**-26**	**-15**	**110**	**230**
31.1	**25.0**	**1.27**	**-23**	**-10**	**112**	**234**
35.8	30.0	1.33	1	34	115	239
43.6	40.0	1.43	15	59	125	257
49.9	50.0	1.53	12	54	142	288

12.4 Dilution of Caustic Solutions

Purchased 20 or 25% caustic solutions should be diluted prior to addition to hot desalted crude to assure uniform dispersion and avoid corrosion and fouling problems downstream of the injection point. One of the following methods can be used to make a 3-5% NaOH solution suitable for injection: batch mixing in a "day tank" with water, continuous injection into a slip stream of the same type of water used as desalter water (but not the desalter water itself going to the desalter), and continuous injection into a dedicated dilution water line.

12.5 Caustic Addition to Crude

Caustic should be added to crude if the salt content of desalted crude remains consistently above 7mg/L or 2PTB despite all other efforts to reduce it further with desalting. Alternatively, caustic should be added if the chloride content of overhead water condensate from the crude column remains consistently above 20mg/L, and periodically above 40mg/L. The following factors should be considered when a caustic addition system is designed.

12.5.1 Location of Addition Point. Diluted caustic should always be injected downstream of the desalter. When caustic is added upstream of the desalter, severe crude emulsification can occur at the mix valve and interfere with brine separation in the desalter vessel. Also the caustic leaves the desalter with the brine effluent and cannot react with hydrogen chloride (HCl) evolved in the downstream HPEs and crude heater. For the same reasons, caustic injection between stages of two-stage desalters is not recommended. Where this has been used, any benefits reported may have been due to caustic carryover to the crude heater in the form of entrained caustic droplets or dissolved sodium naphthenate.

12.5.2 Procedure for Adding Caustic. The usual procedure is to inject a 2.5-3.0% caustic solution into a slip stream of desalted crude diverted from just downstream of the desalter. The temperature of the slipstream is normally below 150 °C (300 °F) and avoids excessive solution bumping and splattering. The caustic addition nozzle or tee connection should be located approximately 20-30 pipe diameters upstream of any elbows, tees, and valves in the slip stream line to ensure proper mixing and prevent impingement by highly corrosive caustic droplets. Following caustic injection, the slip stream of desalted crude is returned to the hot crude line downstream of the last hot preheat exchanger (HPE). This prevents possible fouling of the preheat exchanger tubes by caustic deposits. Some operators inject caustic solution into a slipstream of cold raw crude to provide more time for the caustic to disperse before the slipstream is added to hot desalted crude.

12.5.3 Materials of Construction (Metallurgy). Caustic addition nozzles should be fabricated from Alloy C-276 or a similar high-nickel alloy to minimize potential caustic attack and caustic embrittlement or stress corrosion cracking (caustic SCC). Carbon steel piping can normally be used for caustic solutions containing up to 25% NaOH below approximately 65 °C (150 °F) temperature. However, carbon steel piping requires postweld heat treatment (PWHT) for caustic service above approximately 65 °C (150 °F), with turbulent flow, or when caustic lines are heat traced. An example of preferential caustic attack at welds and heat affected zones (HAZs) in carbon steel piping is shown in Figure 12-1. Also visible is accelerated corrosion caused by turbulence downstream of a reducer fitting. The line carried 15% NaOH solution at 38 °C (100 °F).

12.5.4 Dilution Water. Ample dilution water should be available to ensure that a caustic solution of uniform strength is injected at all times. Piping needs to be designed to prevent backup of caustic into the desalter or of crude into the caustic tank if pump stoppages or pressure surges occur.

12.5.5 Caustic Injection Pumps. Positive-displacement pumps are normally used to deliver a constant volume of dilute caustic solution into the slipstream line. These pumps produce a pulsating delivery pattern and some type of pressure regulator may have to be installed to ensure smooth flow. Injecting the slipstream into the suction side of the desalted crude booster pump contributes to good caustic dispersion and dampens the flow pulsations, but may cause fouling in any downstream preheat exchangers (PHEs).

FIGURE 12-1 CORROSION OF CARBON STEEL PIPING IN 15% CAUSTIC SOLUTION AT 38 °C (100 °F)

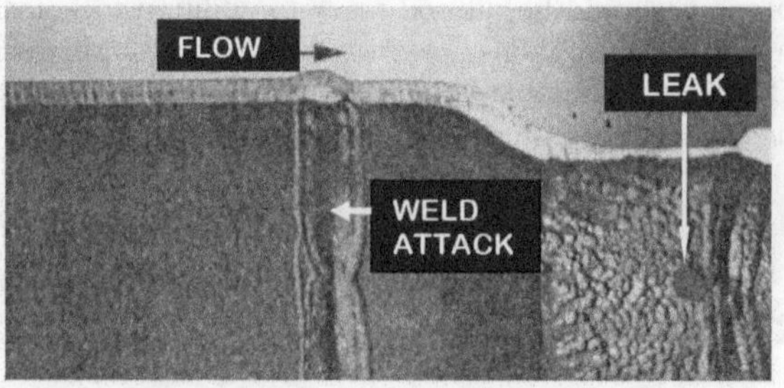

12.5.6 Caustic Supply and Storage. A typical 13,000TPD (100,000BPD) crude unit may require approximately 136kg (300lb) caustic per day when injected at the rate of 10mg/L or approximately 3PTB. This means that appropriate tankage and equipment must be provided for not only storing purchased caustic solution, but also for storing, blending, and injecting approximately 4,300L/d (1,150GPD) diluted caustic solution.

12.6 Optimum Caustic Treatment Level

No more than 10mg/L or 3PTB caustic should be added to desalted crude to minimize coke formation in reduced crude (atmospheric resid) lines and heater tubes and potentially other problems downstream. Larger amounts of caustic may be required, however, when sour crudes containing hydrogen sulfide (H_2S) are charged to the unit. If necessary, caustic addition may have to be reduced to prevent unreacted (excess) caustic from entering the crude column. Caustic addition should be discontinued entirely when the chloride content of overhead water condensate remains consistently below 20mg/L and no desalter upsets are expected to occur. The actual caustic addition rate can be determined by one of the following methods.

12.6.1 Without Sodium Limits. Increasing amounts of caustic are gradually added until the desired chloride level in overhead water condensate from the crude column is reached and maintained. This method is used primarily when the sodium (Na) content of reduced crude is of no concern.

12.6.2 With Strict Sodium Limits. A small amount of caustic is continuously added to smooth out any chloride peaks in overhead water condensate from the crude column. This method is used when strict sodium (Na) limits for reduced crude are called for.

12.6.3 Proportional to Salt Content of Desalted Crude. Approximately 1.7mg/L or 0.5PTB caustic is added for every 3.5mg/L or 1PTB salt in desalted crude, up to a maximum of 10mg/L or 3PTB. This method is less reliable because it does not take into account the chloride level in overhead water condensate from the atmospheric column.

12.7 Caustic Corrosion

When dilute caustic is added to desalted crude at desalter effluent temperatures near 150 °C (300 °F), the water evaporates and the dispersed caustic droplets become increasingly more concentrated, as shown in Figure 12-2. However, the caustic droplets still contain approximately 10% water and are highly corrosive at typical crude heater inlet temperatures near 316 °C (600 °F). Aside from the BP and MP curves for caustic, Figure 12-2 also shows the extent of the stress cracking domain for 300 series stainless steels (hatched area) and carbon steel (area below and left of solid curve). Also shown are three dashed lines defining the approximate corrosion limits for carbon steel. In all cases, the effects of heat tracing on corrosion need to be considered:

- At temperatures between approximately 100 °C (212 °F) and 65 °C (150 °F) depending on concentration, corrosion is less than 0.025mm/y (1mpy).

- At temperatures up to approximately 150 °C (300 °F) and 80% NaOH, corrosion ranges from 0.025 to 0.75mm/y (1 to 30mpy).
- At higher temperatures and concentrations, corrosion exceeds 0.75mm/y (30mpy) in the form of molten salt corrosion.

FIGURE 12-2 CORROSION AND SCC ZONES FOR CARBON STEEL AND SCC ZONE FOR 300 SERIES STAINLESS STEELS

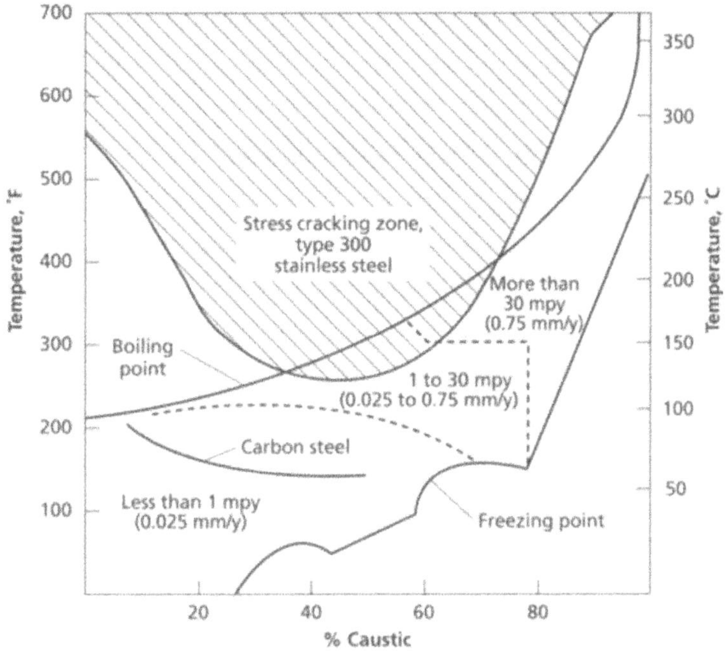

Adapted from C. P. Dillon, *Corrosion Control in the Chemical Process Industries* (New York, NY: McGraw-Hill, 1986).

12.8 Caustic Reaction Mechanism

In the absence of actual salt analyses, it is usually assumed that salt in crude consists of 15% magnesium chloride ($MgCl_2$), 5% calcium chloride ($CaCl_2$), and 80% other salts including sodium chloride (NaCl). At typical crude heater inlet temperatures of 316 °C (600 °F), approximately 90% of the magnesium chloride and 10% of the calcium chloride in crude are hydrolyzed. In contrast, sodium chloride (NaCl) is not hydrolyzed to any significant degree. The amount of hydrogen chloride (HCl) evolved per unit weight of salt in the crude heater is approximately 0.1kg HCl/kg (0.1lb HCl/lb) $MgCl_2$ and 0.003kg HCl/kg (0.003lb HCl/lb) $CaCl_2$ for a total of approximately 0.103kg HCl/kg (0.103lb HCl/lb) salt.

Caustic reacts with hydrogen chloride to form sodium chloride (NaCl) by two mechanisms: Caustic droplets react directly with hydrogen chloride (HCl) vapor as it evolves in the hot preheat exchangers (HPEs) and fired heater in accordance to Equation 12-1:

$$NaOH + HCl \Rightarrow NaCl + H_2O \quad \text{(Eq.12-1)}$$

Or, caustic droplets first react with naphthenic acids to form soluble sodium naphthenates that, in turn, react with hydrogen chloride vapor in accordance to Equations 12-2 and 12-3:

$$NaOH + R\text{-}(CH_2)_n\text{-} COOH \Rightarrow R\text{-}(CH_2)_n\text{-} COONa + H_2O \qquad \text{(Eq.12-2)}$$

$$R\text{-}(CH_2)_n\text{-} COONa + HCl \Rightarrow R\text{-}(CH_2)_n\text{-} COOH + NaCl \qquad \text{(Eq.12-3)}$$

Some amount of hydrogen chloride (HCl) evolves by the time crude enters the preflash column at approximately 220 °C (425 °F) upstream of the fired heater and accounts for overhead corrosion on that column. This means that with caustic addition, neutralization reactions similar to those at the fired heater also occur at the preflash column and downstream at the vacuum heater.

12.9 Potential Problems with Caustic Addition

Most problems with caustic addition occur when too much caustic is added, caustic is added without proper dilution, or spent (waste) caustic is used. The following are some examples of issues that can occur and have occurred. These need to be considered, if caustic addition is to provide the expected benefits.

12.9.1 Improper Addition or Insufficient Mixing. Improper caustic addition or insufficient mixing cause fouling, caustic attack, and caustic embrittlement (caustic SCC) of components downstream of the caustic addition point, including piping, preheat exchanger tubes, and fired heater tubes. Also fouling is likely to occur by other inorganic salts (carbonates, sulfates, silicates, oxides) in desalted crude due to co-precipitation and entrained high-pH water droplets.

12.9.2 Excess Caustic Addition. Excess caustic causes foaming in the preflash drum, preflash column, atmospheric column, and coke drums due to the formation of long-chain sodium naphthenate "soaps." Excess caustic addition has caused coking of the disengaging grids in the vacuum column, increased flash-zone pressure, and resulted in premature coking of coker lines and heater tubes, as shown in Figure 12-3. In this line, a lop-sided coke layer up to 75mm (3in.) thick formed after only three weeks of operation following previous mechanical cleaning. The nominal pipe wall thickness was 12.5mm (0.5in.), but was reduced to approximately 6mm (0.25in.) at the bulged area. Thermal stresses resulted in a tight 100mm (4in.) long crack not visible in photograph.

12.9.3 Fuel Ash Corrosion. When excess caustic reacts with vanadium in crude to form sodium vanadate compounds having vanadium/sodium (V/Na) ratios of less than 10, extremely tenacious deposits are formed on super-heater tubes in boilers that use reduced crude (atmospheric resid) as "Bunker C" fuel oil. The deposits lead to accelerated fouling and corrosion on the fireside of the heater tubes and also on turbine blades in resid-fired gas turbines when the sodium (Na) levels exceed 1mg/L.

FIGURE 12-3 COKE LAY DOWN IN 200MM (8IN.) OD x SCH. 80 COKE DRUM CHARGE LINE DUE TO EXCESS CAUSTIC IN DESALTED CRUDE

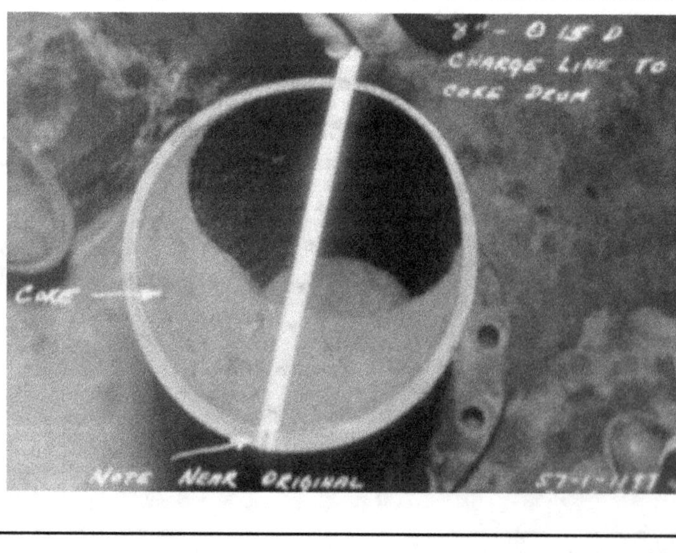

12.9.4 Downstream Sodium (Na) Issues. A decreasing sodium tolerance of many downstream resid processes has discouraged the use of caustic at some crude units. While sodium intake is normally limited to 10mg/L (5PTB) on crude to avoid coke blockages, concentrations quadruple by the time sodium reaches the downstream thermal units. Sodium has promoted coke formation in the fired heaters of visbreakers. Not only do coke deposits form sooner, they are also harder to remove. Another problem has been reduced catalyst life when reduced crude containing sodium is charged to a fluid catalytic cracking unit (FCCU). Sodium has also been shown to deactivate some types of hydrotreating catalysts when the sodium content exceeded 1-2Mg/L in the hydrocarbon feed. As a general rule, the above issues were strictly limited to sodium (Na) or potassium (K) contamination and did not apply to calcium (Ca) contamination.

12.10 Monitoring Frequency

As a general rule, the plant is usually responsible for supplying and maintaining the storage and mixing tanks for caustic solutions; providing and maintaining piping, pumps, and other components to facilitate caustic injection; and stocking sufficient caustic. The process additives supplier is responsible for seeing to it that the caustic addition system is operating properly since caustic addition, like crude desalting, is critical to overhead corrosion control. This means the supplier should coordinate caustic addition with the desalter operation and neutralizer injection into the overhead systems; monitor the caustic pump settings, flow meter readings, and the day tank volume at least once/shift; and ensure heat-tracing for caustic piping and heaters for the caustic supply tank are operational where needed.

12.11 Chapter Summary

Caustic addition to desalted crude offers a convenient and relatively low-cost alternative to two-stage desalting. Caustic addition should be considered, if overhead corrosion cannot be controlled with single-stage desalting, or if the desalter capacity has become too small for the volume or type of crude now being charged. Caustic solutions containing 20 or 25% NaOH in water are readily available in most parts of the country. The relatively high caustic content minimizes shipping, handling, and storing costs, yet does not require alloying to prevent corrosion. The relatively low freezing point avoids the need for special storage and handling facilities (heated tanks, heat-traced lines, etc.) in plants located in the northern part of the United States, but not necessarily in Northern Canada.

Commercial 20 or 25% NaOH solutions should be diluted prior to addition to hot desalted crude to assure uniform dispersion throughout. No more than 10mg/L (3PTB) caustic should be added to minimize coke formation in reduced crude (atmospheric resid) and downstream fired heaters. Caustic droplets react directly with hydrogen chloride as it evolves in the crude heater. Alternatively, caustic droplets first react with naphthenic acids in crude to form soluble sodium naphthenates which, in turn, react with hydrogen chloride. Most issues with caustic addition occur when too much caustic is added, caustic is added without proper dilution, or spent (waste) caustic is used. The process additives supplier is usually responsible for seeing to it that the caustic addition system is operating properly, since caustic addition is critical to overhead corrosion control.

12.12 Read More

1. H. L. Bedell and V. W. Hatchett, "The Use of Chemicals for Retarding Corrosion," *Proc. API* 28M (III), (1948).
2. V. E. Cornich, "Crude Unit Corrosion Control Case History 72 Months Continuous Operation," paper no. 48, CORROSION/82 (Houston, TX: NACE, 1982).
3. E. Q. Camp and C. Phillips, "Neutralization as a Means of Controlling Corrosion of Refinery Equipment," *Corrosion* 6, 2 (1950).
4. J. Gutzeit, "Caustic Injection Revisited – An Old Method for Controlling Crude Tower Overhead Corrosion," *MP* 39, 8 (2000).
5. International Nickel Company, *Corrosion Resistance of Nickel and Nickel-Containing Alloys in Caustic Soda and Other Alkalis*, Corrosion Engineering Bulletin CEB-2 (New York, NY: International Nickel Co., 1973).
6. W. S. Letzsch and D. N. Wallace, "FCC Catalysts Sensitive to Alkali Contaminants," *Oil & Gas J.* 80, Nov. 29 (1982).
7. NPRA Q&A, "Crude Distillation Issues and How to Solve Them," *Oil & Gas J.*, March 30 (1970).

8. C. J. Scherrer and C. R. Baumann, "Update of the Desalted Crude Neutralization Process," paper no. 101, CORROSION/82 (Houston, TX: NACE, 1982).
9. A. S. Al-Omari, et al., paper no. 08551, CORROSION/2008 (Houston, TX: NACE, 2008).
10. NACE Standard Practice SP0403-2015, "Avoiding Caustic Stress Corrosion Cracking of Carbon Steel Refinery Equipment and Piping" (Houston, TX: NACE, 2015).
11. G. Veerabhadraiah et al., "Spent Caustic Management: Remediation Review," presentation at the International Refining and Petrochemical Conference-Asia, July 19–22, 2011, Singapore.
12. G. Egloff and J. C. Morrell, "The Use of Chemical Reagents in the Protection of Atmospheric and Pressure Distillation Equipment Against Corrosion," presentation at ASChE Meeting, St. Louis, MO, Dec. 5-8, 1927.
13. R. D. Merrick and T. Auerbach, "Crude Unit Overhead Corrosion Control," *MP* 22, 9 (1983).
14. C. P. Dillon, *Corrosion Control in the Chemical Process Industries* (New York, NY: McGraw-Hill, 1986).
15. G. R. Lobley et al., "Caustic Cracking: Incidents in Refineries," paper no. 80, Corrosion Control 2007, Sydney, Australia.
16. R. K. Singh Raman, "Role of Caustic Concentration and Electrochemical Potentials in Caustic Cracking Of Steels," *Materials Science and Engineering A,* 441 (2006).
17. Ara Bagdasarian et al., "Crude Unit Corrosion and Corrosion Control," paper no. 615, CORROSION/96 (Houston, TX: NACE, 1996).
18. J. Woodtli and R. Kieselbach, "Damage Due To Hydrogen Embrittlement and Stress Corrosion Cracking," *Engineering Failure Analysis* 7, 427 (2000).
19. R. Baboian, *NACE Corrosion Engineer's Reference Book,* 3rd ed. (Houston, TX: NACE international, 2002).
20. R. B. Rebak, "Environmental Cracking of Corrosion Resistant Alloys In the Chemical Process Industry – A Review," paper no. 07480, CORROSION/2007 (Houston, TX: NACE, 2007).
21. L. L. Shreir et al., *Corrosion Volume 1: Metal/Environment Reactions*, 3rd ed. (Butterworth-Heinemann, 1993).
22. L. L. Shreir et al., *Corrosion Volume 2: Corrosion Control,* 3rd ed. (Butterworth-Heinemann, 1993).
23. P. R. Roberge, *Handbook of Corrosion Engineering* (New York, NY: McGraw-Hill, 2000).
24. J. R. Rue and J. G. Edmondson, "Control of Salt-Initiated Corrosion in Crude Unit Overhead Systems," paper no. 01538, CORROSION/2001 (Houston, TX: NACE, 2001).
25. D. W. Alley and N. D. Coble, "Corrosion Inhibitors for Crude Distillation Columns," *MP* 42, 5 (2003).
26. C. M. Schillmoller, "Select the Right Alloys for Caustic Soda Service," *Chem. Eng. Progress,* May (1996).
27. NACE publication no. 3410, "Refinery Injection and Process Mixing Points" (Houston, TX: NACE, 2001).

13 Preheat Exchanger Fouling

In this Chapter

13.1 First Glance

Preheat exchanger fouling is caused by certain types of crude, by sediment and sludge in crude, and by desalter upsets. The severity and extent of fouling often depends on the crude rate, exchanger geometry, metallurgy, tube temperature, and heat flux. Caustic can contribute to preheat exchanger fouling, if injected at the wrong location or diluted insufficiently prior to injection. Similarly, slop oil can contribute to preheat exchanger fouling, if injected at the wrong location or at too large a volume at any one time.

13.2 Fouling Composition

A number of factors contribute to preheat exchanger fouling, including decreased crude throughput; low flow velocities; high salt, brine, or solids contents of crude; chemical instability and low API gravity of crude; and high temperature gradients across the preheat exchanger tube walls. The principal components and sources of preheat exchanger fouling include one or more of the following categories of compounds present in crude, slop oil, or desalter water:

- Inorganic salts, including chlorides, sulfates, carbonates, and silicates.

- Organic compounds, including asphaltenes, polymers, gums, and waxes.
- Sediment, including sand, silt, and particulate matter of various types.
- Corrosion deposits, including sulfides, oxides, and hydroxides from upstream and midstream production, treating, transportation, and storage operations.
- Corrosion deposits and neutralizer hydrochloride salts in overhead water condensate recycled for use as desalter water.
- Caustic particles due to incorrect caustic injection procedures.

13.3 Fouling Mechanism

Preheat exchanger fouling is usually caused by physical and/or mechanical rather than chemical processes. While hydrocarbon cracking reactions are unlikely to occur at the relatively low temperatures in preheat exchangers ranging from 150 °C (300 °F) to 230 °C (450 °F), crude undergoes a number of changes upon heating. Entrained water droplets dissolve in the heated crude causing dissolved salts in the water droplets to precipitate on hot tube surfaces. Crude destabilization breaks the chemical bonds between asphaltenes and resins and causes asphaltenes to coagulate on the hot tube surfaces. Suspended particles, including salt crystals, rust, sand, silt, and tar, separate out as the crude is heated further and the crude viscosity decreases. While not part of the fouling mechanism as such, corrosion can be expected to occur under accumulated fouling deposits, and the resulting corrosion deposits contribute to the bulk of the fouling deposits.

13.4 Fouling Costs

The cost of preheat exchanger fouling is reflected primarily in increased fuel cost at the fired heater, increased costs due to reduced heat transfer, and the costs of cleaning operations. As a general rule, most fouling occurs on the crude side (usually the tube side) and not on the product side (usually the shell side) of preheat exchangers (PEs).[1]

13.4.1 Direct Fouling Costs. Preheat exchanger fouling decreases the crude inlet temperature at the fired heater. This means the firing rate of the heater will have to be increased to ensure that the crude outlet temperatures are maintained. Therefore one way of measuring the cost of preheat exchanger fouling is to determine the cost of extra fuel gas burned at the fired heater. Nomograms are available to enable unit operators to readily determine this cost. With one nomogram, the basic information required for its use includes the API gravity and the specific heat of the crude charge; the change in outlet temperature of the preheat exchanger (or preheat exchanger train) due to fouling; and the cost of fuel gas, as shown in Figure 13-1. Since the cost of fuel gas and, hence, the cost of extra fuel burned, has varied significantly in recent years, a suitable multiplier may have to be applied. The nomogram assumes that no vaporization occurs in the preheat exchanger, the fired heater efficiency is 75%, and the fuel gas heating value is 37M J/m^3 (1,050BTU/SCF). The cost of preheat exchanger fouling can then be obtained in terms of US$/1,000BBL crude charged.

13.4.2 Indirect Fouling Costs. Aside from increased fuel costs, preheat exchanger fouling increases operating costs for a number of other reasons. Crude throughput may have to be decreased, if the unit is running near capacity. Higher firing temperatures at the heater can be expected to shorten heater tube life to the point where the tube metallurgy has to be upgraded. Extra exchanger capacity may have to be provided to properly cool the product streams on the hot side of the preheat exchangers. Costs associated with periodic bypassing and cleaning of preheat exchangers can be expected to increase significantly. Also preheat exchanger tube life is shortened due to underdeposit corrosion.

13.4.3 Effect of Cleaning Cycles on Fouling Costs. Fouling of preheat exchanger tubes carrying crude usually occurs in several distinct stages. During the initial stages, a thin layer of deposits forms inside the tubes and interferes with heat conductivity through the tube wall. As a result, overall heat transfer Q decreases with operating time. At the same time, the crude outlet temperature decreases, while the outlet temperature of various hot product streams (or "hot oil" where used) on the shell side increases. During the second and final stages, fouling progresses to the point where the pressure drop across the tube side of the exchanger increases and crude flow decreases. As a general rule, Q continues to decrease with operating time. Even when tube side fouling has not yet reached the final stage, flow may have to be

1. Crude is preferentially placed on the tube side to limit the cost of cleaning and/or alloy upgrading to the tubes instead of both the shell and tubes.

cut back intentionally in order that crude outlet temperatures can be maintained. Heat transfer can be partially, if not completely, restored by cleaning one or both sides of the heat exchanger. However, after each cleaning operation, fouling will start all over again. Therefore, heat transfer in the typical preheat exchanger consists of one or more fouling/cleaning cycles per unit operating time, as shown in Figure 13-2.

FIGURE 13-1 NOMOGRAM FOR DETERMINING THE COST OF PREHEAT EXCHANGER FOULING

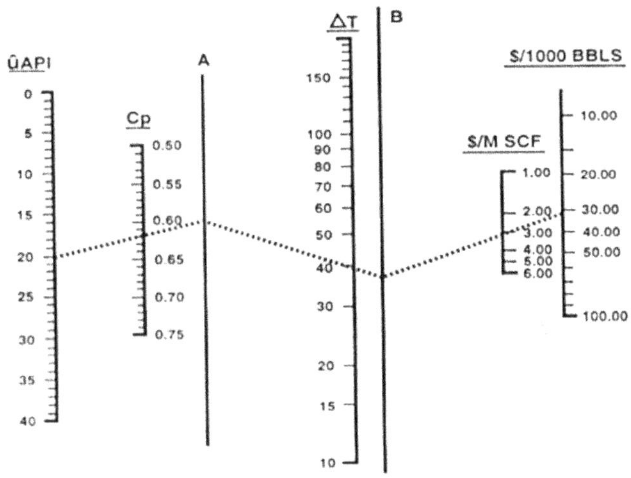

Adapted from P.P. Corneli, "Practical Ideas on Refinery Fouling," *Hydroc. Proc.* 43, 6 (1964).

FIGURE 13-2 EFFECT OF FOULING/CLEANING CYCLES ON HEAT EXCHANGER DUTY

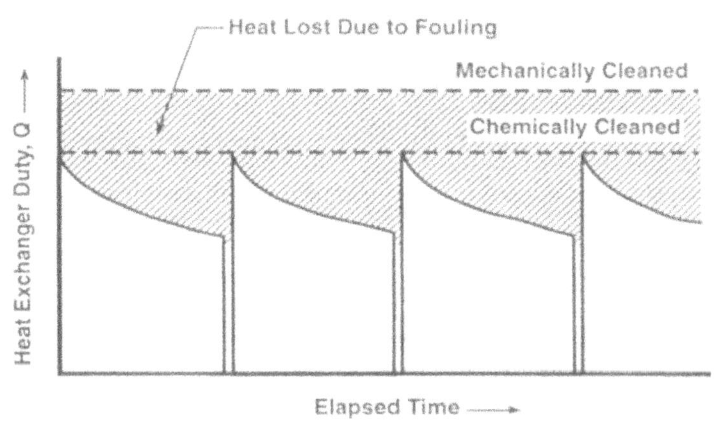

Adapted from J. Gutzeit, *Cleaning of Process Equipment and Piping–A Complete How-To Manual*, publication no. 51 (St. Louis, MO: Materials Technology Institute, 1997).

The hatched area in Figure 13-2 represents the decrease in heat transfer during each fouling cycle. Note that chemical cleaning can only partially restore heat transfer compared to mechanical cleaning, such as hydro jetting[2]. The total annual fouling cost, and the various interrelated cost factors which make up this total cost, can be obtained as follows:

$$C_F = (Q_F \cdot C_E + P + C) \cdot N \quad \text{(Eq.13-1)}$$

2. High-pressure water blasting.

where:

C_F = total annual cost of fouling, US$
Q_F = decrease in heat transfer during each fouling cycle, J (BTU)
C_E = unit energy cost, US$
P = cost of reduced crude throughput during each fouling cycle, US$
C = cleaning cost, US$
N = number of fouling/cleaning) cycles per year

For the sake of simplicity, the costs of reduced heat transfer and crude throughput during the cleaning operation have been ignored, but there is no reason why they cannot be included in the analysis. However, in many applications these costs are zero because the heat exchanger can be bypassed during the cleaning operation without significantly affecting the operation of the rest of the unit. In other cases, any losses can be made up elsewhere without affecting overall costs. The decrease in heat transfer due to fouling, and, hence, the associated energy costs increase with the length of each fouling cycle. Fouling of preheat exchangers usually requires that crude throughput be reduced in order that outlet temperatures for downstream operations can be maintained. The associated cost, P, also increases with the length of each fouling cycle. In contrast, cleaning cost decreases with the length of each fouling cycle because the fewer times a heat exchanger is cleaned, the lower its annual cleaning cost will be.

The various cost factors for typical crude preheat exchangers are shown graphically as a function of a fouling/cleaning cycle in Figure 13-3. Also shown is the U-shaped curve that represents the total annual cost of fouling. This curve reaches a minimum point where the cost of fouling, $Q_F \cdot C_F + P$, may or may not be equal to the cost of cleaning, depending on the exact shape and location of the various cost curves. In this particular example, the minimum point occurs when the fouling cycle is approximately 4 months. If the heat exchanger is cleaned earlier than that, the cleaning costs become excessively high compared to the fouling costs. If cleaned later, the fouling costs become excessively high compared to the cleaning costs. Therefore, the minimum point represents the optimum cleaning interval. The ratio $(Q_F \cdot C_E + P)/C$ represents the benefit/cost ratio and can be used to calculate the return on investment (ROI). While the heat transfer and fouling costs of a single heat exchanger or an entire heat exchanger train can be easily calculated, this is not the case when only one or two heat exchangers of a heat exchanger train are involved. Therefore, several refining companies have developed in-house software programs to handle such problems, including simulation of various process scenarios and graphic analyses. Some suppliers of process antifoulants also have in-house software programs to support their antifoulant marketing efforts.

13.5 Monitoring Exchanger Performance

Preheat exchangers (PEs) probably top the list of crude unit and upgrader components that are cleaned on an ongoing basis because fouling of one or more exchangers has an immediate effect on plant operations. When preheat exchanger outlet or heater inlet temperatures are monitored and recorded on the control board, any changes become immediately apparent to the operators. Fouling increases the pressure drop and decreases the flow through a problem heat exchanger. Even though the crude charge pumps are located upstream of the exchangers, flow decreases as fouling increases, because the pumping rate is more or less fixed. As flow decreases and temperatures change as a result of fouling, heat transfer duty and the heat transfer coefficient decrease. Therefore, fouling can be measured indirectly by measuring the pressure drop across a problem heat exchanger or measuring the overall heat transfer coefficient, U, for a problem heat exchanger.

FIGURE 13-3 EFFECT OF LENGTH OF FOULING CYCLE ON FOULING COST

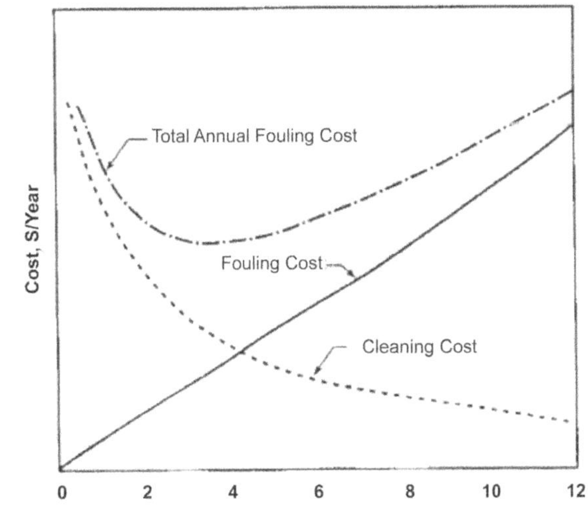

Adapted from J. Gutzeit, *Cleaning of Process Equipment and Piping–A Complete How-To Manual*, publication no. 51 (St. Louis, MO: Materials Technology Institute, 1997).

13.5.1 Monitoring Pressure Drop. Pressure drop should be routinely monitored, if a preheat exchanger is expected to foul on a regular basis. Changes in pressure drop over time are a good fouling indicator and are easily measured, but difficult to interpret in terms of fouling and cleaning costs. Pressure drop can increase precipitously in a heat exchanger that still provides some measure of heat transfer. Excessive tube side pressure drop across a preheat exchanger can cause the inlet channel baffles to collapse because they are not designed to withstand the full pumping pressure when flow is blocked. Collapse of the inlet channel baffles eventually causes the crude to bypass the affected exchanger. To monitor pressure drop on a routine basis, pressure-sensing devices need to be installed just upstream and downstream of the problem heat exchanger. Appropriate data recorders need to be mounted on the unit's control board. Alternatively, pressure gages upstream and downstream of the exchanger can be used to measure pressure drop changes and the resultant data can be hand-plotted against operating time. As a general rule, tube side pressure drop tends to range from approximately 0.5-0.8bar (7-12PSI) across a new or just-cleaned heat exchanger and from 1.4-7bar (20-100PSI) across a badly fouled heat exchanger. Shell side pressure drop is usually lower and ranges from approximately 0.07-0.7bar (1-10PSI) across a new or just-cleaned heat exchanger and from 0.7-3bar (10-45PSI) across a badly fouled heat exchanger.

13.5.2 Monitoring Overall Heat Transfer Coefficient. Change in the overall heat transfer coefficient U with time is a good indicator of the rate of fouling in a problem heat exchanger, but it too is difficult to interpret in terms of fouling and cleaning costs. Nevertheless, overall heat transfer coefficients should be measured and plotted at regular intervals to determine the rate of heat-transfer loss due to fouling. This requires accurate temperature and flow readings for both the crude and the hot heating medium (usually a product stream) passing through the exchanger. It also requires additional instrumentation beyond that normally required for process control. Heat transfer coefficients are difficult to calculate accurately, when a partial phase change due to evaporation or condensation occurs. Rapid temperature and flow-rate fluctuations also present difficulties, as do changes in the crude mix. While software programs are available that can compensate for these factors, calculated heat transfer coefficients are still not entirely accurate because some input data are usually missing. Accuracy can be improved by measuring consecutive data points at nearly identical flow rates (within +/- 5%) and obtaining enough readings to average out any variations. The overall heat transfer coefficient is obtained from actual temperature and flow rate data, using the following relationship:

$$U = Q / (A \cdot F_T \cdot LMTD) \qquad \text{(Eq.13-2)}$$

where:

U	=	Overall heat transfer coefficient, W/m²·C (BTU/hr·ft²·F)
Q	=	Overall heat transfer duty, W (BTU/hr)
A	=	Heat transfer surface, m² (ft²)
F_T	=	Temperature correction factor
LMTD	=	Log mean temperature difference, C (F)

Ideally, the overall heat transfer duty is identical to the heat transfer duty for the cold and hot streams, respectively. Due to inaccurate measurements on one side or the other, overall duty is usually taken as the average of the two individual heat transfer duties. With insufficient instrumentation on one of the streams, the heat transfer duty of the other stream will often have to suffice. The individual heat transfer duties are obtained, using the following relationships:

$$Q_C = R_C \cdot S_C \cdot (T_{C2} - T_{C1}) \quad \text{for the cold stream} \qquad \text{(Eq.13-3)}$$

$$Q_H = R_H \cdot S_H \cdot (T_{H1} - T_{H2}) \quad \text{for the hot stream} \qquad \text{(Eq.13-4)}$$

where:

R	=	Flow rate, kg/hr (lb/hr)
S	=	Specific heat, J/kg C (BTU/lb F)
T_1	=	Stream inlet temperature, C (F)
T_2	=	Stream outlet temperature, C (F)

Here, the subscript "C" denotes the properties of the cold stream (usually crude) in the heat exchanger, while the subscript "H" denotes the corresponding properties of the heating medium (usually a product side stream). The log mean temperature difference (LMTD) is obtained, as follows:

$$\text{LMTD} = \frac{(T_{H1} - T_{C2}) - (T_{H2} - T_{C1})}{\ln (T_{H1} - T_{C2}) / (T_{H2} - T_{C1})} \qquad \text{(Eq.13-5)}$$

To determine fouling trends, the overall heat transfer coefficient, U, is plotted against operating time, starting with the clean overall heat transfer coefficient U_{Clean} obtained right after startup of the unit, or right after cleaning. If available, the design overall heat transfer coefficient U_{Design} can be shown for comparison. A typical curve is shown below in Figure 13-4. It is apparent that U_{Clean} varied from 44 to 52 BTU/hr·ft²·F. In this case, U_{Design} was 70 BTU/hr·ft²·F, but is not shown. Subsequently, the heat exchanger was bypassed for cleaning whenever U had dropped to between 20 and 25 BTU/hr·ft²·F, or approximately 40% of U_{Clean}.

FIGURE 13-4 EFFECT OF FOULING AND CLEANING CYCLES ON OVERALL HEAT TRANSFER COEFFICIENT, U, OF A PREHEAT EXCHANGER

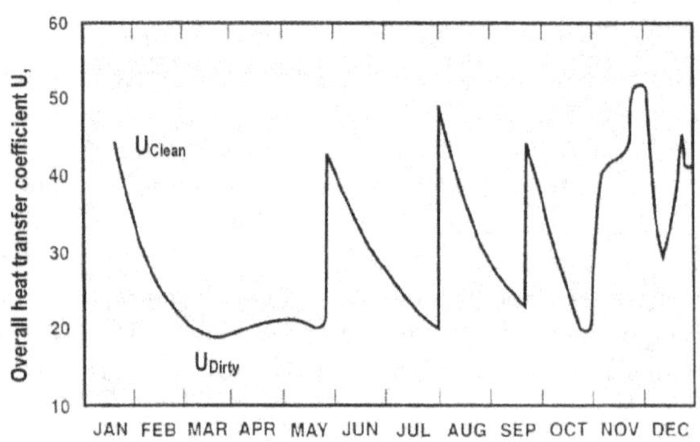

Adapted from J. Gutzeit, *Cleaning of Process Equipment and Piping–A Complete How-To Manual*, publication no. 51 (St. Louis, MO: Materials Technology Institute, 1997).

In general, cleaning restores U to the previous clean values, but the rate of fouling increases with time. For example, at the beginning of the year, the heat exchanger described in Figure 13-4 could be operated for 4 months until bypassing and cleaning became necessary. Subsequent cleaning was required after 2, 1.5, and 1 months, respectively. It is also apparent that once the plant operators recognized the benefits of bypassing the heat exchanger and having it cleaned, they ended up doing so while U was still above 30 BTU/hr·ft²·F, or approximately $0.60U_{Clean}$. Eventually, $U = 0.60U_{Clean}$ became an accepted criterion for initiating heat exchanger cleaning. In crude preheat exchangers, U_{Clean} typically is 115-140 W/m²·C (20-25BTU/h·ft²·F). However, clean heat exchanger tubes will not remain clean for very long. They will continue to transfer heat, while U drops asymptotically until it reaches some finite value that is still well above zero. The heat exchanger will have to be bypassed and cleaned because pressure drop reaches the point where flow is seriously impeded.

13.6 Controlling Exchanger Fouling

In most cases, preheat exchanger fouling is limited to one and sometimes several exchangers of the preheat exchanger train and a number of methods can be used to help control fouling. First, crude should be placed on the tube side to facilitate cleaning of what usually becomes the dirtiest side of the tubes. Second, bypass piping or a parallel spare exchanger should be provided for exchangers that foul on a regular basis. This allows rapid bundle changeout and cleaning by hydro jetting or chemical means while the unit remains on stream. Third, in-line mixing elements can be installed upstream of problem exchangers, and the crude rate can be increased by additional pumping capacity to improve dispersion of slugs of sediment.

If a fouled exchanger is upstream of the desalter (i.e., CPE), part or all of the desalter water can be injected on the suction side of the crude charge pumps to help prevent inorganic salt deposits from forming in the exchanger. If a fouled exchanger is downstream of the desalter (i.e., HPE) desalter operation needs to be carefully monitored and controlled to avoid sudden carryover of entrained slugs of brine and mud with the desalted crude. Caustic should be injected downstream of the fouled exchanger to prevent caustic from contributing to the fouling and the caustic solution strength should be decreased to less than 4.5°Baumé (3%) to improve NaOH dispersion in the desalted crude. Slop oil should also be injected downstream of the fouled exchanger and added at a controlled rate of no more than 1% by trickle feeding to prevent slop oil from contributing to the fouling.

13.7 Use of Antifoulants

Most antifoulants for use in the crude charge are dispersants (amine phosphate ester detergents). In many cases, antifoulant injection into incoming crude has been ineffective and could not be justified from a cost point of view. In some cases, however, antifoulant injection has made deposits easier to remove when the tube bundle was pulled for cleaning. The following should be kept in mind, if antifoulant injection is being considered:

- Antifoulants can delay the onset of fouling, but not prevent fouling. They require clean tubes to start out with in order to be effective. Once the overall heat transfer coefficient U has dropped below approximately 60% of the clean heat transfer coefficient, U_{Clean}, antifoulants are no longer effective and the tube bundle will have to be cleaned.
- Preheat exchanger fouling varies from day to day due to changes in crude slate, desalter operation, and temperatures. This makes it often impossible to inject the right amount of antifoulant for maximum effectiveness.
- Fouling often moves from one preheat exchanger to the next because deposits removed from an upstream exchanger are likely to redeposit in a downstream exchanger.
- Fouling often is a function of salt composition, concentration, and solubility limits in the crude. These are not changed by dispersant-type antifoulants.

13.7.1 Location for Antifoulant Injection. Antifoulants should be injected as close to the point of fouling as possible which can make it difficult to find a suitable injection point for the antifoulant. The effectiveness of antifoulants depends on proper timing and injection, and on the type of foulants encountered. For example, antifoulants tend to be ineffective against hard water scale originating with produced water. Ideally, antifoulants are added only after a problem preheat exchanger has been thoroughly cleaned and inspected during a shutdown. In actuality, antifoulants are often tried out during normal operation in an effort to improve heat transfer. The following alternatives should be considered, if antifoulant injection is to be tried out.

New bundles should be immersed in an open tank of neat (undiluted) antifoulant solution prior to installation. This helps maintain heat transfer during the critical startup phase and the tubes are less prone to foul immediately after startup. It also prevents rusting and surface pitting while the bundle is exposed to the elements during the shutdown. Since the largest drop in heat transfer in a clean exchanger occurs within a day or two after startup, antifoulant injection at twice (or more) the normal level over a 2-week period after startup has been beneficial. If antifoulant is to be used with heat exchangers that have not been cleaned, precautions must be taken because of the strong detergency properties of many antifoulants. Injection at normal and especially at high pretreatment levels can loosen large amounts of fouling and corrosion deposits all at once and carry them downstream. To avoid plugging problems, treatment levels may have to be increased in increments over several weeks, starting with approximately 20 percent of the normal level during the first week.

13.7.2 Optimum Injection Rate. The required amount of antifoulant depends primarily on crude flow rate. With standard antifoulant formulations, no more than 18-22mg/L is usually injected. This corresponds to approximately 285-340L/d (75-90GPD) for a typical 13,000TPD (100,000BPD) crude unit. Lower dosages are usually ineffective, while higher dosages do not necessarily result in significant improvements and are not cost effective.

13.7.3 Antifoulant Evaluation Tests. Antifoulants can be evaluated in the field and in the laboratory. In field trials, the effectiveness of antifoulant injection into the crude charge is often difficult, if not impossible, to evaluate for a number of reasons:

- Heat exchanger tube surfaces should be clean to start with for an antifoulant to work effectively. This is often not the case when a trial starts.
- Fouling rates can fluctuate rapidly due to changes in the crude slate or repeated desalter upsets.
- Fouling may occur on both sides of some heat exchanger tubes, but at different rates.
- Most heat exchangers have insufficient instrumentation for accurate heat transfer or pressure drop measurements. Both are necessary to allow accurate monitoring of the change of fouling with time.
- Two-phase flow and evaporation/condensation phenomena interfere with accurate heat transfer and pressure drop measurements.

In general, a field trial of an unproven product can involve substantial costs to the refinery since most suppliers are unwilling to provide their product free for the test. Human factors often affect the results of field trials. For example, overly optimistic interpretation of initial test data by the antifoulant supplier often leads to disappointment when a heat exchanger is eventually pulled for inspection. There is often a natural reluctance to abort a trial, if there is any kind of chance for eventual success. With long unit runs, trials may then end up lasting indefinitely. While numerous laboratory and pilot plant fouling tests have been devised over the years, most have given inconclusive results for one or more reasons:

- Despite extensive studies, the exact fouling mechanism is not fully understood, and tests results do not duplicate field experience.
- Controlled laboratory fouling tests tend to be accelerated tests and, despite claims to the contrary, often do not accurately represent long-term fouling.
- Controlled field testing with small, parallel, fully instrumented pilot plant heat exchangers has been partially successful, but even then the results are usually open to interpretation.
- Various heat transfer models have been applied to evaluate laboratory or pilot plant fouling data, but despite claims to the contrary, no two usually agree in their findings.

13.7.4 Cost of Antifoulant Injection. Ideally, a complete cost analysis of antifoulant injection includes the cost of alternative fouling control measures, such as appropriate design changes, process changes, or temperature changes. In reality, however, the effectiveness of such changes is difficult if not impossible to predict, given the current need to constantly change crude slates and the general lack of interest in funding fouling research. Plant experience has shown that antifoulant injection is expensive and usually does not justify its cost, even when it seems to work. On a typical crude unit, the annual antifoulant cost (in U.S. $) can easily reach three to four times the daily throughput rate (in BPD). As an example, an antifoulant cost of U.S. $300,000-400,000 per year can be expected for a typical 13,000TPD (100,000BPD) crude unit. For that amount of money, a lot of preheat exchangers can be bypassed and cleaned on a regular basis, assuming a typical cost of less than U.S. $5,000 to 10,000 for bypassing and hydro jetting a heat exchanger on the run.

In the past, some refineries have developed "break-even cost" curves for antifoulant usage at the crude unit. The idea is to use antifoulant only, if the run length with antifoulant injection can be expected to exceed the run length without antifoulant injection by a certain number of days. This approach assumes, however, that the unit is running near capacity, exchangers are clean at the start of a run, and the unit runs are relatively short (1-2yr). It also assumes that crude slates remain pretty much the same over the duration of the run, which is no longer the case. It ignores the fact that with proper planning and preparation, many plants no longer have to come down for preheat exchanger cleaning, because the exchangers can be bypassed and cleaned on the run.

13.8 Keeping Exchangers Clean

Hydro jetting remains the most efficient and preferred offstream mechanical method of cleaning the tube side of preheat exchangers. With the exchanger bypassed and opened, the bundle is either kept in place during cleaning, or pulled and exchanged for a clean bundle. The fouled bundle is eventually cleaned on the ground, as shown in Figure 13-5. Pulling the bundle for cleaning on the ground offers the advantage that the shell side of the bundle can be cleaned as well as the tube side. In either case, hydro jetting should be done by a qualified cleaning contractor because of safety and environmental concerns.

FIGURE 13-5 HYDRO JETTING A PREHEAT EXCHANGER BUNDLE ON THE GROUND

Adapted from J. Gutzeit, *Cleaning of Process Equipment and Piping–A Complete How-To Manual*, publication no. 51 (St. Louis, MO: Materials Technology Institute, 1997).

Aside from offstream mechanical methods for cleaning heat exchangers, five systems for keeping tubes free of fouling deposits during normal operations have been used with preheat exchangers and elsewhere. Except for twisted-tube exchangers, all require that the foulant (for example, crude) is placed on the tube side and that the installation start with clean exchangers. In twisted-tube exchangers, either side can be used.

13.8.1 Koch Fintube™ Twisted Tube Bundles. Foremost among various heat exchanger designs are exchangers using Fintube™ twisted tubes. Twisted tube heat exchangers offer a number of advantages over bare and standard finned tube exchangers, including increased heat transfer, lack of shell baffles, elimination of tube vibration, reduced fouling, and extended run time between cleanings. An example is shown in Figure 13-6. The lack of baffles lowers the pressure drop during operation, allows the bundles to stay cleaner during operation, and makes the bundles easier to clean by hydro jetting on both tube side and shell side during shutdowns. Twisted tube bundles are available purpose build from a number of heat exchanger manufacturers in a variety of materials of construction, including carbon steel, copper alloys, and stainless steels.

FIGURE 13-6 FINTUBE™ TWISTED TUBE HEAT EXCHANGER BUNDLE

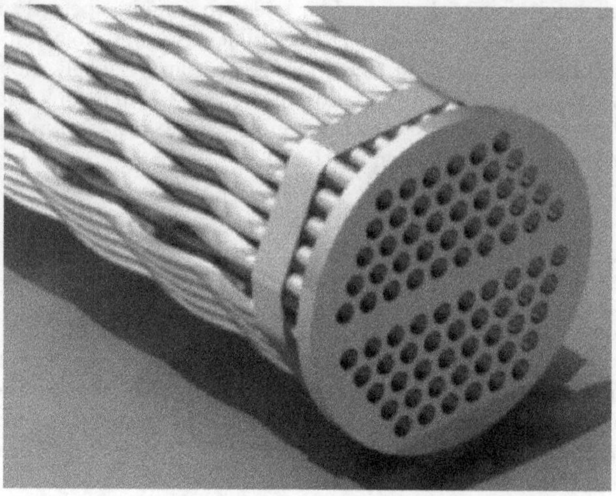

Courtesy Koch Heat Transfer Company, Houston, TX.

13.8.2 Modified Bottle-Brush System™ by WSA[3]. With this system, a small cylindrical stainless steel wire brush is permanently held in a cage or catch basket at one end of each tube. Another catch basket is mounted at the other end of each tube. The brush is propelled through the tube once or twice a shift by reversing normal crude flow. A special channel and piping modifications are required to allow automatic flow reversal at the elevated temperatures seen in preheat exchangers. Originally developed for industrial air conditioning chiller heat exchangers, this system has worked quite well keeping crude preheat exchanger tubes clean in several trial installations. However, eventually more and more brushes get stuck in their catch baskets, or leaks occur at the automatic flow reversal valve. Because of the danger of spilling hot crude, most if not all installations were eventually abandoned.

13.8.3 SPIRELF™ Spiral Spring Cleaning System[4]. With this onstream mechanical cleaning system, shown in Figure 13-7, a stainless steel spring coil is suspended along the length of each tube. Flow-induced vibration of each coil prevents fouling from depositing inside the tubes. The vibrating springs also remove any fouling that has already deposited on the tube walls. This system too has worked quite well and a number of installations throughout the world are in place. Reportedly, the spring cleaning system is able to double the time between hydro jetting or chemical cleaning. However, unless the system is carefully designed and maintained, corrosion, abrasion, and metal fatigue can cause an excessive rate of spring failures. All springs should be replaced whenever the exchanger comes down for cleaning.

FIGURE 13-7 SPIRELF™ CLEANING SYSTEM FOR HEAT EXCHANGER TUBES

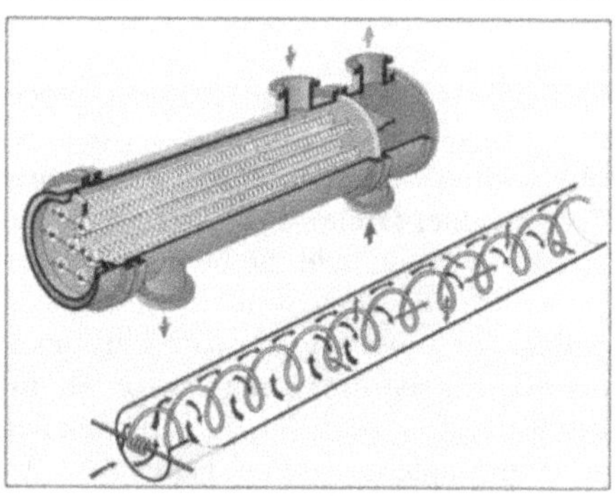

Courtesy Petroval LP, Houston, TX.

3. Trademark of Water Services of America.
4. Trademark of ELF Aquitaine.

13.8.4 TURBOTAL™ Spring Cleaning System.[5] With this system, a spring is held at the inlet end of the tubes by a fixing device that allows the other end to rotate freely around its axis as a result of fluid flow. This rotation causes high flow turbulence and improves the tube side heat transfer coefficient. As the boundary layer is continuously renewed, the wall temperature is lowered and fouling is reduced. Combined with the mechanical effect of the spring rotation, the TURBOTAL™ system reduces fouling by preventing deposit formation and by breaking up the boundary layer along the tube wall. The rotational speed can adapt itself to the operating conditions, such as variations in flow rate, viscosity, density, or vaporization of slugs of water.The fixing device of the coil is located outside each tube. The coil can adapt itself to the geometry of the tubes and the specific process conditions. The TURBOTAL™ system reduces fouling inside heat exchanger tubes and the turbulent effect improves the heat transfer rate (the tube side heat transfer coefficient is multiplied by up to 1.8). Overall, the sum of these two effects results in a reduction of the apparent fouling factor by as much as 75%. The SPIRELF™ and TURBOTAL™ devices can be used in all shell-and-tube heat exchangers with straight tubes as opposed to U-tubes. Specifically, both systems can be used with most fluids up to 360 °C (680 °F) and at velocities of 0.6-3.0m/s (2-6FPS). Typical tubing sizes are 12-50mm (0.5-2in.) OD and 3.3-10m (10-33ft) in length. The methods have been used in various applications aimed at improving energy efficiency and reducing maintenance costs where the main fouling layer was formed by asphaltenes.

13.8.5 FIXOTAL™ Spring Cleaning System.[6] With this system, a wire spring is inserted in each tube and held permanently in place as it expands against the tube wall. In contrast to the SPIRELF™ and TURBOTAL™ devices, the spring does not rotate or vibrate, but simply acts as a turbulence promoter and increases tube side heat transfer. At the same time, tube side fouling is reduced. The wire spring can adapt itself to the geometry of the tubes and the specific process conditions. As a result, the FIXOTAL™ system can be used in all shell-and-tube heat exchangers including those with U-tubes. The springs are available in a large variety of materials and shapes to suit the application. According to the manufacturer, cleaning intervals for preheat exchanger tubes average 2.5 times longer compared to bare tubes. The pressure drop per pass is 0.15-0.2bar (2-3psi).

13.9 Chapter Summary

Preheat exchanger fouling is caused by sediment and sludge in certain crudes and by desalter upsets. The severity and extent of fouling depends on crude throughput, heat exchanger geometry, tube metallurgy, metal temperature, and heat flux. Preheat exchanger fouling decreases the crude inlet temperature at the downstream fired heater. As a result, the heater has to be fired harder to make up for lost heat. Aside from the increased fuel costs and decreased heater tube life, preheat exchanger fouling increases other operating costs, including the costs of bypassing and cleaning the affected exchangers on a more-or-less ongoing basis.

Fouling increases the pressure drop across a problem exchanger and eventually decreases both throughput and heat transfer duty. Pressure drop can increase precipitously in an exchanger and cause the inlet channel baffles to collapse. For this reason, pressure drop should be routinely monitored, if an exchanger is expected to foul on a regular basis. Overall heat transfer coefficients should be monitored routinely to determine the rate of heat transfer loss. The change in overall heat transfer coefficient with time is a good indicator of fouling, but requires accurate temperature and flow readings for both sides of the exchanger. Antifoulants can delay, but not prevent the onset of fouling in preheat exchangers. Because of the large volume of crude passing through preheat exchangers, antifoulant use is expensive and often cannot be justified in terms of energy savings. Moreover, antifoulants are only effective with new or just-cleaned exchangers. Once the overall heat transfer coefficient has dropped to below 60% of the clean coefficient, antifoulants are no longer effective and the tube bundle will have to be pulled for cleaning. Offstream hydro jetting remains the preferred method for cleaning crude preheat exchangers, although several onstream mechanical methods have been developed and used by various petroleum companies.

5. Trademark of Total Oil Company.
6. Trademark of Total Oil Company.

13.10 Read More

1. F. Pouponnot and A. W. Krueger, "Heat Exchanger Tube Inserts - An Update with New Applications in Crude Distillation Unit, Vacuum Applications, and Reboilers," paper presented at the AIChE Spring Meeting, Orlando, FL (2006).
2. A. W. Krueger and F. Pouponnot, "Heat Exchanger Performance Enhancement through the Use of Tube Inserts in Refineries and Chemical Plants – Successful Applications: Spirelf, Turbotal, and Fixotal Systems," Proc. Int. Conf. on Heat Exchanger Fouling and Cleaning, VIII (2009).
3. J. Gutzeit, *Cleaning of Process Equipment and Piping – A Complete How-To Manual*, MTI publication no. 51. (St. Louis, MO: Materials Technology Institute of the Chemical Process Industries, Inc., 1997).
4. P. P. Corneli, "Practical Ideas on Refinery Fouling," *Hydroc. Proc.* 43, 6 (1964).
5. J. D. Crawford and R.M. Miller, "Refinery Process Fouling – Effect of Trace Metals and Remedial Techniques," *Proc. API* 43, III (1963).
6. H. M. Joshi, "Mitigate Fouling to Improve Heat Exchanger Reliability," *Hydroc. Proc.* 78, 1 (1999).
7. L. L. Ludwig, "Are Your Crude Unit Antifoulants Saving You Money," *Hydroc. Proc.* 64, 1 (1985).
8. NACE Publ. T8170, "Process Side Antifoulants in Petroleum Refineries," *MP* 9, 6 (1970).
9. NPRA Q&A, "Crude Distillation Issues and How to Solve Them," *Oil & Gas J.* March 30 (1970).
10. P. Van der Wee and P. A. Tritsmans, "What to do About Crude Unit Preheat Exchanger Fouling," *Hydroc. Proc.* 45, 8 (1966).
11. J. G. Speight, *Fouling in Refineries* (Houston, TX: Gulf Professional Publishing, 2015).

14 Corrosion in Fired Heaters

In this Chapter

14.1 First Glance

Fired heaters (furnaces) are the principal heat source for most high-temperature processes and represent a substantial investment for the plant. Crude units have at least one fired heater in the atmospheric section to heat the desalted crude charge to the atmospheric column and at least one fired heater in the vacuum section to heat the reduced crude (atmospheric resid) charge to the vacuum column. The same applies to the DRU and VDU of bitumen upgraders. Depending on past revamps and upgrades, there may be additional fired heaters to increase unit throughput. Most fired heaters contain steam coils to supply low- or medium-pressure process steam. On some units, a fired heater is used to provide circulating "hot oil" for heat exchangers and reboilers on the unit. All fired heaters use either gas or oil as fuel.

14.2 Basic Heater Construction

Several configurations of fired heaters are used, depending on age, size, design, operating parameters, type of crude or reduced crude, and refinery preferences. A typical example is the double-radiant section cabin heater with horizontal tube rows (coils) and an overhead convection section shown in Figure 14-1. Fired heaters are purpose-designed and field-erected by the heater manufacturer who guarantees the heater's overall performance. Heaters are designed around their burners. Many have burners located along the floor (bottom) of the firebox, while others have burners located in one or both end walls. Most fired heaters contain radiant and convection coils to heat the charge. They can also contain separate steam coils to provide process steam.

When heaters with horizontal tube rows are fired from the floor, a large number of small burners along the length of the firebox can be used. With this design, the flames are relatively short and clearances between flames and tubes are not critical. When heaters with horizontal tube rows are fired from one or both end walls of the firebox, only 2 to 4 burners per end wall can be used. The firebox must be wide enough to provided adequate clearance between the flames and tubes along the entire length of the firebox to prevent flame impingement, "hot spots," and early tube failures due to stress rupture.

FIGURE 14-1 FIRED HEATER WITH TWIN RADIANT SECTIONS AND OVERHEAD CONVECTION SECTION

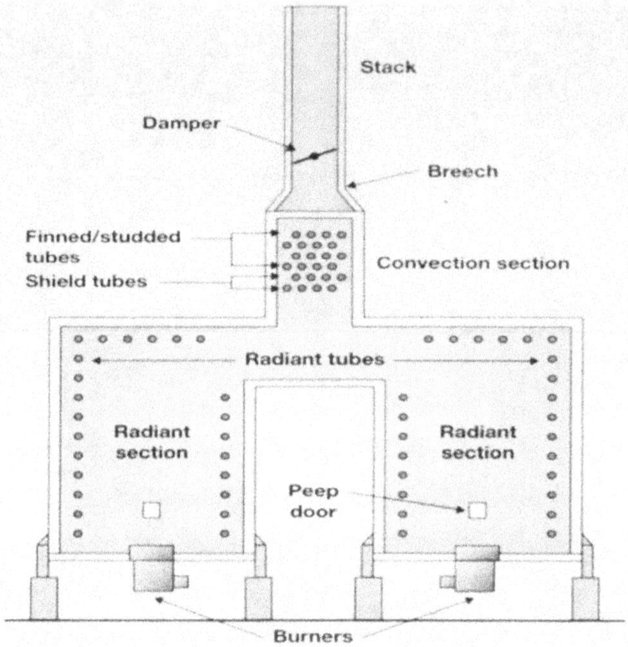

Adapted from A. Garg, "Improve Vacuum Heater Reliability," *Hydroc. Proc.* 78, 3 (1999).

14.3 Heater Tubes

As a general rule, heater tubes have a nominal outside diameter of 100-150mm (4-6in.). Tube size in vacuum heaters often increases toward the heater outlet to keep vapor velocities below sonic values. Tube length ranges from 12-18m (40-60ft) and usually depends on available tube-mill capacity. The length of convection tubes is based on a convection heat transfer coefficient that, in turn, depends on flue gas velocity and/or size of the convection section. The bottom two tube rows in the convection section (shield tubes) are often subject to overheating and accelerated corrosion when finned or studded tubes are substituted for bare tubes to increase heat transfer.

14.3.1 Heat Flux. High average radiant heat-flux values lower the construction cost of a fired heater for a given duty, but decrease tube life and increase maintenance costs. As a general rule, average radiant heat flux values for heater tubes should not exceed 30-45KW/m^2 (10,000-15,000BTU/h·ft^2) in atmospheric heaters, and 15-30KW/m^2 (5,000-10,000BTU/h·ft^2) in vacuum heaters. Average radiant heat flux above these values increases coke deposition to the point where hot spots form that eventually lead to early tube failure. Average convection heat flux values should not exceed 10-30KW/m^2 (3,300-10,000BTU/h·ft^2) for bare tubes, and 15-40KW/m^2 (5,000-13,000 BTU/h·ft^2) for finned tubes.

14.3.2 Mass Flow Velocity. The minimum mass flow velocities through heater tubes should be 1,000kg/m²·sec (200lb/ft²·sec) in atmospheric crude heaters and 1,250kg/m²·sec (250lb/ft²·sec) in vacuum heaters. Mass flow velocities below these values increase coking to the point where hot spots form. If necessary, enough velocity (dilution) steam should be added to reach and exceed these values.

14.4 Heater Tube Problems

Tubes, return bends, and headers represent the most expensive components of a fired heater and are also the most frequently inspected, repaired, or replaced. Heater tube issues include certain metallurgy requirements. Since heater tubes operate with metal temperatures well above 230 °C (450 °F), alloying against HTSC is required. Even in the absence of naphthenic acids, low-alloy Cr-Mo steels up to and including 9Cr-1Mo are required for acceptable tube life in atmospheric heaters. While 12Cr stainless steel has also been used for this purpose, it is usually avoided, because of the possibility of 475 °C (885 °F) embrittlement occurring. Austenitic Type 316 or Type 317 stainless steel tubes are required in vacuum heaters with acidic crudes.

14.5 Transfer Lines

Transfer lines connect the heater outlet manifold to a distillation column. In the atmospheric section of crude units, the transfer line carries a mixture of crude and crude vapors at temperatures between 315-370 °C (600-700 °F); in the vacuum section, the transfer line carries a mixture of reduced crude (atmospheric resid) and vapors at temperatures between 350-370 °C (670F-700 °F). In the DRU section of upgraders, the transfer line carries a mixture of preflashed dilbit at approximately 300-315 °C (580-600 °F); in the VDU, the transfer line carries a mixture of bitumen resid and vapors at temperatures between 350-370 °C (670-700 °F). In all cases, two-phase vapor velocities can be very high and can reach 70–80% of sonic values.

Because of the low pressure and large vapor volume, vacuum heater transfer lines must be sized to avoid having to raise heater outlet temperatures to maintain flow. Otherwise, coke formation in the heater tubes increases and causes run length between decoking to decrease. As far as corrosion is concerned, vapor velocities normally should not exceed 60m/sec (200fps) to minimize HTSC, and 50m/sec (165fps) to minimize NAC under two-phase flow conditions.

14.6 Chapter Summary

Fired heaters are the principal heat source for most high-temperature processes. Crude units include at least two fired heaters to heat incoming and reduced crude, respectively. Depending on past revamps and upgrades, there can be additional heaters to increase throughput. Most fired heaters also contain steam coils to supply low or medium-pressure process steam. All fired heaters use either gas or oil as fuel. The same applies to the DRU and VDU of bitumen upgraders. Fired heaters are purpose-designed and field-erected by the builder who guarantees the heater's overall performance. Heaters are designed around their burners. Many have burners located along the floor (bottom) of the firebox, while others have burners located in one or both end walls.

Tubes, return bends, and headers represent the most expensive components of a fired heater and are also the most frequently inspected, repaired, or replaced. Flame impingement often shortens tube life. Since heater tubes operate with metal temperatures well above 230 °C (450 °F), alloying against HTSC and NAC is usually required. In the absence of naphthenic acids, low-alloy Cr-Mo steels up to and including 9Cr-1Mo are required for acceptable tube life. Austenitic Type 316 or Type 317 stainless steel tubes are usually required with acidic crudes in vacuum heaters. Transfer lines have similar requirements. Vapor velocities in heater tubes and transfer lines normally should not exceed 60m/sec (200fps) to minimize HTSC, and 50m/sec (165fps) to minimize NAC under two-phase flow conditions.

14.7 Read More

1. API Division of Refining, *Guide for Inspection of Refinery Equipment, Chapter II – Conditions Causing Deterioration or Failures,* 2nd ed. (Washington, DC: API, 1973).
2. W.L. Nelson, *Petroleum Refinery Engineering,* 4th ed. (New York, Toronto, London: McGraw-Hill, 1958).
3. D. A. Hansen and R. B. Puyear, *Materials Selection for Hydrocarbon and Chemical Plants* (New York, NY: Marcel Dekker, 1996).
4. E. L. Hildebrand, "Materials Selection for Petroleum Refineries and Petrochemical Plants," *MP* 11, 7 (1972).
5. M. J. Humphries and G. Sorell, "Corrosion Control in Crude Oil Distillation Units," *MP* 15, 2 (1976).
6. T. M. Krebs, "Selection of Furnace Tubing - Part1: Mechanical Considerations, Types of Tubing Used," *Petroleum Engineer* 30, 2 (1958).
7. NACE RP0170-2004, "Protection of Austenitic Stainless Steels and Other Austenitic Alloys from Polythionic Acid Stress Corrosion Cracking During Shutdown of Refinery Equipment" (Houston, TX: NACE, 2004).
8. L. M. Wyatt et al., *An Atlas of Corrosion and Related Failures,* publication no. 18 (St. Louis, MO: Materials Technology Institute, 1987).
9. A. Garg, "Improve Vacuum Heater Reliability," *Hydroc. Proc.* 78, 3 (1999).
10. L. Engel, "Korrosion in Mineralolraffinerien," *Erdöl und Kohle* 27, 6 (1974).
11. J. L. Hau and A. Seijas, "Furnace Tube Life Assessment as Impacted by the Methodology Used," paper no. 08547, CORROSION/2008 (Houston, TX: NACE, 2008).
12. ASM Committee on Petroleum Refinery Applications, "Selection of Steel for High-Temperature Service in Petroleum Refinery Applications," *Metals Handbook,* 8th ed., Vol. I (Metals Park, OH: ASM, 1961).
13. C. H. Samans, "Which Steel for Refinery Service? Part 1: Steels for Service below 650 °F, *Hydroc. Proc.* 42, 10 (1963).
14. C. H. Samans, "Which Steel for Refinery Service? Part 2: Steels for Service above 650 °F, *Hydroc. Proc.* 42, 11 (1963).
15. J. Gutzeit, R. D. Merrick, and L. R. Scharfstein, "Corrosion in Petroleum Refining and Petrochemical Operations," *Metals Handbook,* 9th ed., Vol. 13 – Corrosion (Metals Park, OH: ASM, 1987).

15 Upper Column Corrosion

In this Chapter

15.1 First Glance

Corrosion in upper crude columns occurs primarily in columns that operate at relatively low top temperatures and is usually referred to as "cold tower corrosion." Low-temperature column operation can be the result of careless equipment changes or a sudden decrease in the stripping steam rate for the column. As a general rule, however, low-temperature column operation is used by choice for one or more of the following reasons:

- To extract more heat from the overhead system.
- To increase the production of kerosene (jet fuel).
- To allow a greater portion of C_5/C_6 fractions to enter the naphtha side cuts for hydrotreating and isomerization to a higher-octane product.
- To allow a greater portion of C_{6+} fractions to enter the naphtha side cuts for conversion to aromatics with reformer feed.
- To reduce the volume of light ends that has to be send to a gas plant.

Disadvantages of low-temperature column operation include:

- Steam condensation in the top of a column instead of the overhead system.
- General temperature reduction throughout the entire overhead system.
- Various forms of aqueous corrosion due to hydrogen chloride (HCl) in overhead vapors.
- Increased risk of acid corrosion in previously dry components.
- Increased risk of $NH_3 \bullet HCl$ formation and resulting corrosion and fouling problems.

Increasing the upper column temperature is one way of avoiding cold tower corrosion. As a general rule, however, column temperatures are not easily increased without major operational concessions no matter how attractive this may be from a corrosion viewpoint. While this chapter addresses primarily corrosion in the top of atmospheric columns, similar corrosion issues apply to preflash columns that normally operate at lower top temperatures. Similar corrosion issues also apply to DR columns in bitumen upgraders that operate at lower top temperatures compared to crude unit columns.

15.2 Cause of Upper Column Corrosion

Upper column corrosion is not caused by naphtha vapors or condensate because naphtha is not particularly corrosive at typical column temperatures. Instead, the great bulk of corrosion issues is caused by a combination of hydrogen chloride (HCl) vapor, water in one form or another, and a few other problem chemicals. While each of these contributes to corrosion, the root cause of corrosion is the hydrolysis of residual chloride salts in desalted crude or dilbit in the hot preheat exchangers (HPEs) and fired heater to form hydrogen chloride (HCl) vapor. Without moisture or water condensate, HCl vapor as such is not corrosive at upper column temperatures. For corrosion, HCl vapor has to dissolve in water condensate and become dilute hydrochloric (HCl) acid. Moreover, HCl vapor has to penetrate and react with existing corrosion deposits on metal surfaces. To minimize the amount of HCl vapor rising in a column, the full regimen of preventive measures developed by the industry over many years should be implemented, starting with proper gravity settling and blending of crude receipts in tankage and ending with caustic addition to desalted crude. These measures are further discussed in Chapters 8 to 10 and 12 of the *Guide*:

15.2.1 Hydrogen Chloride and Hydrochloric Acid. Corrosion in the top of crude unit columns is caused by the same chemical compounds that cause corrosion in overhead systems. They include primarily hydrogen chloride (HCl) vapor formed when desalted crude is heated. The acid vapor travels up various columns, forms dilute hydrochloric acid (HCl) in the presence of water, and reacts with ammonia (NH_3) to form ammonium hydrochloride ($NH_3 \cdot HCl$) salt. Hydrogen chloride (HCl) evolves by hydrolysis of residual magnesium chloride ($MgCl_2$) and calcium chloride ($CaCl_2$) in the desalted crude feed. When crude is heated in the hot preheat exchangers (HPEs) and fired heater prior to distillation, these salts partially react with dissolved water, entrained water, and dilution steam in the crude to form hydrogen chloride (HCl) vapor.

At typical crude heater temperatures of 316 °C (600 °F), approximately 90% of the $MgCl_2$ and 10% of the $CaCl_2$ are hydrolyzed, as shown in Figure 15-1. In contrast, sodium chloride (NaCl) is not hydrolyzed to any significant degree and stays with atmospheric resid (reduced crude) going to the vacuum section, or elsewhere. The remaining 10% $MgCl_2$ and 90% $CaCl_2$ also stay with reduced crude going to the vacuum section. A small amount of HCl vapors then evolves at the higher temperatures of the vacuum heater and accounts for corrosion issues in the vacuum overhead system, as shown in Figure 15-2. Regardless of how or where it is formed, hydrochloric (HCl) acid can corrode carbon steel and most common alloys in the upper columns. Certain other chemical compounds also contribute to upper column corrosion. In some cases, acetic acid (CH_3COOH) and sulfuric acid (H_2SO_4) have been identified in overhead water condensate; however, their exact source and role in upper column corrosion has never been convincingly established.

Reported corrosion data are often obtained in the laboratory to help explain a possible, but not necessarily probable, connection with the corrosion seen in a particular column. It can then be counter-productive to try to extrapolate and apply these data to all columns. In other cases, it is more realistic to concentrate on improving the desalter operation rather than blaming unexpected corrosion issues on yet another candidate chemical compound. Desalter issues of one kind or another still remain a major stumbling block on the road toward effective corrosion control on many units. For this reason, Chapter 9 of the *Guide* is devoted to desalter theory, design, and operation.

FIGURE 15-1 HYDROLYSIS OF MAGNISIUM AND CALCIUM CHLORIDE SALTS IN CRUDE AS A FUNCTION OF TEMPERATURE

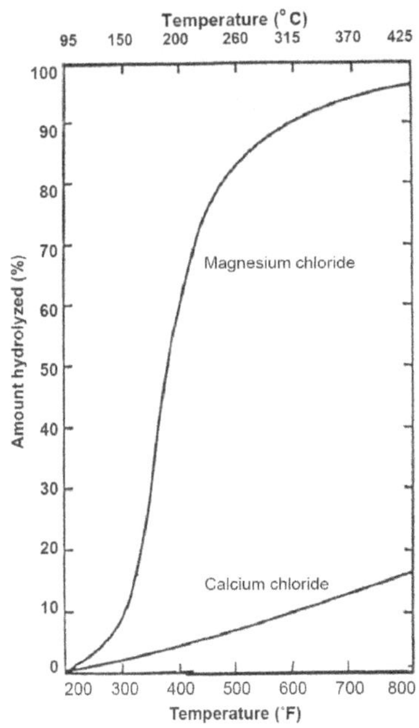

Adapted from L.E. Fisher, G. C. Hall, and R. W. Stenzel, "Crude Oil Desalting to Reduce Refinery Corrosion Problems," *MP* 1, 5 (1962)

15.2.2 Other Problem Chemicals. In addition to hydrogen chloride (HCl) vapor, and regardless of the above caveats, other chemical compounds are known to have contributed to upper column corrosion and their detrimental effects are documented in some detail. Examples of problem chemicals are listed below in order of decreasing importance.

15.2.2.1 Ammonia. Ammonia (NH_3) is formed by dissociation of organic nitrogen compounds in desalted crude in the hot preheat exchangers (HPEs) and fired heater. Ammonia is also introduced into crude units when stripped sour water is used as desalter water. Ideally, stripped sour water contains no ammonia, but 40mg/L or more ammonia is seen in water coming from many sour water strippers. Another source of ammonia is overhead water condensate from the atmospheric column that is subsequently used as desalter water. Therefore, even without intentional ammonia injection in the overhead system, the ammonia levels in overhead water condensate can build up to anywhere from 10 up to 3,000mg/L.[1]

The total amount of ammonia going overhead depends on the desalter water pH. Ammonia levels in overhead water can be reduced by keeping the desalter water pH as low as possible without causing desalter corrosion problems. On many crude units, however, the overhead water pH is difficult to control by relying solely on desalter water pH; ammonia or a liquid organic neutralizer has to be injected as well.

1. M. R. Beychok, *Aqueous Wastes from Petroleum and Petrochemical Plants* (John Wiley & Sons, 1967).

FIGURE 15-2 DISTRIBUTION OF HCl IN TYPICAL CRUDE UNIT

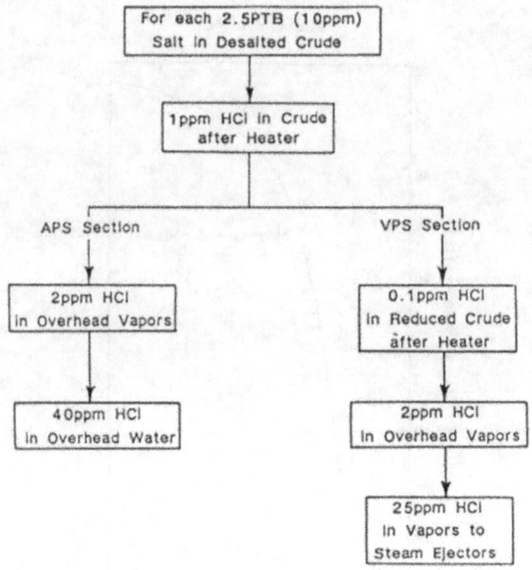

15.2.2.2 Ammonium Hydrochloride Deposits. A mist of ammonium hydrochloride (NH_3·HCl) particles forms whenever hydrogen chloride (HCl) vapor comes in contact with ammonia (NH_3) vapor. For all practical purposes, this occurs already in the top of most columns. It also occurs immediately downstream of any ammonia injection point when ammonia is intentionally injected into an overhead vapor line for pH control. The mist sublimes and forms corrosive deposits when the partial pressure of ammonium hydrochloride (NH_3·HCl) exceeds a limiting value. Ammonium hydrochloride is completely soluble in overhead water condensate, but not in naphtha condensate.

15.2.2.3 Neutralizer Hydrochloride Deposits. A corresponding mist of liquid organic neutralizer hydrochloride (R-NH_2·HCl) particles forms whenever liquid organic neutralizers are injected into an overhead vapor line for pH control and contact hydrogen chloride (HCl) vapor. Depending on the particular chemicals involved, the mist sublimes to form corrosive deposits when the partial pressure of the neutralizer hydrochlorides exceeds a limiting value. Some liquid organic neutralizer hydrochlorides, however, are so hygroscopic that they disperse in naphtha condensate instead of forming corrosive deposits. These hydrochlorides return to the column with reflux and are known to cause fouling on column trays partly down the column. Like ammonium hydrochloride, all liquid organic neutralizer hydrochlorides are completely soluble in overhead water condensate, but insoluble in naphtha condensate. At one time, a supplier of process additives advocated injecting a liquid organic neutralizer into the reflux stream returning to the atmospheric column. The idea was to neutralize hydrogen chloride (HCl) vapor in the upper column and thus keep HCl out of the overhead system. Unfortunately, no one bothered to ask what happens to the resultant hydrochloride salt until the upper column trays became completely fouled and the column had to be washed out.

15.2.2.4 Corrosion Deposits. Corrosion deposits in upper columns consist primarily of mixed iron sulfides/hydroxides/oxides and form porous deposits at pH above 6-7. The deposits usually contain entrapped hydrogen chloride (HCl), ammonium hydrochloride (NH_3·HCl), or liquid organic neutralizer hydrochlorides (R-NH_2·HCl) and become sticky, acidic, and highly corrosive. Most corrosion deposits are insoluble in water condensate that forms in upper columns depending on the column temperature.

15.2.2.5 Corrosion Inhibitor Deposits. Concentrated (neat) inhibitor forms sticky, gooey deposits when diluted incorrectly prior to injection into an overhead vapor line for corrosion control. The deposits can also end up in upper columns when the injection nozzle is too close to the vapor outlet elbow from the top of a column. Inhibitor droplets can fall back into the column when pressure surges occur. Inhibitor has also been known to back up into a column when someone forgot to open the block valve for the reflux slipstream used to dilute the inhibitor prior to injection into the overhead line. Inhibitor deposits become acidic and highly corrosive at temperatures above approximately 120 °C (250 °F). Once formed, inhibitor deposits are insoluble in both hydrocarbon and water condensate and cannot be simply removed by water washing. Fouled column trays will have to be removed and cleaned by hydro jetting during shutdown.

15.2.2.6 Hydrogen Sulfide. Dissolved hydrogen sulfide (H_2S) is present in sour crude and also is formed in the fired heater from organic sulfur compounds. Like HCl, hydrogen sulfide rises in the preflash, atmospheric, and vacuum columns and combines with ammonia to form an ammonium bisulfide ($NH_3{\cdot}H_2S$) mist. Compared to ammonium chloride ($NH_3{\cdot}HCl$), ammonium bisulfide ($NH_3{\cdot}H_2S$) does not sublime at the relatively high temperatures in the upper columns. It does sublime, however, in overhead condensers when the partial pressure of ammonium bisulfide exceeds a limiting value. Ammonium bisulfide is soluble in overhead water condensate and causes sour-water corrosion at pH above 7.5. In the presence of oxygen (air) contamination, yellow sulfur (Sn) deposits are formed.

15.2.2.7 Organic Chlorides. Most organic chlorides originate with crudes contaminated with various organic solvents and surfactants used during well completion, workover, and stimulation operations. Organic chlorides can also enter crude with slop oil, chemical byproducts, and waste streams at the refinery. They are not removed by desalting, because they are insoluble in water. Organic chlorides partially hydrolyze in the crude heater to form hydrogen chloride (HCl) that contributes to column top corrosion. Various other issues with organic chlorides are discussed in Chapter 11 of the *Guide.*

15.2.2.8 Carbon Dioxide. Carbon dioxide (CO_2) is used in enhanced recovery operations and remains in some crude sent to the crude unit. Carbon dioxide can also originate in dilution steam added at the crude heater and stripping steam added at the bottom of the atmospheric column and at sidestream columns. Carbon dioxide rises in the preflash and atmospheric columns and dissolves in water condensate to form dilute carbonic (H_2CO_3) acid, which can cause severe vapor-space corrosion in upper columns.

15.2.2.9 Oxygen (Air). While not corrosive as such, oxygen reacts with dissolved hydrogen sulfide and ammonia to form elemental sulfur, inorganic polysulfides [$(NH_4)_2S_3$], and mixed sulfur oxides (SO_2, SO_3, and others). Oxygen can enter atmospheric columns with air intrusions at leaking crude pump seals, with fresh makeup water for the desalter, with cooling tower water from leaking tube bundles, or with poorly deaerated steam. Sulfur-containing deposits are readily apparent by their yellow color. Polysulfides have been blamed for accelerated corrosion in the form of pitting, but tend to form protective iron polysulfide films in some sour-water services.

15.2.2.10 Acetic Acid. Light organic acids, such as acetic acid (CH_3-COOH), proprionic acid (CH_3CH_2-COOH), and to a lesser degree formic acid (H-COOH), are formed in the crude heater due to decomposition of naphthenic acids present in most crudes. Acetic acid has been identified in overhead water condensate at levels as high as 250 mg/L and can cause severe corrosion even in the absence of HCl. Proprionic acid has been identified in overhead water condensate at levels as high as 100 mg/L.

15.2.2.11 Sulfuric Acid. Sulfuric acid (H_2SO_4) is used for formation fracturing (acidizing) operations during upstream operations. When oil wells are not properly purged after acidizing, initial crude shipments can contain high levels of sulfuric acid for a short period of time. Sulfuric acid has also been added to desalter water to lower the desalter pH, but normally should not be used for that purpose to avoid forming hard sulfate scales. Hydrochloric acid (HCl) would be a better choice despite the fact that a small amount of chloride ions is introduced. In either case, some residual acid is likely to stay with desalted crude, vaporize in the fired heater, and travel up the atmospheric column to dissolve in overhead water condensate. While sulfuric acid has been identified in overhead water condensate, it has not been shown for certain that sulfuric acid was actually responsible for increased corrosion in the upper column seen at the time.

15.3 Upper Column Metallurgy

Materials of construction (metallurgy) for crude units in general are reviewed in Chapter 4 of the *Guide.* Despite greater use of alloys in recent years, carbon steel remains the basic material of construction for the upper section and dome areas of various columns at temperatures below approximately 230 °C (450 °F). Bubble cap trays, shown in Figure 15-3, are still found in older designs and usually have bubble caps made of carbon steel or 12Cr stainless steel (Type 405, 410, or 410S), while the tray decks are made of carbon steel. Compared to valve trays, bubble-cap trays have high maintenance costs and are rarely specified for new construction. Trays are fabricated of circular steel plates and usually installed inside the column at intervals of about 0.6-0.75m (24-30in.) up the height of the column. The spacing is chosen primarily for ease of installation and access because it allows climbing through the column for future repair or maintenance. The principal areas where corrosion occurs include the inside, top surface, and slots of the bubble caps;

the tray decks at the downcomers; and the shell areas at the downcomers. The following lists alternative metallurgy options for components in the top of various columns. The alloys are shown in order of increasing corrosion resistance.

15.3.1 Low-alloy Cr-Mo Steels. The corrosion resistance is similar to that of carbon steel. These steels are sometimes used instead of carbon steel because of ready availability, but may require postweld heat treatment (PWHT) to prevent hydrogen stress cracking (H_2S cracking) at hard welds.

FIGURE 15-3 VAPOR/LIQUID FLOW IN COLUMN WITH BUBBLE-CAP TRAYS

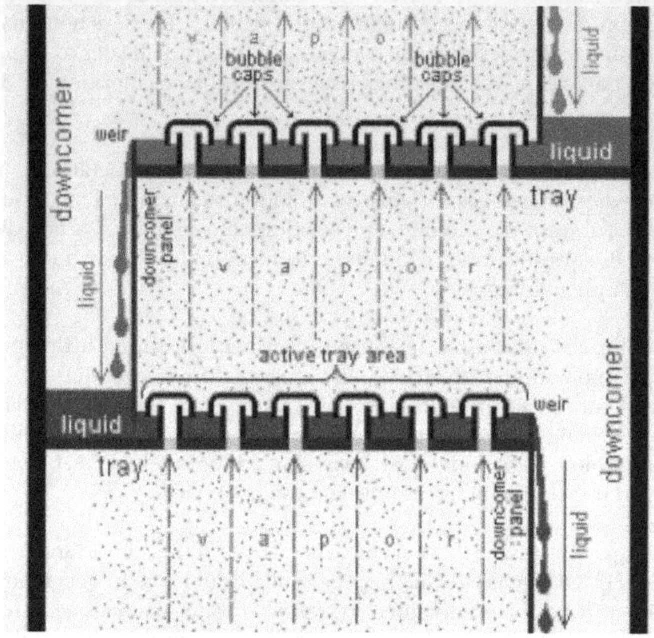

Adapted from H. Padleckas, "Fractionating Column," *Wikimedia*, November (2006).

15.3.2 12Cr Stainless Steel. Clad 12Cr stainless steel (Types 405, 410, or 410S) is the standard metallurgy for columns and trays against high-temperature sulfidic corrosion (HTSC) at temperatures above 230 °C (450 °F), i.e. in the lower sections of columns. It is often used for the entire length of columns, but may be subject to underdeposit corrosion in the upper column sections. Affected carbon steel shell areas are either weld overlaid or clad with 12Cr stainless steel, as needed. Valve trays are used in most current column designs and are usually made from carbon steel, but require tray valves made of 12Cr stainless steel. Carbon steel tray valves corrode through in short order because attack occurs from both sides of a relatively thin piece of metal. It should be noted that all 12Cr stainless steels are likely to pit under chloride containing deposits. The martensitic grade (Type 410) requires postweld heat treatment (PWHT) to prevent cracking at hard welds.

15.3.3 Austenitic Stainless Steel. Austenitic stainless steel (Types 304, 316, or 317) overlays or cladding may pit and stress-corrosion crack (SCC) in upper columns sections due to chlorides, but have been used successfully in some columns when other alloys were unavailable. They may have to be protected against SCC during shutdown in accordance with NACE RP0170[2].

15.3.4 Alloy 400. Alloy 400 (66Ni-32Cu) was often used in the upper column sections, usually in the form of cladding or weld overlays of the dome and shell areas down to the fourth tray from the top. The upper 3 to 4 trays too were made from Alloy 400. The alloy has good corrosion resistance to most chloride-containing deposits and to dilute hydrochloric acid condensate, provided temperatures are limited to below 120 °C (250 °F). At higher temperatures and

2. NACE Standard RP0170-2004: *Protection of Austenitic Stainless Steels and Other Austenitic Alloys from Polythionic Acid Stress Corrosion Cracking During Shutdown of Refinery Equipment.*

in the presence of hydrogen sulfide (H_2S), Alloy 400 can corrode at higher rates than carbon steel, as confirmed by the results of accelerated laboratory tests listed in Table 15-1. Alloy 400 can also corrode under acid ammonium chloride ($NH_3 \cdot HCl$) deposits in vapor areas in the presence of hydrogen sulfide (H_2S).

TABLE 15-1 CORROSION IN NAPHTHA/WATER MIXTURES AT 177 °C (350 °F)[3]

Alloy	Corrosion Rate, mm/y (mpy)	
	H_2S Saturated	H_2S Saturated + 1,500mg/L Cl
Carbon Steel	0.37 (15)	0.7 (28)
Alloy 400	2.8 (112)	2.5 (99)
12Cr Stainless Steel	0.25 (11)	0.025 (1)
Type 304 SS	0.1 (4)	0.05 (2)

15.3.5 Alloy 625. Alloy 625 (60Ni-21Cr-9Mo-5Fe) is used for selective alloying in the top of columns regardless of temperature, but is prone to pitting under chloride containing deposits.

15.3.6 Alloy C-276. Alloy C-276 (58Ni-15Cr-16Mo-6Fe-2Co-3W) is preferred over Alloy 400 and Alloy 625 for selective alloying of column trays, upper shell areas, domes, and overhead vapor lines exiting the column. Despite some isolated claims to the contrary, the alloy is fully resistant to pitting under tray deposits under the reducing conditions encountered in crude and DR units.

15.4 Corrosion Mechanism

Corrosion in upper columns is an electrochemical process that requires the presence of water in one form or another as the electrolyte. Metal loss occurs at anodic sites of the corroding metal, while hydrogen gas (H_2) evolves at cathodic sites. Two basic types of corrosion reactions can occur, depending on the absence or presence of hydrogen sulfide (H_2S). In most column tops and overhead systems, both probably take place at the same time or alternate with changes in operating parameters. In the absence of hydrogen sulfide, HCl reacts with bare metal surfaces to form ferrous ions (Fe^{+2}) at anodic sites and atomic hydrogen (H_0) at cathodic sites. The overall reaction is the formation of soluble iron chlorides and evolution of molecular hydrogen (H_2). The electrolyte is overhead water condensate or crystalline water in tray deposits. Expressed in most simplistic terms, the partial corrosion reactions are:

$$Fe \Rightarrow Fe^{+2} + 2e^- \quad \text{(at anodic sites)} \qquad \text{(Eq.15-1)}$$
$$2H^+ + 2e^- \Rightarrow 2H_0 \Rightarrow H_2\uparrow \text{ (at cathodic sites)} \qquad \text{(Eq.15-2)}$$

The overall corrosion reaction is:

$$Fe + 2HCl \Rightarrow FeCl_2 + H_2\uparrow \qquad \text{(Eq.15-3)}$$

In the presence of hydrogen sulfide, H_2S reacts with bare metal surfaces to form ferrous ions (Fe^{+2}) at anodic sites and sulfide ions (S^{-2}) and atomic hydrogen (H_0) at cathodic sites. The overall reaction is the formation of complex iron sulfide (Fe_xS_y) deposits and evolution of molecular hydrogen (H_2). The electrolyte is water condensate or crystalline water in tray deposits. Expressed in most simplistic terms, the partial corrosion reactions are:

3. J. Gutzeit, Unpublished results of laboratory tests in naphtha/H_2O mixtures saturated with H_2S at pH 4.3.

$$Fe \Rightarrow Fe^{+2} + 2e^- \quad \text{(at anodic sites)} \qquad \text{(Eq.15-4)}$$

$$2HS^- + 2e^- \Rightarrow S^{-2} + 2H_0 \Rightarrow H_2\uparrow \quad \text{(at cathodic sites)} \qquad \text{(Eq.15-5)}$$

The overall corrosion reaction is:

$$Fe + H_2S \Rightarrow FeS\downarrow + H_2\uparrow \qquad \text{(Eq.15-6)}$$

When hygroscopic chloride salts form sticky deposits in the top of columns, underdeposit corrosion occurs even in the absence of a free water phase. In that case, the basic corrosion reactions are still the same, but the electrolyte now is water bound to the crystal structure of the salts. Corrosion can also be caused by shock condensation of dilute hydrochloric acid (HCl) on the upper shell and dome sites where external insulation is missing[4], and on the top tray when cold naphtha reflux enters from the overhead receiver drum. In the presence of hydrogen sulfide (H_2S), hydrogen stress cracking (HSC) can occur in the heat-affected zones (HAZ) of hard welds. These may have to be postweld heat treated (PWHT) to prevent cracking. With proper overhead pH control and in the absence of hydrogen cyanide (HCN) contamination, hydrogen blistering normally is not a problem in the upper columns of crude units and upgrader DRUs.

15.5 Components Subject to Corrosion

Corrosion in the top of columns occurs at temperatures between approximately 93 °C (200 °F) and 175 °C (350 °F) or lower, depending on the unit. The affected columns include the preflash column, atmospheric column, vacuum column, and the corresponding upgrader columns.

15.5.1 Preflash Column. Corrosion occurs in the upper preflash column, including the shell areas, dome, top 3 or 4 trays, and the outlet elbow for the preflash column overhead vapor line.

15.5.2 Atmospheric Column. In the top of "hot columns," corrosion is less prevalent due to localized dewpoint depression, but can occur at temperatures as high as 175 °C (350 °F) as various types of underdeposit corrosion. Corrosion occurs primarily in the following sections of "cold columns" operating between approximately 93 °C (200 °F) and 120 °C (250 °F):

- Upper column including the shell areas, dome, and the top 3 or 4 trays. Here, corrosion damage and fouling can affect column operation and can be severe enough to cause leaks in a matter of 2 or 3 years and require extensive repairs or alloy cladding.
- Upper pump-around circuits where corrosion can cause pipe leaks that may have to be clamped.
- Upper heavy naphtha (HN) side-draws where corrosion and fouling can affect downstream operations and compromise product quality.
- Outlet elbow for the overhead vapor line.

15.5.3 Vacuum Column. Corrosion occurs in the upper vacuum column including the shell areas, dome, upper trays, and the outlet elbow for the overhead vapor line.

15.6 Underdeposit Corrosion in Upper Columns

No reliable data are available for predicting the corrosion of carbon steel under neutralizer hydrochloride salt or corrosion deposits in the top of columns. As a general rule, however, underdeposit corrosion in the form of localized corrosion and pitting occurs at much higher rates than general corrosion. Once initiated, corrosion depends on the pH under the deposits or within the pits, instead of the pH of water condensate that may or may not be present. Some idea of the effect of moisture on carbon steel corrosion under ammonium chloride deposits can be gleaned from limited data obtained in air at 150 °C (300 °F) and presented in Figure 15-4.

4. Unless required for personnel protection, the upper sections of columns are not normally insulated even in cold climates. Since most columns are internally alloy clad or weld overlaid, lack of insulation is rarely a problem.

FIGURE 15-4 EFFECT OF MOISTURE ON CORROSION OF CARBON STEEL UNDER $NH_3 \cdot HCl$ DEPOSITS AT 150 °C (300 °F)

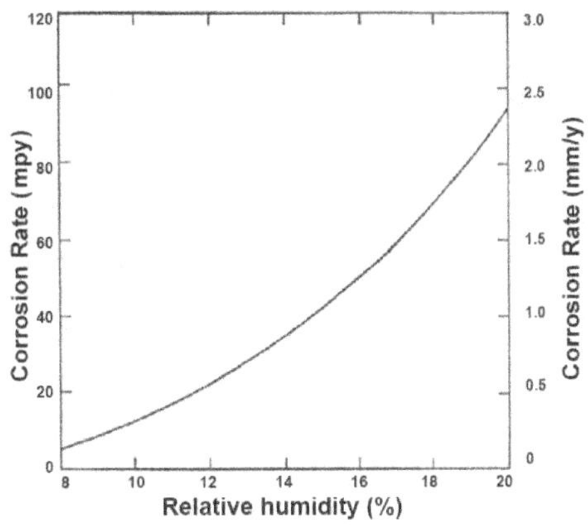

In the absence of a free water phase, the concept of pH is meaningless except in the sense that pH decreases within the deposits as hydrogen ions (H^+) are formed due to corrosion. The rate of pitting continues unabated and most likely increases with time, even after the pH of bulk overhead water is raised by neutralizer injection. Few data are available to predict corrosion of alloys other than carbon steel in the upper columns and domes. At one crude unit, a 5mm (0.125in.) thick Alloy 400 liner in the top of the atmospheric column lasted only 5 years. In another unit, Alloy 400 trays corroded at 0.075-0.1mm/y (3-4mpy). For all practical purposes, only Alloy C-276 is fully resistant to corrosion in the upper section of columns.

15.7 Dew Point in Column Top

Dew point is an indication whether water condenses on or above the top column tray of atmospheric columns that operate at an absolute pressure of approximately 4.5bar (65psi). Assuming the vapors have a water vapor content of 5%, the water vapor partial pressure is approximately 1.24bar (18psi). Without suitable process software, however, the dew point is difficult to calculate when accompanied by naphtha condensation. In many columns, the dew point should be approximately 104 °C (220 °F). Therefore, no water would be expected to condense in the top of "hot columns" that operate well above 120 °C (250 °F). Conversely, water would be expected to condense in the top of "cold columns" that operate between approximately 93 °C (200 °F) and 120 °C (250 °F) where most corrosion takes place. Moreover, incoming cold reflux can lower local tray temperatures well below the dew point and cause shock condensation and resulting acid formation.

15.8 Location of Upper Column Corrosion

Sudden and unexpected spikes in the salt, free water, and BS&W contents of incoming crude interfere with proper desalting, produce desalter upsets, and increase the amount of hydrogen chloride (HCl) rising to the top of columns. Unless desalter upsets are avoided, corrosion of the upper shell and dome sites by ammonium chloride ($NH_3 \cdot HCl$) deposits and wet hydrogen chloride (HCl) and hydrogen sulfide (H_2S) vapors is likely to occur. When the upper section of a column is opened and entered for inspection, the three principal types and locations of corrosion found include corrosion under tray deposits, corrosion due to shock condensation, and general corrosion of the top tray, as shown in Figure 15-5.

15.8.1 Corrosion under Tray Deposits. Most tray deposits consist of one or more of the following deposit categories, including neutralizer salts, corrosion deposits, and corrosion inhibitor residues. Corrosion under neutralizer salt and corrosion deposits is probably the biggest problem on the upper 3 or 4 trays of most columns. The principal areas where corrosion occurs on bubble-cap trays include the inside, top surface, and slots of bubble caps; the tray decks at the downcomers; the shell areas at tray downcomers; and the areas between adjacent caps. The specific locations on valve tray decks include the areas underneath the valve tray openings where tubercles tend to form; the areas between the valve disks and the openings where "doughnut" shaped deposits tend to form; the tray decks proper; and the shell areas above the normal liquid level on the trays. Corrosion is also seen on the shell areas above the upper four trays, the column dome, and the overhead vapor outlet nozzle. Severe underdeposit corrosion can be found under valve tray deck deposits even when the trays are made from Alloy 400. Examples of tray corrosion damage are:

- Valve disks that are loose, stuck open, have legs bent or broken off, or missing entirely, as shown in Figure 15-6. Missing valve disks tend to collect in a pile at low spots in the column.
- The underside of bubble caps, at risers, and around sidewall slots of the caps, as shown earlier in Figure 15-7. On one column, nearly 25% of 250 12Cr bubble cap-and-riser assemblies on each of Trays No. 3 through 5 counting from the top of the column were found to be severely pitted or holed through.
- Tray-deck corrosion at riser openings resulted in loose risers and was concentrated at the top flange of the tray support beam and at tray section joints, as shown earlier in Figure 15-8. This type of corrosion is probably caused by a combination of ammonium hydrochloride ($NH_3 \cdot HCl$), corrosion deposits, and water entering on with reflux.

15.8.2 Corrosion under Neutralizer Salt Deposits. Contact between ammonia (NH_3) and hydrogen chloride (HCl) vapors produces a mist of ammonium chloride ($NH_3 \cdot HCl$) particles that sublimes in the upper column. The usual source of ammonia is decomposition of organic nitrogen compounds in the crude heater. Another source of ammonia is direct injection into the overhead system for pH control: Any excess ammonia then enters the column with reflux. Ammonium chloride ($NH_3 \cdot HCl$) can enter the upper column as water droplets entrained in reflux. Inevitably, these salt droplets evaporate and leave deposits behind, as the reflux stream flows down to hotter tray levels. The same applies to liquid organic neutralizer hydrochlorides ($R\text{-}NH_2 \cdot HCl$) when liquid organic neutralizers are injected into an overhead system for pH control. Most if not all neutralizer hydrochloride salt deposits are sticky and acidic. They are corrosive even in the absence of free water condensate. Some liquid organic neutralizer hydrochloride salts, like MOPA·HCl[5], are so hygroscopic that they readily "dissolve" in naphtha overhead condensate, return to the column with reflux, and work their way down the column to cause underdeposit corrosion of midlevel trays. Since most salt deposits are partially removed when a column is cleared by steam-out and water washing prior to personnel entry, salt deposits are often difficult to find, quantify, and analyze.

FIGURE 15-5 PRINCIPAL LOCATIONS FOR CORROSION IN THE TOP OF COLUMNS

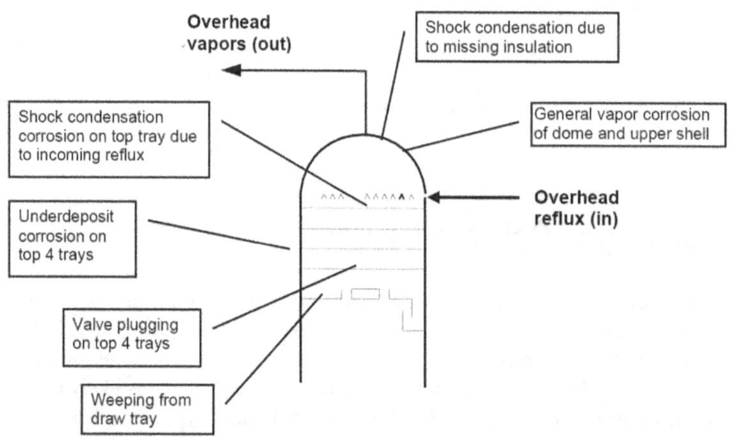

5. The hydrochloride salt of methoxypropylamine (MOPA) has a melting point of 100 °C (212 °F).

FIGURE 15-6 CORRODED 12CR VALVE DISK FROM NO. 4 TRAY OF ATMOSPHERIC COLUMN

FIGURE 15-7 CORRODED 12CR BUBBLE CAPS AND CARBON STEEL DECK FROM NO. 3 TRAY IN ATMOSPHERIC COLUMN

FIGURE 15-8 CORRODED CARBON STEEL DECK FROM NO. 3 TRAY IN ATMOSPHERIC COLUMN

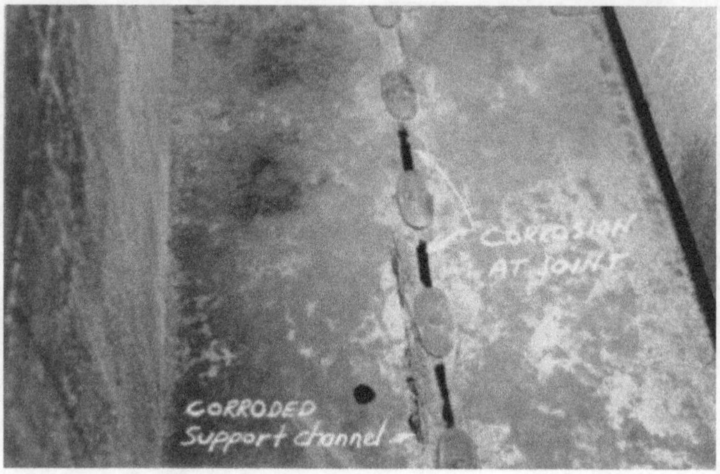

15.8.3 Corrosion under Corrosion Deposits. Corrosion deposits form in situ as a result of corrosion in the top of the column. Corrosion deposits can also enter a column with reflux and lay down on the upper trays of a column. The amount or volume of corrosion deposits entering with reflux depends on the severity of overhead corrosion. All corrosion deposits tend to absorb hydrogen chloride (HCl), ammonium chloride (NH_3·HCl), or liquid organic neutralizer hydrochloride (R-NH_2·HCl) salts and cause severe underdeposit corrosion. Most corrosion deposits also tend to be sticky and acidic. Like neutralizer hydrochloride salt deposits, they are corrosive even in the absence of free water condensate. Corrosion deposits can be so voluminous that trays are plugged to the point where they become flooded. Once corrosion deposits have formed, they can only be partially removed by mechanical cleaning (hydro jetting) from the upper manholes after the column is opened. In the worst cases, the tray assemblies will have to be removed for proper cleaning on the ground and replaced with new trays.

15.8.4 Corrosion under Inhibitor Residues. Excess filming-amine corrosion inhibitor can enter a column with reflux and lay down on the upper trays. Small amounts of inhibitor tend to bind corrosion deposits to the tray areas. Large amounts of inhibitor can completely foul the upper trays. For example, in one unit, severe fouling occurred when the reflux slipstream check valve at the overhead vapor line was stuck in the closed position and inhibitor was continuously pumped backwards into the reflux line instead of forward into the overhead vapor line. Corrosion inhibitor residues form sticky and "gooey" deposits when solvents and diluents in the product evaporate. They become corrosive even in the absence of free water condensate. As in the case of corrosion deposits, inhibitor residue deposits can only be removed by mechanical cleaning (hydro jetting) after the column is opened. In the worst cases, the tray assemblies will have to be removed for proper cleaning on the ground and replaced with new trays.

15.8.5 Corrosion Due to Shock Condensation. Shock condensation corrosion is a special case of initial condensate corrosion. Shock condensation on the vessel wall and dome occurs when water droplets condense on cold metal surfaces even though process temperatures are still above the dew point. The water droplets absorb hydrogen chloride (HCl) and form dilute hydrochloric acid that can cause severe corrosion at the point of shock condensation. Shell surfaces stay relatively free of deposits, because the acid condensate dissolves prior corrosion deposits. The condensate droplets are then swept away by hot vapors, but usually contact adjacent metal surfaces or the column vapor outlet nozzle before evaporating and exiting the column with overhead vapors. When column tops are insulated for personnel protection, shock condensation corrosion is found primarily in the upper crude column and preflash column at temperatures within 14 °C (25 °F) of the dew point. Suspect areas include uninsulated areas of the upper shell of the crude column. With pieces of exterior insulation missing, these metal surfaces can be expected to be much colder than insulated surfaces during periods of cold weather and high winds. Corrosion damage due to shock condensation corrosion has certain characteristic features described below:

- Localized etching, surface roughening, and metal loss at the point of shock condensation.
- Narrow regions of sharp-edged pitting, striations, and shallow grooving in the condensate flow direction.
- Sharp and angular edges along corroded areas.
- Lack of corrosion deposits due to flow turbulence.

Shock condensation corrosion can also occur when cold reflux at 38-52 °C (100-125 °F) is discharged on the top tray at 115-120 °C (240-250 °F) and not distributed evenly through a distributer pipe. As a result, water carried over with the reflux stream collects at low spots on the tray and any deposits on the subcooled tray areas become wet and corrosive. As the water evaporates again, solids stay behind. To reduce water carryover with reflux, suitable baffles can be installed in the reflux drum. On some units, a separate water knockout (K.O.) drum or remote boot is placed between the reflux drum and the column.

15.8.6 General Corrosion of the Upper Trays. General corrosion in the top of columns is often referred to as acid corrosion, overall corrosion, metal wasting, and probably various other designations. Regardless of what it is called, general corrosion is caused by hydrogen chloride (HCl) dissolved in water condensate. General corrosion requires water condensate to be present, either as a continuous or intermittent water layer, puddles, or water droplets dispersed throughout the vapor and naphtha phases. Water containing dissolved HCl becomes dilute hydrochloric acid with a pH between 2 and 4, and lower. In contrast to other types of upper column corrosion, general corrosion is fairly predictable. Once a suitable location has been identified (for example, a specific shell area near the top tray), periodic ultrasonic thickness (UT) measurements can be used to predict the corrosion rate reasonably well. Alternatively, a retractable electrical resistance (Corrosometer™) probe with flexible sensing element can be inserted through a nozzle and placed on a tray surface to monitor corrosion continuously, as shown in Figure 15-9.

Semipermanent coupon racks can also be placed on trays, but are only practical when a column will be opened for maintenance on a regular schedule. With extended runs of 5 or more years, coupon racks are often forgotten and left in place indefinitely. Even when they are eventually recovered, the data obtained is usually no longer pertinent. General corrosion can be found especially at the discharge area below the reflux inlet nozzle, on the upper shell and dome areas of a column operating near the dew point, and in the top pump-around circuit where used. Corrosion damage due to general corrosion can be in the form of overall surface etching, more or less uniform surface roughening and metal loss, sharp-edged pitting all over, and sharp and angular edges along the corroded areas. It is often characterized by complete absence of corrosion deposits at the corroded areas.

15.9 Methods for Controlling Corrosion

Corrosion and fouling issues are far more critical in columns than in overhead systems, because a unit may have to come down when column trays and other internals are blocked and require thorough cleaning, repairs, or replacement. It cannot be stated often enough that control of upper column corrosion requires effective crude desalting upstream of the column to reduce the amount of hydrogen chloride (HCl) going up a column. This may require upgrading the electrical system of an existing desalter, upgrading an existing desalter to two-stage desalting, increasing the size of an existing desalter, optimizing the desalter operation, installing monitoring equipment to alert unit operators when upsets occur and caustic injection into desalted crude is needed. These topics are discussed in some detail in Chapter 9 and Chapter 12 of the *Guide.* Other methods for controlling upper column corrosion and fouling have been used in the past and are briefly summarized below:

FIGURE 15-9 RETRACTABLE PROBE WITH FLEXIBLE SENSING ELEMENT FOR PLACEMENT ON COLUMN TRAY

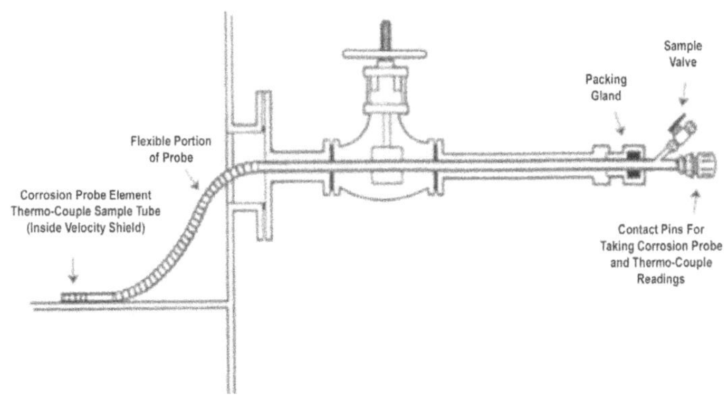

Courtesy of Betz Process Chemicals, The Woodlands, TX.

15.9.1 Tank Settling of Crude Receipts. Tank settling of crude receipts helps decrease the amount of hydrogen chloride (HCl) going up the columns by reducing desalter upsets. Proper tank settling is discussed in Chapter 8 of the *Guide.*

15.9.2 Alloy Upgrading. In the past, the usual remedy for upper column corrosion was cladding or weld overlaying the upper column and dome with Alloy 400 and installing Alloy 400 trays. As noted earlier, Alloy 400 can be unreliable at column temperatures above 120 °C (250 °F) when significant amounts of hydrogen sulfide (H_2S) are present. Alloy C-276 can be substituted for Alloy 400 to reduce corrosion regardless of column temperature, but may not always be cost effective. However, at least one small refinery decided to go that route to eliminate crude column corrosion once and for all. Contrary to some reports, titanium should not be used in upper columns because of possible underdeposit corrosion, hydriding, and embrittlement.

15.9.3 Exterior Insulation. Unless needed for personnel protection, installing, repairing, or replacing exterior insulation to prevent shock condensation probably makes sense only for columns with relatively low top temperatures in the 105-120 °C (220-250 °F) range and then only in northern climates.

15.9.4 Modifying the Reflux Inlet Distributor. Reflux can contain more entrained water than is often appreciated. It is often cost effective to redesign and realign reflux inlet piping to ensure even reflux distribution over the entire top tray area and avoid forming water puddles on the tray.

15.9.5 Installing a Coalescer in the Reflux Line. To reduce the amount of entrained water in reflux going to the upper column, a coalescer can be installed in the reflux return line.

15.9.6 Installing a Coalescer in the Pump-Around Line. A coalescer can also be installed in a pump-around line to reduce the amount of entrained water.

15.9.7 Raising the Column Top Temperature. If practical from a process viewpoint, the column top temperature should be raised by at least 10 °C (18 °F) above the calculated dew point to avoid acid condensation.

15.9.8 Keeping Neutralizers and Inhibitors out of Dry Stages of an Overhead System. Ammonia, liquid organic neutralizers, and filming-amine corrosion inhibitors should not be injected into the dry stage of a two-stage overhead system to avoid sending excess neutralizer, neutralizer salts, and inhibitor into the upper column with reflux.

15.10 Ineffective Methods for Controlling Corrosion

Some methods for controlling column top corrosion and fouling have turned out to be counterproductive over the long run by creating new problems including fouling problems. Typical examples, all related to process additives, are briefly summarized below:

15.10.1. Injecting a Neutralizer into Crude Feed. In the past, a process additives supplier proposed injecting a liquid organic neutralizer into the bottom of the atmospheric column to neutralize hydrogen chloride (HCl) before it reaches the upper column and overhead system. This practice was quickly abandoned when rapid tray fouling and underdeposit corrosion occurred throughout the column. In some cases, sidestream strippers also were full of deposits. Even when fouling was not a problem, liquid organic neutralizers in column feed reacted with naphthenic acids and caused foaming in the column. Organic neutralizer hydrochloride salts then end up in one or more sidestreams and cause issues in various downstream units. For example, salts in reformer naphtha feed can cause run-away conversion, coking of catalysts, and promote chloride pitting and stress-corrosion cracking of 300 series stainless steel components.

15.10.2 Injecting a Neutralizer into the Reflux Line. One process additives supplier injected ammonia and liquid organic neutralizers continuously into the reflux line in an attempt to neutralize hydrogen chloride (HCl) in the upper column before it enters the overhead system. This practice, too, caused rapid tray fouling and underdeposit corrosion throughout the column, but may still be used on some units.

15.10.3 Injecting a Neutralizer into the Column Top. In the past, several process additives suppliers injected inorganic and liquid organic neutralizers directly into the upper column or top pump-around (where used) to neutralize hydrogen chloride (HCl) before it enters the overhead system proper. Rapid tray fouling and underdeposit corrosion usually occurred, forcing the column to be shut down for cleaning.

15.10.4 Injecting a Corrosion Inhibitor into the Column Top. Most upper column issues are caused by salt deposits and not by acid attack. Aside from causing tray fouling, filming-amine corrosion inhibitors should not be injected, because they are unable to penetrate deposits and reach the underlying metal surface. Where filming-amine corrosion inhibitors were injected in the past, they ended up in the heavy naphtha or kerosene cuts and caused issues with WSIM[6] ratings when jet fuel was run.

15.10.5 Injecting a Corrosion Inhibitor into the Reflux Line. One process additives supplier injected a filming-amine corrosion inhibitor continuously into the reflux line to use the inhibitor's inherent detergent properties to loosen and move ammonium hydrochloride (NH_3-HCl) tray deposits further down the column. In essence, the inhibitor was used as an antifoulant. As indicated in Chapter 13 of the *Guide,* if an antifoulant is to be used with columns that have not been thoroughly cleaned, tray fouling and underdeposit corrosion can be moved elsewhere.

15.11 Water Washing the Top Trays

While often ineffective, water has been injected into the reflux line to the upper column in an attempt to dislodge neutralizer hydrochloride deposits on the top trays. However, without suitable piping and equipment in place, the unit may have to come down for water washing. Alternatively, on-stream water washing to remove salts off the top trays has been done by closing off the water draw from the overhead receiver drum and sending wet reflux to the top of the column. At the same time, the upper column temperature was dropped low enough to remove the water at the top sidecut draw. In all cases, the effectiveness of water washing should be carefully monitored with pH readings of the discharge water.

15.12 Repair Methods

Corroded carbon steel tray components can be repaired or replaced in kind, depending on the severity of corrosion. In many applications, replacement or overlaying with Alloy 400 does not provide sufficient long-term improvement to justify its cost. For greater corrosion resistance, corroded column internals can be selectively replaced or overlaid with Alloy C-276. As noted in Chapter 6 of the *Guide*, a number of precautions apply when a column is first opened for inspection or maintenance. Sulfide deposits evolve poisonous hydrogen sulfide (H_2S) gas when subjected to water and cleaning solutions. H_2S gas is heavier than air and can flow out of the manholes of an unvented column with deadly consequences when unprotected personnel try to enter.

15.13 Chapter Summary

Corrosion in the top of preflash and crude columns occurs in columns that operate at top temperatures between approximately 93 °C (200 °F) and 120 °C (250 °F). Low-temperature operation is often used by choice to obtain certain product slates; top temperatures cannot be simply raised no matter how attractive this may be from a corrosion viewpoint.

Corrosion is caused by the same chemical compounds that cause corrosion in overhead systems. They include primarily hydrogen chloride (HCl), ammonia (NH_3), and hydrogen sulfide (H_2S) entering with desalted crude. In addition, ammonia, liquid organic neutralizers, various neutralizer hydrochloride salts, filming-amine corrosion inhibitors, corrosion deposits, and water can enter with overhead reflux or via desalter water.

In atmospheric columns, the affected internals include primarily the top four trays, the upper vessel shell and dome, the upper pump-around circuit, and the upper naphtha side-draw.

6. Water Separometer Index, Modified, per ASTM D2550-69 (latest revision), ASTM, West Conshohocken, PA.

Carbon steel remains the standard material of construction. Alloy 400 is often used as a replacement for carbon steel or 12Cr stainless steel on the upper four trays. Alloy 400 is also used extensively to clad or weld overlay the dome, and the shell down to the fourth tray from the top. However, at temperatures above 120 °C (250 °F), Alloy 400 often corrodes worse than carbon steel. Alloy C-276 can be substituted for Alloy 400 to reduce corrosion regardless of column temperature.

Various types of corrosion including corrosion under salt and corrosion deposits; corrosion due to shock condensation on upper shell, dome, and tray areas; and general corrosion of the top tray are found in upper columns and elsewhere. Corrosion of column shell areas is usually monitored with UT thickness readings. Special types of corrosion probes are available for monitoring corrosion on column trays.

15.14 Read More

1. S. H. Bernhardt et al., "Condensation of Immiscible Mixtures, Heat Transfer-Tulsa," *AIChE Symposium Series* 68, 118 (1970?): pgs. 21-37.
2. S. G. Ewing, "Electrochemical Studies of the Hydrogen Sulfide Corrosion Mechanism," *Corrosion* 11, 1 (1955).
3. R. T. Foley, "The Role of the Chloride Ion in Iron Corrosion," *Corrosion* 26, 2 (1970).
4. R.D. Merrick and T. Auerbach, "Crude Unit Overhead Corrosion Control," *MP* 22, 9 (1983).
5. E. Mohsen, "Crude Unit Operation / Corrosion are Linked," *Oil & Gas J.*, April 17 (1978).
6. K. Reiser, "The Control of Corrosion in Refinery Distillation Units," *J. Institute of Petroleum* 53, 527 (1967).
7. J. R. Richmond and G. S. Browne, "Automatic pH Control of Crude Column Overhead Streams," *MP* 21, 11 (1982).
8. J. R. Rue, "Cold Tower Aqueous Corrosion: Causes and Prevention," paper no. 211, CORROSION/90 (Houston, TX: NACE, 1990).
9. C. J. Scherrer et al., "Crude Units: Focusing on Corrosion of Initial Condensation Equipment," paper no. 66, CORROSION/81 (Houston, TX: NACE, 1981).
10. H. Z. Kister, "What Caused Tower Malfunctions in the Last 50 Years," Trans I. Chem. E., 81, Part A, Institution of Chemical Engineers, January 2003.
11. H. Z. Kister, *Distillation Operation* (New York, NY: McGraw-Hill, 1990).
12. H. Z. Kister, "Are Column Malfunctions Becoming Extinct or Will They Persist in the 21st Century," Trans. I. Chem. E., Part A, Chem. Eng. Res. 75: Institution of Chemical Engineers, 1997.
13. W. R. Ellingsen, "Diagnosing and Preventing Tray Damage in Distillation Columns," DYCORD/86, IFAC Proceedings of International Symposium on Dynamics and Control of Chemical Reactors and Distillation Columns (Bournemouth, UK: 1986).
14. J. J. France, "Troubleshooting Distillation Columns," paper presented at the AIChE Spring Meeting (Houston, TX: 1993).
15. N. P, Lieberman, *Process Design for Reliable Operation*, 2nd ed. (Houston, TX: Gulf Publishing, 1988).
16. N.P. Lieberman, *Troubleshooting Process Operations*, 3rd ed. (Tulsa, OK: PennWell Books, 1991).
17. N. P. Lieberman and E. T. Lieberman, *A Working Guide to Process Equipment* (New York, NY: McGraw-Hill, 1997).
18. G. H. Shieveler, "Use Heavy-Duty Trays for Severe Services," *Chem. Eng. Progr.*, August 1997.

16 Overhead Corrosion

In this Chapter

16.1 First Glance

Overhead corrosion has been well recognized and understood since the 1940s, but continues to be a major operating, maintenance, and expense issue, if not a nightmare, on many units. By definition, the term "overhead corrosion" includes various forms of low-temperature aqueous corrosion in overhead systems of distillation columns. While overhead corrosion is also considered to occur in the upper section and dome of these columns, it is sufficiently different to justify a separate discussion. Therefore, this chapter addresses primarily the why's, where's, when's, and how's of corrosion and fouling in overhead systems of the atmospheric and vacuum section of crude units and the DRU and VDU of bitumen upgraders.

16.2 Cause of Overhead Corrosion

The bulk of overhead corrosion issues is caused by hydrogen chloride (HCl) in combination with water (steam) condensate and other chemical compounds. HCl vapor evolves in the hot preheat exchangers (HPEs) and fired heater by hydrolysis of residual chloride salts in desalted crude. For overhead corrosion and fouling to occur, HCl vapor has to:

- Dissolve in water condensate and form hydrochloric (HCl) acid.
- React with ammonia (NH_3) or ammonium hydroxide ($NH_3 \cdot H_2O$) vapor and form ammonium hydrochloride ($NH_3 \cdot HCl$) salt.
- React with liquid organic neutralizers ($R\text{-}NH_2 \cdot H_2O$) and form neutralizer hydrochloride ($R\text{-}NH_2 \cdot HCl$) salts.
- Penetrate existing corrosion deposits on metal surfaces.
- Penetrate existing corrosion inhibitor films.

To limit the amount of HCl vapor going overhead, crude receipts are first gravity settled in tankage and then desalted. Slop oil and caustic should be added to the desalted crude. These measures are discussed in greater detail in Chapters 8, 9, 10 and 12 of the *Guide.* Other measures to further reduce corrosion and fouling include upgrading the metallurgy of some components, limiting overhead vapor velocities, injecting neutralizers and corrosion inhibitors, and water washing. The latter three measures are covered in Chapters 17, 18, and 19 of the *Guide.* Since the chemistry and mechanism of HCl evolution were covered in some detail in Chapter 15 on upper column corrosion, they are not repeated here. But as in the case of upper column corrosion, certain problem chemicals other than HCl contribute to overhead corrosion and are listed below in approximate order of decreasing importance:

16.2.1 Ammonia. Ammonia (NH_3) vapors are formed by dissociation of organic nitrogen compounds in the crude preheat exchangers (PEs) and fired heater. Ammonia is also introduced into crude units when stripped sour water is used as desalter water. Ideally, stripped sour water contains no ammonia, but 40mg/L NH_3 or more is seen with many sour water strippers. Another source of ammonia is ammonium hydroxide ($NH_3 \cdot H_20$) in water condensate[1] recycled from the atmospheric overhead receiver to the desalter for use as desalter water. Therefore, even without intentional ammonia injection in an overhead system, ammonia levels in overhead water condensate can range anywhere from 10 to 3,000mg/L[2]. The total amount of ammonia going overhead depends on the desalter water pH. Low ammonia levels can be maintained in overhead water by keeping the desalter water pH as low as possible without causing desalter corrosion issues that typically occur at pH below 5.0. However, on many crude units the overhead pH is difficult to control by relying solely on desalter water pH and ammonia or a liquid organic neutralizer has to be injected.

16.2.2 Ammonium Hydrochloride Deposits. A mist of ammonium hydrochloride ($NH_3 \cdot HCl$) particles forms whenever NH_3 vapor comes in contact with HCl vapors. Ammonium chloride mist forms already in the top of most columns. It also forms immediately downstream of an ammonia injection point when ammonia or ammonium hydroxide ($NH_2 \cdot H_2O$) is intentionally injected into an overhead vapor line for pH control. The mist sublimes and forms corrosive deposits when the partial pressure of ammonium chloride exceeds a limiting value, as shown in Figure 15-10 below. Ammonium hydrochloride is completely soluble in overhead water condensate, but insoluble in naphtha condensate.

16.2.3 Organic Neutralizer Deposits. Fouling problems are caused by gunk formation when a neat neutralizer and corrosion inhibitor are injected together into a reflux or overhead vapor line. The gunk usually ends up in downstream separator drums. When a neutralizer is injected too close to elbows, tees, and heat exchanger inlet nozzles, impingement corrosion occurs and results in corrosive deposits.

16.2.4 Organic Neutralizer Hydrochloride Deposits. As with ammonia, a mist of neutralizer hydrochloride ($R\text{-}NH_2 \cdot HCl$) particles forms when liquid organic neutralizers are injected intentionally into an overhead vapor line for pH control. Depending on the particular chemical involved, the mist sublimes and forms corrosive deposits when the partial pressure of the neutralizer hydrochloride exceeds a certain limiting value. Some liquid organic neutralizer hydrochlorides are hygroscopic, readily disperse (but not dissolve) in naphtha condensate, do not form corrosive deposits, and return to the column with reflux. Others are known to cause fouling on column trays midway down the column. As a general rule, neutralizer hydrochlorides are completely soluble in overhead water condensate, but insoluble in naphtha condensate.

1. The term water condensate is often preferred over aqueous condensate.
2. M. R. Beychok, *Aqueous Wastes from Petroleum and Petrochemical Plants* (John Wiley & Sons, 1967).

16.2.5 Corrosion Scale. Corrosion deposits in overhead systems consist primarily of mixed iron hydroxides/oxides/sulfides that form porous deposits or scale at pH above 6.0 to 7.0. Based on thermodynamic considerations and pH, the key sulfides are mackinawite (FeS), greigite (Fe_3S_4), pyrrhotite ($Fe_{1-X}S$), and pyrite (FeS_2) while the key iron hydroxides/oxides are ferric oxide (Fe_2O_3), magnetite (Fe_3O_4), and ferrous hydroxide [$Fe(OH)_2$].[3] The deposits also contain entrapped HCl, ammonium hydrochloride ($NH_3 \cdot HCl$), or neutralizer hydrochlorides ($R\text{-}NH_2 \cdot HCl$) and often become sticky, acidic, and highly corrosive. Corrosion deposits are insoluble in overhead water and naphtha condensate, but can be prevented from depositing by continuous water washing.

16.2.6 Corrosion Inhibitor Deposits. Neat corrosion inhibitor droplets can form sticky and gooey deposits in vapor lines when the inhibitor is insufficiently or incorrectly diluted prior to injection. They can also end up in column tops due to issues with the injection system. Most inhibitor deposits become acidic and highly corrosive at temperatures above approximately 120 °C (250 °F). Once formed, these deposits are insoluble in both hydrocarbon and water condensate. They cannot be removed simply by water washing, and a fouled heat exchanger may have to be bypassed for hydro jetting during shutdown.

16.2.7 Hydrogen Sulfide. Dissolved hydrogen sulfide (H_2S) is present in sour crude and is also formed in the fired heater from organic sulfur compounds. Like HCl, H_2S travels up the preflash, atmospheric, and vacuum columns and combines with ammonia to form an ammonium bisulfide ($NH_3 \cdot H_2S$) mist. This mist sublimes and forms corrosive deposits when the partial pressure of $NH_3 \cdot H_2S$ exceeds a certain limiting value. $NH_3 \cdot H_2S$ dissolves in overhead water condensate and causes sour-water corrosion at pH above 7.5-8.0. In the presence of oxygen from air contamination, yellow sulfur deposits are often formed.

16.2.8 Organic Chlorides. Organic chlorides can enter raw crude with slop oil, chemical by-products, and waste streams originating upstream, midstream, and at the refinery. They are not removed by desalting, because they are insoluble in water. Organic chlorides partially hydrolyze in the crude heater to form HCl that can dramatically increase overhead corrosion. Various issues with organic chlorides are discussed in detail in Chapter 11 of the *Guide*.

16.2.9 Carbon Dioxide. Carbon dioxide (CO_2) originates with upstream operations and is present as a dissolved gas in crude charged to the unit. CO_2 can also originate with steam added at the crude heater, crude column, and sidestream columns. CO_2 travels up the preflash and atmospheric columns and dissolves in overhead water condensate to form dilute carbonic (H_2CO_3) acid. At the relatively high temperatures and low pH values of overhead water condensate, CO_2 can cause severe vapor-space corrosion in dead areas of heat exchanger shells and separator drums.

16.2.10 Oxygen (Air). While not corrosive as such, oxygen reacts with dissolved hydrogen sulfide and ammonia to forms elemental sulfur (S_n), inorganic polysulfides [$(NH_4)_2S_3$], and sulfur oxides (SO_2, SO_3, and others). Oxygen usually enters overhead systems with air intrusions at leaking pump seals, makeup water for the desalter, cooling tower water at leaking tube bundles, or poorly de-aerated steam. Sulfur deposits in heat exchanger bundles are readily apparent because of their yellow color. Polysulfides have been blamed for accelerated overhead corrosion in the form of pitting, but tend to form protective iron polysulfide films in some sour-water services. Oxygen leaks into the vacuum section can cause severe operating and corrosion problems throughout.

16.2.11 Light Organic Acids. Light organic acids including formic acid (H-COOH), acetic acid (CH_3-COOH), and proprionic acid (CH_3CH_2-COOH) can form in the fired crude heater due to decomposition of naphthenic acids present in most types of crude. Acetic acid (CH_3COOH) has been positively identified in overhead water condensate at levels as high as 250mg/L and can cause severe corrosion even in the absence of HCl. Proprionic acid has been identified in overhead water condensate at levels as high as 100mg/L. Formic acid has not been identified, but is considered to be the most corrosive of the three.[4]

16.2.12 Sulfuric Acid. Sulfuric acid (H_2SO_4) is used for formation fracturing (acidizing) operations during crude production. When oil wells are not properly purged after acidizing, subsequent crude shipments can contain high levels of residual sulfuric acid for short periods of time. H_2SO_4 has also been added to desalter water to adjust pH, but normally should not be used for that purpose to avoid the formation of hard sulfate scales. In either case, some residual acid is likely to stay with desalted crude, vaporize in the fired crude heater, and travel up the atmospheric column and enter the overhead system. While sulfuric acid has been identified in overhead water condensate, it has not been shown for certain that sulfuric acid was actually responsible for any increased corrosion seen at the time.

3. J. Ning, Y. Zheng, D. Young, B. Brown, and S. Nesic, "Thermodynamic Study of Hydrogen Sulfide Corrosion of Mild Steel," *Corrosion* 70, 4 (2014).
4. G.B. Elder, "Materials of Construction for Organic Acids," in *Process Industries Corrosion - The Theory and Practice*, B.J. Moniz, W.I. Pollock, eds., pgs.287-296 (Houston, TX: NACE, 1986).

16.3 Overhead System Metallurgy

The general metallurgy (materials of construction) for crude units and the corresponding upgrader units is reviewed in Chapter 4 of the *Guide.* The following lists alternative metallurgy options specifically for components of overhead systems. The listing is shown in approximate order of increasing corrosion resistance and likely cost.

16.3.1 Carbon Steel. Carbon steel remains the standard material of construction for overhead systems. Depending on the chloride content and pH of overhead water condensate, carbon steel can corrode at excessive rates and requires neutralizer, corrosion inhibitor, and wash water injection for corrosion control. Corrosion by dilute hydrochloric (HCl) acid at different pH values can be predicted reasonably well from the published correlations shown in Figure 16-1 and Figure 16-2. These correlations are based on laboratory tests and are best used for predicting changes in corrosion as a result of changes in pH instead of predicting absolute corrosion rates.

Corrosion rates measured in the field are usually lower than those obtained in the laboratory because a large volume of naphtha condensate reduces the statistical likelihood of acid droplets contacting any given steel surface. HCl is a strong acid and completely dissociates in water; the chloride content is directly related to pH, as shown in Figure 16-3. The following guidelines apply to carbon steel corrosion with respect to pH measured in overhead water condensate taken off the receiver boot:

- Minimum corrosion occurs in the pH5.5-6.5 range.
- At pH below 5.0, corrosion rapidly increases with decreasing pH due to increasing HCl content. At the same time, any protective iron sulfide (FeS) deposits loosen, dissolve, and foul downstream components.
- At pH above 6.5, corrosion increases rapidly with increasing pH due to increasing ammonium bisulfide ($NH_3 \cdot H_2S$) content.
- At pH above 8.5, stable deposits of mackinawite (FeS), greigite (Fe_3S_4), pyrrhotite ($Fe_{1-X}S$), and pyrite (FeS_2)[5] stifle further corrosion, but increase the likelihood of fouling.

However, bulk pH does not necessarily represent the pH of all water condensate in an overhead system. For example, the pH of stagnant water in the boot of overhead receiver drums is likely to vary widely depending on the rate of water withdrawal and may be much lower than the bulk pH. Local pH variations affect corrosion or fouling at various points in an overhead system.

FIGURE 16-1 CORROSION OF CARBON STEEL BY AQUEOUS HCl SOLUTIONS AT 57 °C (135 °F)

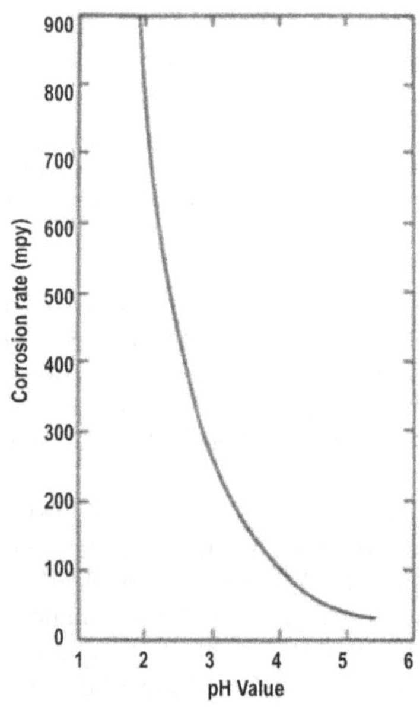

Adapted from J. Gutzeit, "Corrosion of Steel by Sulfides and Cyanides In Refinery Condensate Water," *MP* 7, 12 (1968).

5. J. Ning, Y. Zheng, D. Young, B. Brown, and S. Nesic: *op.cit.*

FIGURE 16-2 CORROSION OF CARBON STEEL BY AQUEOUS $NH_3 \cdot H_2S$ SOLUTIONS AT 57 °C (135 °F)

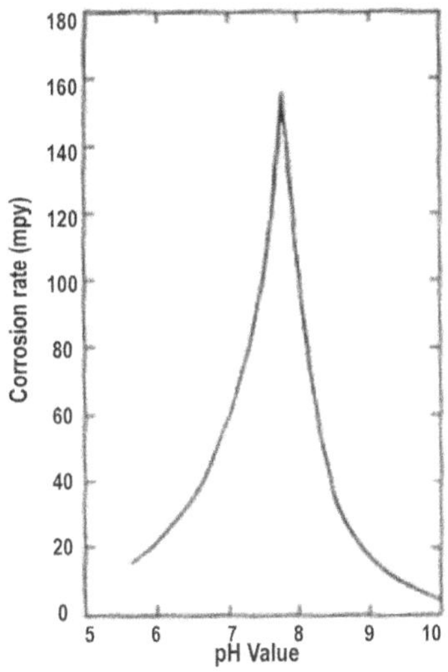

Adapted from J. Gutzeit, "Corrosion of Steel by Sulfides and Cyanides In Refinery Condensate Water," *MP* 7, 12 (1968).

FIGURE 16-3 HCl CONTENT OF OVERHEAD WATER CONDENSATE AS A FUNCTION OF pH

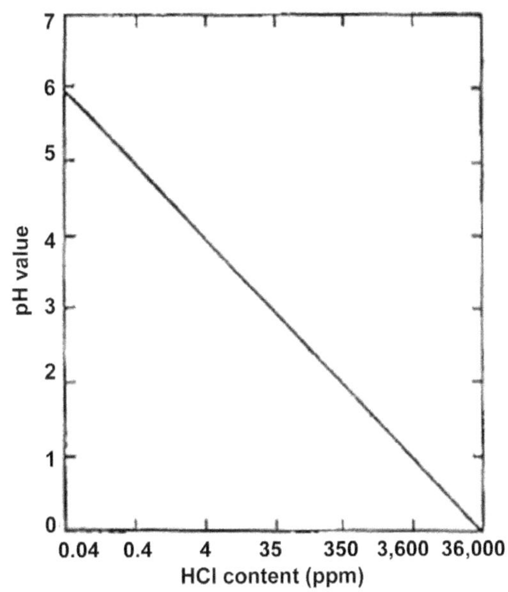

In general, carbon steel in crude unit overhead systems is not subject to hydrogen blistering provided the pH is maintained in the recommended pH5.5-6.5 range. However, at pH below 2.0 in the presence of a "poisoning agent" such as hydrogen sulfide (H_2S) or cyanide (HCN)[6], carbon steel corrodes at high enough rates to introduce atomic hydrogen

6. Poisoning agents such as hydrogen sulfide (H_2S) or cyanide (CN) interfere with (poison) the formation of molecular hydrogen (H_2) at the corroding steel surface and facilitate entry of atomic hydrogen (H_O).

(H_O) into the steel and cause hydrogen blistering and cracking, as shown in Figure 16-4. In this particular case, approximately 80 blisters were found in the lower 1/3rd of the shell of a water-cooled heavy naphtha (HN) condenser after 5yrs of service. HN entered the shell at approximately 140 °C (285 °F) and exited at 96 °C (205 °F). The blisters measured 12-50mm (0.5-2.0in.) in diameter at the steel surface and were 5-8mm (0.2-0.3in.) deep. The affected HN condenser was the upper (hottest) unit of a bank of two in series. The lower unit, operating at an inlet temperature of 96 °C (205 °F) and outlet temperature of 52 °C (125 °F), and the downstream separator drum were free of blisters.

16.3.2 Low-alloy Cr-Mo Steels. The corrosion resistance low-alloy Cr-Mo steels is similar to that of carbon steel. These steels are sometimes used instead of carbon steel because of ready availability, but require postweld heat treatment (PWHT) to prevent hydrogen stress cracking (sulfide cracking) at hard welds.

16.3.3 12Cr Stainless Steel. 12Cr stainless steel (Types 405, 410, or 410S) is rarely used for overhead heat exchanger tubes because it tends to pit under chloride containing deposits. Moreover, it is not readily available in tubular form. The martensitic grade (Type 410) requires PWHT to prevent hydrogen stress cracking at hard welds.

FIGURE 16-4 HCI CORROSION AND HYDROGEN BLISTERING OF HEAVY NAPHTHA CONDENSER SHELL

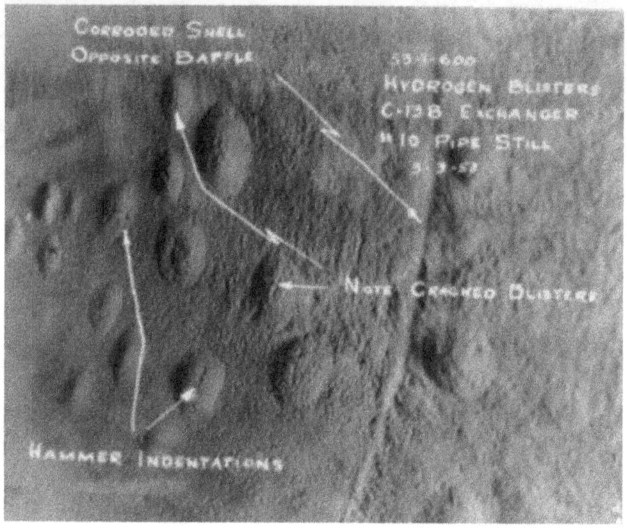

16.3.4 Austenitic Stainless Steel. Austenitic stainless steel (Types 304, 316, or 317) may pit in the presence of chloride deposits. They are likely to chloride stress-corrosion crack (CSCC) at temperatures above approximately 65 °C (150 °F) upon exposure to air and components should be protected against CSCC during shutdown in accordance with NACE RP0170[7].

16.3.5 Alloy 400. Alloy 400 (66Ni-32Cu) is resistant to dilute HCl acid condensate in overhead systems. It is often used for cladding or weld overlaying elbows and tees in overhead vapor lines to prevent impingement corrosion, and for lining the bottom of separator drums to prevent pitting corrosion. The alloy has good corrosion resistance to most chloride-containing deposits, but can corrode under acidic ammonium hydrochloride ($NH_3{\cdot}HCl$) deposits in the presence of H_2S. As shown in Table 4-3 in Chapter 4 of the *Guide*, Alloy 400 can corrode at higher rates than carbon steel at temperatures above approximately 120 °C (250 °F) when used for cladding and weld overlaying elbows and tees of overhead vapor lines. The expected service life is only 5-10yr under these conditions and Alloy C-276 would be a better choice.

16.3.6. Alloy 625. Alloy 625 (60Ni-21Cr-9Mo-5Fe) is often used for heat exchanger tubes and selective alloying of overhead vapor lines and drums regardless of temperature. While Alloy 625 is often considered to be fully resistant to hydrochloric acid (HCl) corrosion, the alloy is prone to pit under chloride containing deposits. For example, severe tube corrosion occurred when a combination of liquid organic neutralizer and filming-amine corrosion inhibitor was

7. NACE Standard RP0170-2004, *Protection of Austenitic Stainless Steels and other Austenitic Alloys from Polythionic Acid Stress Corrosion Cracking during Shutdown of Refinery Equipment.*

injected upstream of the first heat exchanger in the first stage of a multistage overhead system. While the vapor temperatures were only 110-115 °C (230-240 °F), the dewpoint was 102 °C (215 °F) and water condensation took place. No wash water was used. Carbon steel tubes in this heat exchanger failed after only 0.5-2.0yr and eventually were replaced with Inconel 625 tubes. Despite their high alloy content, the latter tubes were also severely attacked under neutralizer hydrochloride salt deposits after only 2 years, as shown in Figure 16-5.

FIGURE 16-5 CORROSION OF ALLOY 625 TUBES IN OVERHEAD EXCHANGER UNDER NEUTRALIZER HYDROCHLORIDE DEPOSITS AT 110 °C (230 °F)

16.3.7 Alloy C-276. Alloy C-276 (58Ni-15Cr-16Mo-6Fe-2Co-3W) is preferred for selective alloying of overhead vapor lines and fittings, such as elbows and tees. Despite some isolated claims to the contrary, the alloy is fully resistant to impingement corrosion in vapor lines under the reducing conditions encountered in most overhead systems. The alloy is less suitable, however, under oxidizing conditions, including applications where aerated steam, cooling water, or air leaks can occur. Alloy C-276 is highly resistant to acidic chlorides, such as ammonium hydro chloride ($NH_3{\cdot}HCl$) or ammonium sulfate [$(NH_4)_2{\cdot}SO_4$].

16.4 Cooling Water Metallurgy

Cooling water corrosion (CWC) can present special issues for plants, because copious amounts of water are required for steam generation and for cooling various process streams. Not surprisingly, plant locations are chosen wherever water in the form of surface waters, river water, or seawater is readily available. On average, a typical 13,900TPD (100,000BPD) refinery requires up to 189,000m^3/d (1,193,000BPD) water or approximately 12 times the amount of crude for cooling and steam generation[8]. Other data indicate that refineries use 4-6L (1-2.5Gal) water for every 4L (1Gal) of product[9]. Refineries using untreated once-through river water, brackish water, or seawater for cooling often end up switching, in turn, from carbon steel to Admiralty brass to copper nickel 70-30 to titanium in order to find a tube material sufficiently resistant to CWC. Titanium tubes are often required for brackish or seawater. Refineries using treated recirculating cooling water often can use carbon steel and Admiralty brass tubes, provided excessive process temperatures are avoided. Even then some plants have switched to titanium tubes, given the excellent resistance of titanium against corrosion on the process side.

16.4.1 Admiralty Brass. Admiralty brass (70Cu-28Zn-1Sn) is used extensively for condenser tubes in cooling water services. The high zinc (Zn) content of this alloy protects it against corrosion by hydrogen sulfide (H_2S) on the process side by forming protective zinc sulfide (ZnS) films. However, most brasses have limited strength at elevated temperatures above approximately 230 °C (450 °F). They also have limited resistance against HTSC and should never be used under

8. L.E. Otts, Jr. "Water Requirements of the Petroleum Refining Industry, Geological Survey Water-Supply," paper 1330-G (Washington, DC: U.S. Government Printing Office, 1963).
9. U.S. Environmental Protection Agency, 2014.

these conditions, even if the strength requirements are met. Current production of Admiralty brass in the United States normally is inhibited against preferential dezincification in cooling water service by alloying with 0.04% phosphorous (P) or arsenic (As). Like other brass alloys, however, the alloy may stress-corrosion crack (SCC) during shutdown when even traces of ammonia residues are exposed to the atmosphere during cleaning and maintenance operations. To prevent cracking, Admiralty brass tube bundles should be washed with a dilute acid solution immediately after pulling from the shell.

16.4.2 Naval Brass. Naval brass (60Cu-39Zn) is used in cooling water service when a high-strength corrosion resistant material is required for condenser tube sheets. Like Admiralty brass tubes, naval brass has reasonably good resistance against process-side corrosion by hydrogen sulfide (H_2S) because protective zinc sulfide films are formed. Most brasses have limited strength at elevated temperatures above 230 °C (450 °F). They also have limited resistance against HTSC and should never be used under these conditions, even if the strength requirements are met. Like Admiralty brass, naval brass is subject to stress-corrosion cracking in the presence of residual ammonia, moisture, and air when tube bundles are opened to the atmosphere prior to cleaning during shutdown. A dilute sulfuric acid wash has been used to prevent cracking. While all currently available naval brass production is inhibited against preferential dezincification through the addition of 0.7% Sn (tin), uninhibited naval brass tubesheets may still be found in older plants.

16.4.3 Copper Nickel 70-30. This alloy is also used for condenser tubes in cooling water service. In contrast to Admiralty brass, however, copper nickel 70-30 does not stress-corrosion crack during shutdowns when exposed to air. The copper nickel 90-10 variety should not be used because the alloy is subject to intergranular corrosion by wet H_2S in overhead systems.

16.4.4 Specialty Stainless Steels. Specialty stainless steels (2250, 3RE60) are used for condenser tubes in cooling water service to prevent pitting. However, purchasing these tubes often requires long lead times and orders are subject to minimum quantity requirements.

16.4.5 Titanium (Various Grades). Titanium is used in tubular form for water-cooled condensers using brackish or seawater. Experience has shown that unalloyed titanium, such as Grade 2, can fail due to pitting, hydriding, and cracking under chloride-containing deposits on the process side near carbon steel baffle holes and tubesheets. These deposits will have to be removed continuously by water washing to prevent premature tube failures due to pitting and embrittlement. Titanium tubes for overhead coolers and condensers should meet BWG18 dimensions with a wall thickness of 1.25mm (0.049in.). Welded tubes with BWG18 (average) dimensions are preferred over tubes with BWG18 (minimum) dimensions in order to be price competitive with Admiralty brass or specialty stainless steel tubes. When substituted for the thicker BWG12 carbon steel or BWG14 Admiralty brass tubes, titanium tubes can be subject to severe vibration damage, unless the baffle spacing is decreased.[10] For the same reason, BWG20 or BWG22 titanium tubes should not be used. Most titanium tube bundles are easily damaged by careless handling during shutdowns.

16.5 Expected Tube Life

In many overhead systems, tube life is limited by shell side corrosion instead of tube side corrosion. Because of the high cost of tube materials other than carbon steel, it is sometimes more cost-effective to stay with carbon steel tubes despite a relatively short tube life. With suitable bypass piping in place and spare tube bundles on hand, a heat exchanger can be bypassed on the run; the bundle pulled, rotated, cleaned, repaired, or replaced; and the exchanger returned to service without major delays. The same applies to tube bundles that foul regularly and have to be pulled for cleaning by hydro blasting to improve heat transfer.

Shell side salt and corrosion deposits have lead to early tube failures regardless of metallurgy, including titanium and Alloy 625 tubes. Continuous water washing is usually required to prevent deposit formation and control the resultant underdeposit corrosion. In many overhead systems, Admiralty brass, copper nickel 70-30, or titanium tubes can be substituted for carbon steel to obtain acceptable tube life. Specialty stainless steel tubes and Alloy 625 tubes have been used on a limited basis, but sometimes with questionable results considering the higher cost. The following guidelines apply especially to water-cooled coolers and condensers in overhead systems of crude units and upgraders:

10. As noted earlier, new titanium tubes have failed due to vibration damage within hours from installation.

- Expected tube bundle life is 1-3yr with BWG12 carbon steel tubes, 3-5yr with BWG14 Admiralty brass tubes, 7-12yr with BWG16 copper nickel 70-30 tubes, and 10+yr with BWG18 titanium tubes.
- When titanium tubes are used, bundle life is no longer determined by tube corrosion, but often depends on the life of carbon steel tubesheets, baffles, and other hardware where used. All-titanium bundles are preferable, but expensive.

16.6 Types of Overhead Systems

On most crude units, either single or two-stage systems are used with preflash and atmospheric columns, depending on crude slate, product slate, column top temperatures, overhead vapor volume, and other operating parameters. Aside from the vacuum ejector system, single-stage systems are used with vacuum columns.

16.6.1 Single-Stage Type 1 Overhead Systems[11]. Type 1 systems shown in Figure 16-6 (left side) are used with "cold" columns where overhead vapors at 120 °C (250 °F) are cooled and condensed in a set of coolers and condensers and send to a single receiver drum for separation of fuel gas and naphtha condensate from water condensate. At the relatively low overhead temperatures, water-cooled condensers are often used for this purpose; a fin-fan air cooler may be used as a trim cooler following the water-cooled condensers. A portion of naphtha condensate is returned to the column as reflux; the remainder becomes product for further processing or goes to storage. Water condensate may be reinjected upstream of the overhead exchangers as wash water. Excess water may also be used as desalter water or sent to a sour water stripper. Fuel gas is further processed or flared, as needed.

16.6.2 Two-Stage Type 2 Overhead Systems[12]. Type 2 systems shown in Figure 16-6 (right side) are used with "hot" columns where only a portion of overhead vapors at 150 °C (300 °F) is cooled and condensed in the first stage. At the relatively high overhead temperatures, heat exchangers using incoming cold crude as the heating medium are often used for this purpose. Only naphtha condensate is separated from the vapors in the first-stage receiver drum to be returned to the column as reflux. The balance of vapors is sent to the second stage where naphtha is further cooled, condensed, and separated from fuel gas and water condensate in the second receiver drum. Fin-fan air coolers and water-cooled condensers may be used in the second stage. The water condensate may be reinjected upstream of the second-stage coolers and condensers as wash water. Any excess water is either used as desalter water or sent to a sour water stripper for disposal. The advantage of the two-stage over the single-stage system is a reduction of fuel costs at the fired crude heater because the hot overhead vapors are better utilized to preheat the incoming charge instead of being wasted in water-cooled condensers.

FIGURE 16-6 TYPE 1 VERSUS TYPE 2 OVERHEAD SYSTEMS

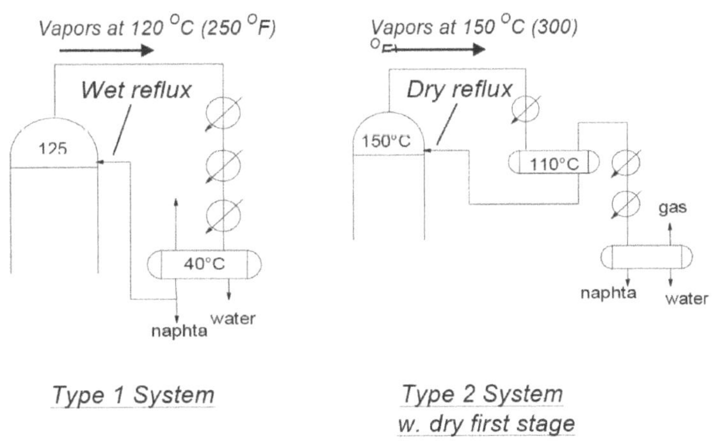

11. Single-stage systems are referred to as single-drum systems in some references and reports.
12. Two-stage systems are referred to as two-drum systems in some references and reports.

16.6.3 Two-Stage Type 3 Overhead Systems. In a Type 3 system, the temperature of the first receiver drum is kept lower compared to that of the Type 2 system. As a result, water condensate is withdrawn from both the first and second drum along with naphtha condensate. In general, naphtha condenses first, followed by ammonium hydrochloride and neutralizer hydrochloride salts, and finally water. Depending on the dew point, the first stage of a 2-stage system can operate either dry (Type 2) or wet (Type 3), as shown schematically in Figure 16-7:

- At overhead vapor temperatures in the 150-180 °C (300-350 °F) range (hot column), no water condenses in the first stage and the first drum simply operates as a reflux drum. The first stage is considered to be dry.
- At relatively low overhead vapor temperatures in the 105-120 °C (220-250 °F) range (cold column), water condenses in both stages and is withdrawn from both receiver drums. The first stage is considered to be wet.
- At intermediate overhead vapor temperatures in the 120-150 °C (250-300 °F) range, water may well condense in some, but not all components of the first stage. The first stage may have to be considered to operate wet, even though relatively little or no water is withdrawn from the first-stage receiver.

When overhead temperatures shift the dry/wet cutoff point into the upstream or downstream direction during normal operation, corrosion patterns and tendencies in two-stage overhead systems change correspondingly. The same applies to single-stage systems where the cutoff point can shift from one component to the next, or even within a single component (for example, a condenser). Also important are the effects of sudden changes in the steam partial pressure due to water surges (upsets) from the desalter.

16.6.4 Other Overhead Systems. Under special conditions, atmospheric overhead systems may consist of three or more condensing stages. While there may be advantages from a process point of view, corrosion in these condensing systems is more difficult to monitor and control than in single-stage or 2-stage systems.

FIGURE 16-7 CONDENSATION PROFILES FOR SINGLE AND TWO-STAGE OVERHEAD SYSTEMS

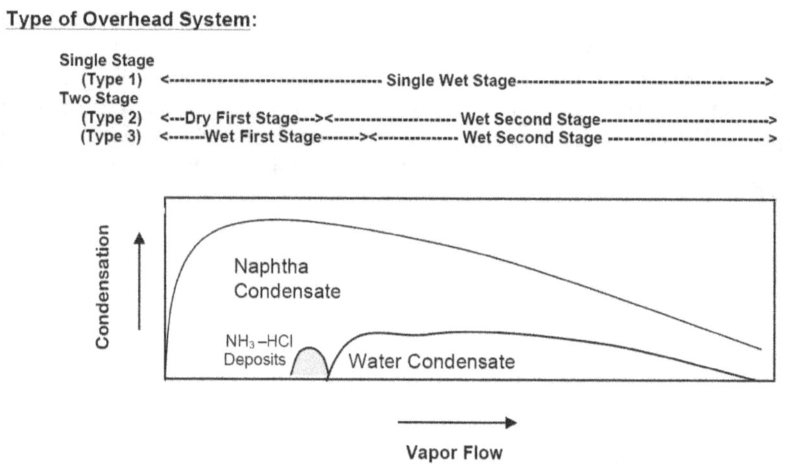

16.6.5 Dry versus Wet Operation of Overhead Systems. From a corrosion point of view, the difference between dry and wet operation of the first stage of Type 2 and Type 3 overhead systems and dry or wet components of Type 1 overhead systems is not appreciated by some unit operators and process chemicals suppliers and therefore ignored. As a consequence, corrosion issues continue unabated because the wrong control measures are implemented. Specifically, the following general guidelines apply:

- In dry components or dry sections of overhead systems, corrosion is caused primarily by ammonium hydrochloride ($NH_3{\cdot}HCl$), neutralizer hydrochloride ($R\text{-}NH_2{\cdot}HCl$), or corrosion deposits. Without free water condensate, these deposits are trapped in place and their concentration increases to the point where tubes become partially plugged, corrode, and start to leak.
- In wet components or wet sections of overhead systems, deposits are washed away by condensing steam and any corrosion is caused primarily by dilute HCl acid.

16.6.6 Vacuum Ejector Systems. Vacuum for the overhead system of vacuum columns is pulled by one or more steam jet ejectors. Most vacuum overhead systems are similar to the single-stage, Type 1 overhead systems used in the atmospheric section except that steam jet ejectors and intercondensers are tacked onto the vapor line. In older systems, vacuum is pulled after the bulk of overhead vapors is first condensed in a water-cooled barometric or tubular precondenser. Vapors leaving the precondenser then go to the first ejector stage, followed by its own water-cooled ejector condenser. In newer systems, no precondenser is used and the overhead vapors go directly to the first ejector stage. In either case, 2 to 4 ejector/condenser stages in parallel and/or series are used depending on the required degree of vacuum, as shown in Figure 16-8. Water condensate drains directly from each condenser shell via a standpipe into a large seal drum (hot well or vacuum sump). On some systems, a pump sends hydrocarbon condensables for further processing while residual vapors go to the fuel gas system. During startup, all ejectors are operational, but during steady operation fewer ejectors have to be engaged. Ejectors normally operate with superheated, medium pressure steam (motive steam). Back pressure is reduced by steam condensation and the static head of condensate-filled standpipes. Water from the seal drum is often used as desalter water. In general, corrosion issues are similar to those seen in the overhead systems of preflash and atmospheric columns, but less severe because less HCl evolves in the vacuum section.

FIGURE 16-8 VACUUM COLUMN WITH PRECONDENSER AND TWO EJECTOR/CONDENSER STAGES

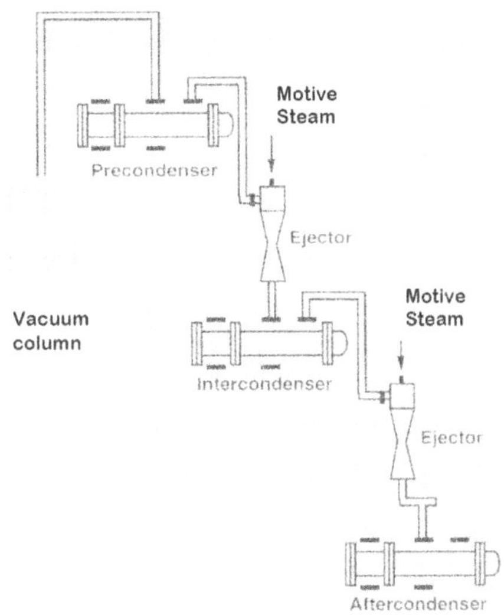

Adapted from J.R. Lines, "Understanding Ejector Systems Necessary to Troubleshoot Vacuum Distillation," *Hydroc. Proc.* 73, 10 (1994).

16.7 Overhead Heat Exchangers, Coolers, and Condensers

Most overhead systems use two or three types of heat exchangers, including crude/overhead heat exchangers, fin-fan air-cooled coolers and condensers, and water-cooled coolers and condensers. Depending on heat transfer duty, flow velocity, and pressure-drop constraints, heat exchangers are configured as single or parallel banks of one or more units in series. Often, but not always, similar corrosion issues are seen in parallel banks of a given unit. In any given bank, more severe corrosion is seen in the hotter upstream units than in the cooler downstream units. The various heat exchanger types can differ with respect to location, metallurgy, and corrosion problems, as follows:

16.7.1 Crude/Overhead Heat Exchangers. The first set of heat exchangers in both single-stage (Type 1) and 2-stage (Type 2 or Type 3) overhead systems of preflash and atmospheric columns often are crude/overhead exchangers. In most applications, straight-tube shell-and-tube heat exchangers are used, as shown in Figure 16-9. For ease of cleaning and to minimize the cost of alloying, crude is usually placed tube side instead of shell side. These are the hottest overhead exchangers and allow hot overhead vapors to be cooled with cold crude. By using cold crude instead of cooling water, tube side corrosion and fouling issues are by and large avoided. Crude temperatures are not high enough for sulfidic (SC) or naphthenic acid corrosion (NAC) and carbon steel or low-alloy Cr-Mo tubes can be used. Corrosion and fouling by entrained brine in crude may cause tube side problems, but with proper design the exchangers operate above the dewpoint on the overhead side and carbon or low-alloy steel tubes present few shell side problems. Typical tube life is over 10 years.

FIGURE 16-9 SHELL-AND-TUBE HEAT EXCHANGER WITH REMOVABLE TUBE BUNDLE AND STRAIGHT TUBES

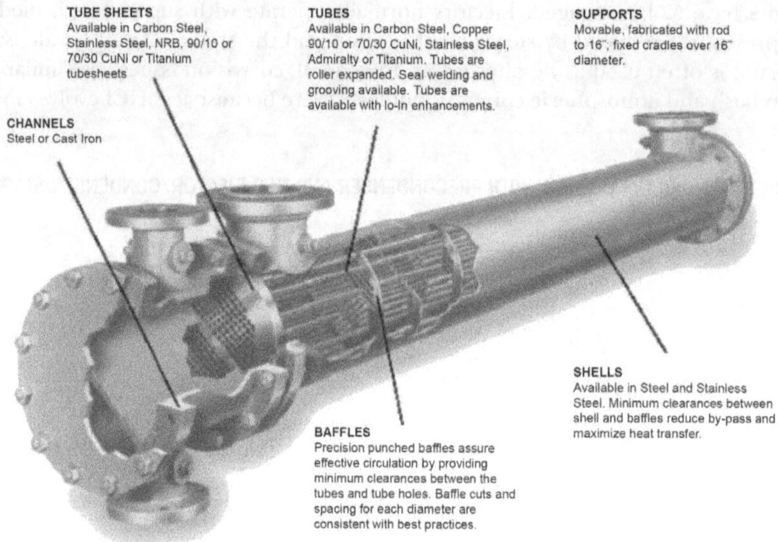

Courtesy of Tubular Exchangers Manufacturers Association (TEMA).

16.7.2 Air-Cooled Heat Exchangers. Air-cooled heat exchangers are an alternative to crude/overhead exchangers and are usually fin-fan types, shown in Figure 16-10. A fin-fan cooler uses one or more "fans" under or over multiple sets of aluminum finned tube bundles to force air to cool the tube side process stream (hence the term "fin-fan cooler"). The horizontal tube bundles are enclosed by a plenum chamber to uniformly distribute the air across all tubes. With air-cooled heat exchangers, overhead heat is wasted to the atmosphere (i.e., not recovered for heating crude or some other process stream). However, like crude/overhead exchangers, air-cooled heat exchangers avoid the corrosion and fouling issues seen with water-cooled heat exchangers in this service. Air-cooled heat exchangers are often designed to operate above the dew point and use aluminum-finned BWG12 carbon steel tubes. Typical tube life is over 10 years. Air-cooled heat exchangers are also used as coolers or condensers in the second stage of two-stage Type 2 and Type 3 overhead systems and then operate at or below the dew point. They can become subject to serious overhead corrosion by dilute HCl acid, unless the pH is controlled between 5 and 6 and a corrosion inhibitor is injected. In wet service, carbon steel tubes last only 3-5 years. Depending on location, air-cooled coolers or condensers can be subject to fouling and underdeposit corrosion on the air side, when air-born dirt, sand, tree pollen, or cooling tower spray collects between the aluminum fins.

FIGURE 16-10 FIN-FAN AIR COOLER WITH MULTIPLE SETS OF FINNED TUBES AND DUAL BOTTOM FANS

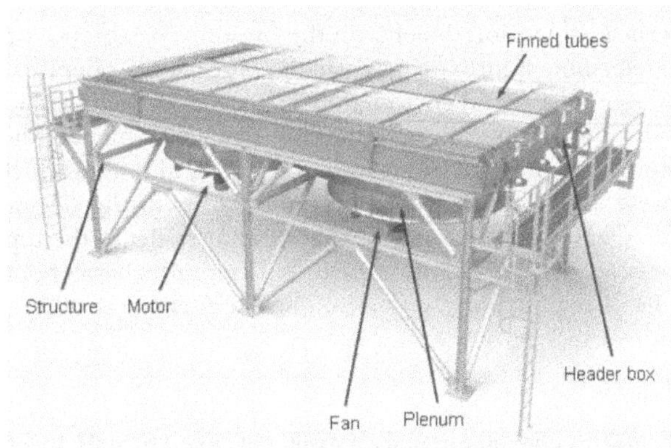

Courtesy of Alfa Laval AB.

16.7.3 Water-Cooled Heat Exchangers. These shell-and-tube exchangers usually follow crude/overhead exchangers or air-cooled coolers or condensers. Provided overhead temperatures are sufficiently low, water-cooled heat exchangers are also used in place of crude/overhead exchangers, air-cooled coolers, or air-cooled condensers. With water-cooled heat exchangers, overhead heat is wasted to the cooling water system (i.e., not recovered for heating crude or some other process stream). Water-cooled coolers or condensers usually carry cooling water on the tube side and have relatively low shell side temperatures to keep the cooling water from boiling in the tubes. To minimize corrosion and fouling, cooling water inlet temperatures should be no higher than 27 °C (80 °F) and outlet temperatures no higher than 52 °C (125 °F)[13]. Consequently, water-cooled coolers or condensers normally operate at or below the dew point on the overhead (shell) side and without pH control and corrosion inhibitor injection can be subject to severe corrosion by dilute HCl acid. Even when used as trim coolers at low overhead temperatures, water-cooled heat exchangers can become subject to cooling water corrosion and fouling problems. In most applications, treated recirculating cooling water is required to prevent excessive corrosion and fouling. As a general rule, carbon steel tubes have a relatively short life, especially when the heat exchanger is undersized and operates at higher-than-normal water temperatures.

Waterside corrosion due to high overhead vapor temperatures often generates excessive amounts of corrosion deposits, as shown in Figure 16-11. Unless an affected heat exchanger can be replaced with an fin-fan cooler, BWG10 or BWG12 carbon steel tubes may have to be upgraded to BWG14 Admiralty brass, BWG16 copper nickel 70-30, or BWG18 titanium tubes. The channel box should be coated with a coal-tar epoxy and protected with Zn/Al sacrificial anodes.

FIGURE 16-11 FOULED CARBON STEEL CHANNEL AFTER 2-YEARS OPERATION AT EXCESSIVE WATER TEMPERATURES

13. J. Gutzeit, "Corrosion and Fouling of Admiralty, Aluminum and Steel Tubes in Open Recirculating Cooling Water Systems," *MP* 4, 7 (1965).

16.8 Dew Point of Overhead Vapors

Since the type and severity of overhead corrosion depend on the presence or absence of water condensate, one should be able to use the dew point to determine approximately which components operate dry and corroded by salt and/or corrosion deposits, and which operate wet and corroded by HCl acid. Moreover, the dew point could then be used to determine the potential for shock condensation in otherwise dry components. Shock condensation in overhead vapor lines can cause droplet impingement at downstream elbows, tees, and inlets of the first bank of overhead heat exchangers. Since dew point represents the steam saturation temperature at the calculated partial pressure, it should be easily obtained from *Steam Tables*[14]. However in contrast to the vapor outlet in the top of the atmospheric and other columns, steam in overhead systems condenses only after the bulk of naphtha has condensed. As shown by the example in Table 16-1, the relative amounts of naphtha and water condensate formed at different overhead temperatures can be calculated from distillation data.

TABLE 16-1 NAPHTHA AND STEAM CONDENSATION IN ATMOSPHERIC OVERHEAD SYSTEM AS A FUNCTION OF TEMPERATURE

Temperature		Amount Condensed	
°C	°F	Naphtha, %	Steam, %
102	**215**	**2**	**0**
100	210	10	0
96	205	18	0
93	200	27	0
90	195	36	0
88	190	45	0
85	185	54	0
83	180	62	0
80	**175**	**70**	**1**
78	173	74	20
78	172	76	30
76	170	80	50
73	165	87	74
71	160	90	80
68	155	91	85
65	150	92	88
60	140	93	93

In this example (in bold type), naphtha starts to condense at 102 °C (215 °F), while steam does not start to condense until the temperature has dropped to 80 °C (175 °F) at which point 70% of the naphtha has already condensed. Therefore under most conditions, the steam partial pressure and dew point become "moving targets" in the sense that the:

- Total pressure decreases continuously as vapors condense.
- Volume, composition, and molecular weight of overhead vapors change continuously as naphtha condenses.
- Volume of steam in overhead vapors changes continuously as steam condenses.
- Total pressure decreases as the pressure drops through various heat exchangers and condensers of an overhead system.

Using *Steam Tables* under these conditions ignores the fact that a major portion of hydrocarbon vapor has already condensed by the time the overhead temperature approaches the dew point and leads to different partial pressures. Since iterative calculations to determine the highest temperature at which liquid water is in equilibrium with steam (i.e., the dew point) are tedious and time-consuming, process software can be used for estimating the dew point. According to the data[15] summarized below in Table 16-2, dew points for six overhead systems predicted by four proprietary software

14. *Properties of Saturated and Superheated Steam*, 3rd ed. of Various Publishers, 1940.
15. Results of round-robin tests reported at STG 34 Information Exchange Meeting at NACE CORROSION 2004.

models agree within 1 °C (3 °F) to 3 °C (6 °F), as they should. In contrast, dew points obtained from *Steam Tables* were 15 °C (28 °F) to 43 °C (77 °F), or an average of 27 °C (48 °F) lower than dew points calculated by the software. However, in light of continuing changes in process conditions on a unit, it remains to be seen how accurate and useful dew points actually need to be for predicting overhead corrosion and fouling.

TABLE 16-2 DEW POINTS PREDICTED BY STEAM TABLES AND FOUR DIFFERENT PROPRIATARY SOFTWARE PROGRAMS

Crude Unit Overhead System	Predicted Dew Point, °C (°F)					
	Software Program					
	No. 1	No. 2	No. 3	No. 4	Spread	Steam Tables
A	115(239)	116(242)	116(242)	115(240)	1(3)	73(163)
B	100(212)	101(214)	101(214)	102(215)	2(3)	84(184)
C	110(229)	108(226)	108(226)	109(227)	2(3)	93(199)
D	99(210)	97(206)	97(206)	99(211)	3(5)	81(177)
E	93(199)	90(193)	90(193)	92(198)	3(6)	51(125)
F	91(197)	89(192)	88(191)	91(197)	3(6)	61(142)

16.9 Velocity Limits for Overhead Lines

In the case of carbon steel piping, the velocity limits listed in Table 15-3 are recommended to reduce or prevent potential impingement attack in vapor lines. Similar velocity limits should be used with carbon steel overhead condenser tubes. No velocity limits apply when vapor lines, inlet lines of overhead condensers and receiver drums, and condenser tubes are fabricated from corrosion-resistant alloys, such as Alloy 400, Alloy 625, or Alloy C-276. Plastic or alloy ferrules can be installed in the inlet of carbon steel condenser tubes to protect the tube ends against impingement attack and increased corrosion due to flow turbulence. To estimate the maximum vapor velocity in piping, the following equation can be used:

$$V_{Sonic} = \sqrt{K \cdot g \cdot R \cdot T} \qquad \text{(Eq.16-1)}$$

where:

V_{Sonic} = sonic velocity, FPS
K = $C_p / C_v \sim 1.4$
g = $32.2 Ft/sec^2$
R = $1{,}544/$ mol. wt $= 1{,}544/67 \sim 23$
T = absolute temperature, $^{O}R \sim 680$

The bulk vapor velocity in overhead lines can be easily estimated to determine if impingement attack is likely to occur at elbows, tees, and inlets of the first set of crude/overhead heat exchangers. Flow areas for common pipe sizes of vapor lines are listed in Table 16-4. It is common to find a wide range of vapor velocities in overhead lines of different plants, ranging from 15m/s (50FPS) to over 90m/s (300FPS), and sometimes even higher.

TABLE 16-3 RECOMMENDED VELOCITY LIMITS FOR CARBON STEEL VAPOR LINES IN OVERHEAD SYSTEMS

Flow Conditions	Bulk Vapor Velocity, m/s (FPS)	Basis
Maximum vapor velocity	0.3 mach	Standard design guidelines in terms of sonic velocity
Dry (0% water)	30 (100)	Standard design guidelines for dry vapor lines
2-Phase flow with 0 - 0.5% water	30 (100)	Estimate based on statistical likelihood of metal/water contact
2-Phase flow with 0.5 - 5.0% water	15 (50)	NACE and API recommendations for typical overhead vapor lines
Presence of injection point	9 (30)	NACE and API recommendations for downstream of injection points
Potential for cooling water leaks	9 (30)	NACE and API recommendations for downstream of water coolers

To calculate vapor velocity in common units:

$$V = \frac{11(T + 460)}{3{,}600\ P \cdot A} \left[(R_{HC}/MW_{HC}) + (R_{Steam}/MW_{Steam})\right] \qquad \text{(Eq.16-2)}$$

where:

- V = flow velocity, FPS
- T = temperature, F
- P = total overhead pressure, psig
- A = flow area of pipe, ft^2
- R_{HC} = hydrocarbon flow rate, lb/h[16]
- R_{Steam} = steam flow rate, lb/h
- MW_{HC} = molecular weight of hydrocarbon vapors (~100)[17]
- MW_{Steam} = molecular weight of steam (=18)

TABLE 16-4 FLOW AREAS FOR STANDARD PIPE DIAMETERS

Nominal Pipe Size	Flow Area, ft^2
6 IPS SCH. 40	0.20
8 IPS SCH. 40	0.35
10 IPS SCH. 40	0.55
12 IPS SCH. 30	0.80
14 IPS SCH. 30	1.0
16 IPS SCH. 30	1.3
18 IPS SCH. 20	1.6
20 IPS SCH. 20	2.0
22 IPS SCH. 20	2.5
24 IPS SCH. 20	3.0

16. R (in lb/hr) / MW = R (in mol/hr).
17. Use actual molecular weight, if known.

16.10 Types of Overhead Corrosion

Depending on temperature and velocity, seven principal types of corrosion are usually identified at different locations of overhead systems. They are listed below in order of decreasing overhead temperature, starting with the column outlet:

- Corrosion under salt deposits.
- Corrosion under corrosion product deposits.
- Initial condensate corrosion.
- Shock condensation corrosion.
- Impingement attack.
- General corrosion.
- Fretting wear/corrosion and midspan collision damage of thin-walled tubes.

16.10.1 Corrosion under Salt Deposits. Hydrogen chloride (HCl) vapor reacts with ammonia (NH_3) vapor and with neutralizers injected intentionally for pH control to form ammonium hydrochloride ($NH_3 \cdot HCl$) and neutralizer hydrochloride ($R\text{-}NH_2 \cdot HCl$) mist. If the chloride concentration is sufficiently high and the temperature sufficiently low, the mist sublimes[18] forming salt deposits in dry areas of the overhead system. The resulting underdeposit corrosion is often misdiagnosed as shock-condensation corrosion, and the neutralizer injection rate is increased which, of course, only increases corrosion. The appearance of corrosion under salt deposits varies. Special features of corrosion under salt deposits are as follows:

- Most salt deposits are sticky and acidic.
- Some deposits are hygroscopic and "dissolve" in naphtha.
- Salt deposits can plug overhead exchangers.
- Initial deposit formation corresponds to the salt point of the neutralizer salt.
- With increasing HCl content of overhead vapors, salt deposits form at increasingly higher temperatures, as shown in Figure 16-12.
- Salt deposits can usually be removed by water washing.

The principal components affected by corrosion under salt deposits include the overhead vapor lines downstream of a neutralizer injection point, heat exchanger tubes, the tube side of air-cooled coolers or condensers, stagnant areas and the bottoms of heat exchanger and receiver shells, and heat exchanger tubes on both sides of baffle holes. Aside from tube failures, localized metal loss and hole through under salt deposits has also been found on the short (inner) radius of elbows at areas of stagnant flow. Typical locations in single-stage Type 1 overhead systems are areas where vapor temperatures are well above the dew point. In multi-stage Type 2 overhead systems, the entire first stage operates dry and salt deposits throughout. In multi-stage Type 3 overhead systems, some components of the first stage where vapor temperatures are above the dew point are likely to be affected.

FIGURE 16-12 NH_3-HCl SALT POINT AS A FUNCTION OF CHLORIDE CONTENT OF OVERHEAD WATER CONDENSATE

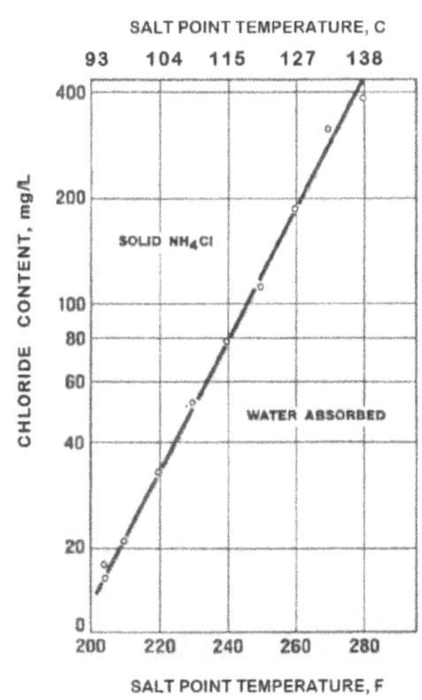

18. Sublime indicates either evaporation of a solid or condensation of a gas without going through a liquid state. Some authors prefer to use the term de-sublime when condensation takes place.

16.10.1.1 Corrosion of Carbon Steel under Salt Deposits. Localized corrosion and voluminous deposits are found on carbon steel tubes when a neat liquid organic neutralizer and a dilute filming-amine corrosion inhibitors were injected in the first stage of a two-stage Type 2 overhead system, as shown in Figure 16-13. Overhead vapor temperatures were 150-177 °C (300-350 °F), the dew point was calculated to have been 110 °C (230 °F), and no water was found in the downstream receiver drum. Positive identification of this type of corrosion is sometimes difficult, because the original salt deposits are often gone by the time an exchanger is cleared by water washing prior to opening and bundle removal. Carbon steel tubes in this heat exchanger failed after only 2 years due to corrosion under salt deposits. Without neutralizer injection, carbon steel tubes should have lasted 10+ years in a dry system such as this.

FIGURE 16-13 CORROSION OF CARBON STEEL TUBES UNDER SALT DEPOSITS IN A CRUDE/OVERHEAD EXCHANGER AT 160 °C (320 °F)

16.10.1.2 Corrosion of Alloy 625 under Salt Deposits. Severe corrosion of Alloy 625 tubes occurred in at least one plant when a combination of liquid organic neutralizer and filmer was injected upstream of a condenser in the first stage of a multistage atmospheric overhead system. While the overhead vapor temperatures were only 110-116 °C (230-240 °F), the dew point was 102 °C (215 °F) and no water condensate formed. No wash water was injected. Carbon steel tubes in this condenser failed repeatedly at 0.5-2yr intervals and were eventually replaced with Alloy 625 tubes. Despite the high alloy content, these tubes too were attacked under salt deposits after only 2yrs, as shown in Figure 16-14. While Alloy 625 is considered fully resistant to hydrochloric acid (HCl) corrosion, it can be prone to pitting attack and crevice corrosion under salt deposits and components should be kept free of deposits by water washing.

FIGURE 16-14 CORROSION OF ALLOY 625 CONDENSER TUBES BY SALT DEPOSITS AT 110 °C (230 °F)

16.10.2 Corrosion under Corrosion Deposits. In wet components, corrosion under corrosion product deposits depends on the pH of the deposits instead of the pH of bulk overhead water that may be present. Often corrosion is likely to continue unabated even after the pH of bulk overhead water was raised by neutralizer injection. Corrosion is often found on tube exterior surfaces under deposits on both sides of baffle holes. Corrosion deposits are also found at the shell exit nozzles for condensate lines. Exposed metal surfaces can corrode even in the absence of water condensate due to formation of a H_2S electrolyte sublayer under the naphtha layer, as shown schematically in Figure 16-15. Therefore, some amount of corrosion products can be expected to deposit in all components and in all stages of overhead systems. The amount of corrosion products depends on corrosion severity and on the pH of water condensate that is present. At pH below 5, corrosion products are dissolved or loosened enough to be picked up by the process stream. At pH above 7.5-8, dissolved iron (Fe^{+2}) combines with hydroxide (OH^-) and disulfide (HS^-) radicals to form more or less stable corrosion products. Constant pH swings not only increase the corrosion rate, but also the amount of corrosion products formed. Wash water injection, mechanical cleaning (hydro jetting), or acid cleaning is usually required for removal.

FIGURE 16-15 CORROSION UNDER NAPHTHA CONDENSATE DUE TO ELECTROLYTE SUBLAYER

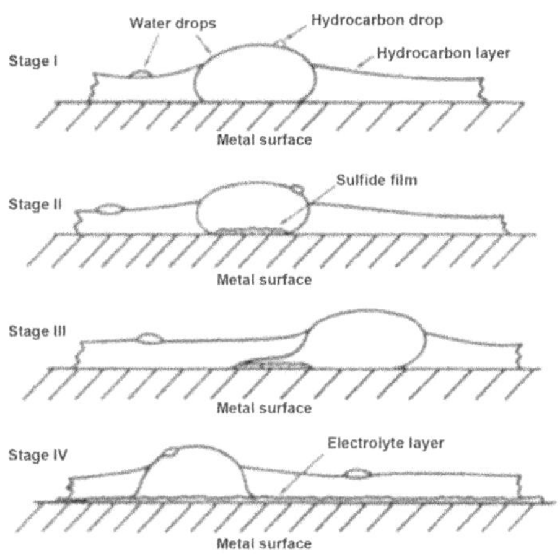

Modified from S. H. Bernhardt et al., "Condensation of Immiscible Mixtures," Heat Transfer Tulsa, *AIChE Symposium Series* 68, 118 (1971?).

Aside from in-situ formation of corrosion deposits, corrosion deposits that originate with upstream corrosion can be deposited in downstream components. For example, corrosion deposits formed in carbon steel overhead lines often end up in downstream heat exchangers, because tube bundles make excellent "collection baskets," as shown in Figure 16-16. Since corrosion deposits tend to absorb HCl, they also cause underdeposit corrosion similar to that caused by salt deposits. The presence of chlorides can be confirmed by chemical analysis of the deposits. Some corrosion deposits can contain water hardness and elemental yellow sulfur (S_n) due to tube leaks in water-cooled condensers.

FIGURE 16-16 LOOSE CHUNKS OF CORROSION DEPOSITS ON TOP ROW OF OVERHEAD HEAT EXCHANGER BUNDLE

The appearance of corrosion damage under corrosion deposits varies depending on location and metallurgy:

- Localized, lake-type corrosion accompanied by voluminous corrosion deposits in the case of carbon steel tubes.
- Localized, shallow pitting and relatively little corrosion in the case of alloy tubes, including titanium.
- Hydriding and cracking of titanium tubes at baffle holes.
- All-over pitting of titanium tubes usually accompanied by hydriding.
- Localized metal loss and hole-through under deposits on the short (inner) radius of elbows due to stagnant flow.
- Localized pitting along the bottom of receiver drums where corrosion deposits tend to collect over time.

For example, 19mm (0.75in.) ODx16BWG titanium grade 2 tubes shown in Figure 16-17 failed due to underdeposit corrosion at carbon steel baffles after only 1.5yrs service. Vapor inlet temperatures were approximately 112-115 °C (235-240 °F). The gray-white deposits were identified to be mixed iron oxides/hydroxides that contained HCl and ammonium chloride ($NH_3{\cdot}HCl$) residues and had a pH of 3.5. The corrosion products probably entered from upstream components. Titanium spacer tubes seen in the top of the photograph corroded at a higher rate than the water-cooled tubes, because of the temperature difference. Several Type 304 stainless steel tie rods inside the spacer tubes failed due to chloride stress-corrosion cracking (CSCC).

The overhead system was routinely treated with a liquid organic neutralizer and a filming-amine corrosion inhibitor. The source of the ammonia was probably stripped sour water used as desalter water. Carbon steel and Admiralty brass tubes had previously lasted 2-3yrs in this water-cooled condenser. After hydro jetting, the titanium tube surfaces were near-original: The inked grade identification could still be clearly read on most tubes. The tubes also showed shallow pitting at tool marks left from the tube rolling and drawing operation. The tube metal was very brittle at baffle locations, resulting in the damage shown in Figure 16-18. Pitting produced atomic hydrogen (H_0) that diffused into the metal and caused it to become hydrided, embrittled, and eventually crack, as shown in Figure 16-19. Continuous water washing would have prevented these failures.

FIGURE 16-17 CORROSION OF TITANIUM TUBES IN OVERHEAD CONDENSER UNDER IRON OXIDE/HYDROXIDE DEPOSITS AT CARBON STEEL BAFFLES

FIGURE 16-18 CRACKED TITANIUM TUBES AT BAFFLE HOLES AFTER HYDRO JETTING

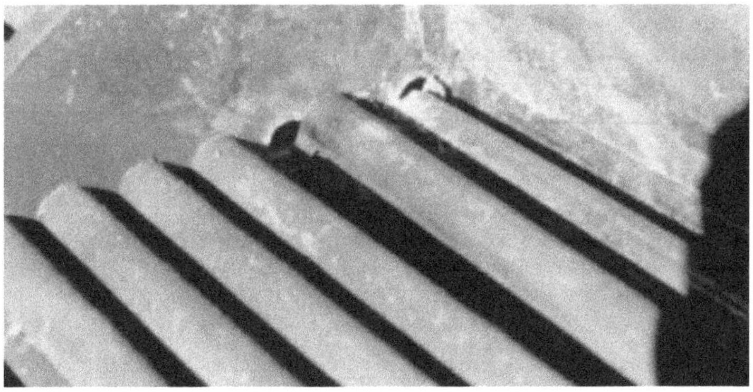

In the shells of heat exchangers and reflux drums, corrosion is usually found in the bottom as a narrow band due to accumulated deposits, as shown in Figure 16-20. The corroded shell area measured 150mm (6in.) wide x 1.5m (60in.) long and contained pits up to approximately 12mm (0.5in.) deep. The damaged area ended at one of the heads of the drum. The remainder of the shell areas had minor pitting to 3mm (1/8in.) deep. After sandblasting, a 10mm (3/8in.) carbon steel patch measuring 400mm (16in.) wide x 2m (80in.) long was welded over the heavily pitted bottom area and the entire bottom half of the vessel was coated with coal tar epoxy.

FIGURE 16-19 PITTING, HYDRIDING, AND CRACKING OF TITANIUM TUBE AT BAFFLE LOCATION IN CONDENSER SHOWN IN FIGURE 16-17 AND FIGURE 16-18

Tube OD surface exposed to overhead vapors

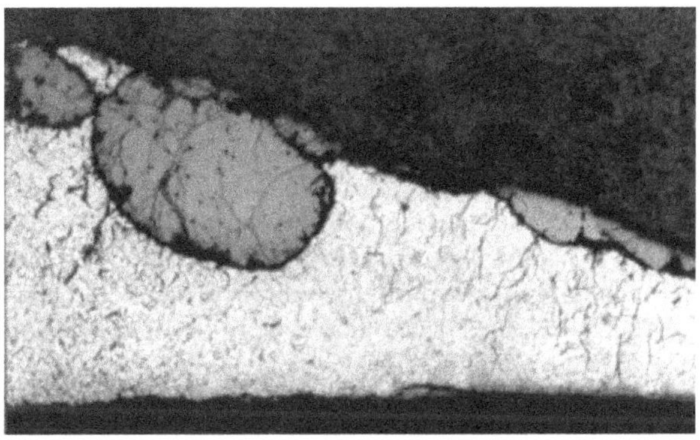

Tube ID surface exposed to cooling water

FIGURE 16-20 PITTING OF CORRODED CARBON STEEL SHELL OF OVERHEAD RECEIVER DRUM UNDER CORROSION PRODUCT DEPOSITS

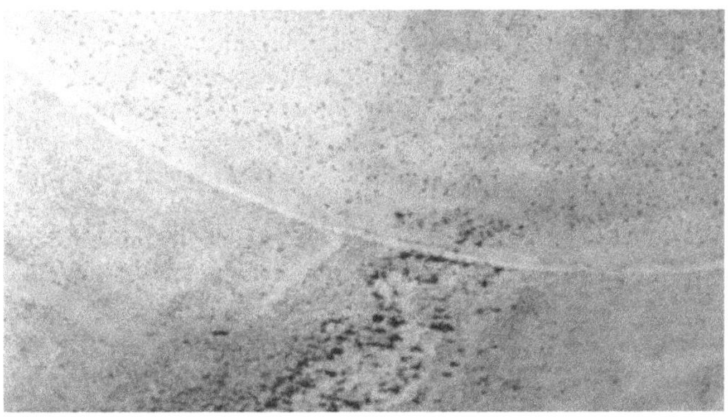

16.10.3 Initial Condensate Corrosion. Water condensation in overhead systems usually takes place after the bulk of hydrocarbons has condensed. Corrosion is confined to the general location where vapor temperatures have dropped to or below the dew point and HCl is absorbed into the condensing water droplets. Initial water condensate is highly acidic, because HCl is much more soluble than ammonia or most liquid organic neutralizers and contacts a relatively small amount of water (droplets). At the point of initial water condensation, the pH is approximately 1.5 units lower than the pH of bulk water condensate sampled off the overhead receiver drum. Initial condensate corrosion can also occur just downstream of water injection points, if the water does not contain a neutralizer. In Type 1 single-stage overhead systems, initial condensate corrosion is seen in components where overhead temperatures first reach or drop below the dew point. In wet stages of Type 2 or Type 3 two-stage overhead systems, initial condensate corrosion is found where steam first condenses. In all cases, corrosion damage is in the form of localized surface roughening, etching, sharp-edged pitting, narrow areas of striations, and shallow grooves in the flow direction. As a rule, no corrosion deposits are found at the corroded areas.

16.10.4 Shock Condensation Corrosion. Shock condensation corrosion is a special case of initial condensate corrosion. Shock condensation occurs when water droplets condense on cold metal surfaces, even though the bulk vapor temperatures are still well above the dew point for the system. The water droplets absorb HCl vapor to form dilute HCl acid droplets that corrode metal surfaces at, or just downstream of, the point of shock condensation. Normally, the acid droplets are immediately swept away by hot vapors, but contact fresh metal surfaces further downstream. The corroding metal surfaces are free of deposits, because deposits would prevent acid droplets from contacting the underlying metal.

In Type 1 single-stage overhead systems, shock condensation can occur at areas where vapor temperatures are still well above the dew point. In Type 2 or Type 3 two-stage overhead systems, shock condensation can occur in the first stage at areas when vapors are still above the dew point. The coldest tubes in heat exchangers probably have to be at least 14 °C (25 °F) degrees below the dew point before shock condensation takes place. Shock condensation also occurs in overhead piping downstream of injection points when a reflux sidestream is used to dilute corrosion inhibitors, or when recycled water condensate is used as wash water. Another location for shock condensation corrosion is the top tray of the atmospheric column when incoming reflux from the overhead receiver drum is unevenly distributed and forms pockets or puddles of corrosive water, as discussed in Chapter 15 of the *Guide*.

Uninsulated segments of overhead piping that operate within 14 °C (25 °F) degrees of the dew point can be subject to shock condensation corrosion during periods of cold weather and high winds. Normally, insulating overhead lines should prevent shock corrosion, but experience at some units indicated that this is not necessarily so. For example, contrary to conventional wisdom, no shock condensation occurred in uninsulated vapor lines of at least one plant in northern Canada with winter temperatures as low as -40 °C (-40 °F). One possible explanation is that high vapor velocities and a large enough temperature difference between hot vapors and cold pipe walls prevents a layer of water condensate to form long enough for corrosion to occur. Another possibility is that a thin layer of ice outside the pipe prevents decreases pipe wall temperature enough to prevent shock condensation. Under normal conditions, uninsulated piping is at least 10 °C (18 °F) degrees colder than adjacent insulated piping and applying outside insulation only makes sense for overhead systems with relatively low vapor temperatures in the 105-120 °C (220-250 °F) range. As with initial condensate corrosion, corrosion damage is in the form of localized surface roughening, etching, sharp-edged pitting, narrow areas of striations, and shallow grooves in the flow direction. As a rule, no corrosion deposits are found at the corroded areas.

16.10.5. Impingement Attack. Impingement attack is a combination of corrosion and erosion that occurs when corrosive acid droplets impinge at high vapor velocities against a solid metal surface and cause metal loss in the form of pitting. Impingement attack removes any protective passive films, inhibitor films, or corrosion deposit layers that were present and eventually forms areas were high flow turbulence further accelerates corrosion and metal loss. Corrosion damage is in the form of localized sharp-edged pitting, narrow areas of striations, and deep grooves in the flow direction. As a rule, no corrosion deposits are found at the corroded areas. Classic examples where attack occurs are the long (outer) radii of elbows and crosses or tee fittings in overhead vapor lines. Other examples are weld beads and backing rings of circumferential pipe welds and the inlets to overhead heat exchangers, coolers, condensers, and receiver drums. When injection points are located too close to an inlet nozzle to a heat exchanger, the underlying baffle and tubes are rapidly corroded. The onset and severity of impingement attack depend on piping geometry, flow regime, and flow velocity. Complex pipe geometries and spray-type flow regimes are most likely to cause impingement attack. Vapor velocities above 15m/s (50FPS), but sometimes as low as 9 m/s (30FPS), can lead to impingement attack. Because of the high flow velocities, filming-amine corrosion inhibitors are ineffective against impingement attack.

Impingement attack can result from shock condensation in hot vapor lines. Injection points for corrosion inhibitors, neutralizers, or wash water can also lead to impingement attack, including hole-through. Injection points should be located at least 10-20 pipe diameters upstream of the nearest elbow, tee, or inlet nozzle to avoid impingement issues. The top tube rows and baffle areas of heat exchangers that carry overhead vapors on the shell side are often subject to impingement corrosion, as shown in Figure 16-21. These tube rows may have to be sacrificed after plugging the tube ends, or a special deflector baffle plate may have to be installed. Impingement attack of the inlet ends of heat exchangers that carry overhead vapors on the tube side is also fairly common. The tube ends can become paper thin at the inlet tubesheet and the ligaments between tubes become severely pitted, as shown in Figure 16-22. The first 75-100mm (3-4in.) tube lengths are usually affected, as shown in Figure 16-23. Unless vapor velocities can be reduced, the usual approach is to fit 150mm (6in.) long alloy ferrules into each tube inlet. In some cases, however, corrosion simply moves downstream of the ferrule due to flow turbulence. Since fin-fan air coolers carry overhead vapors on the tube side, impingement attack can occur on tube end plugs, as shown in Figure 16-24.

FIGURE 16-21 IMPINGEMENT ATTACK OF ADMIRALTY BRASS TUBES AND CARBON STEEL BAFFLES UNDER VAPOR INLET NOZZLE

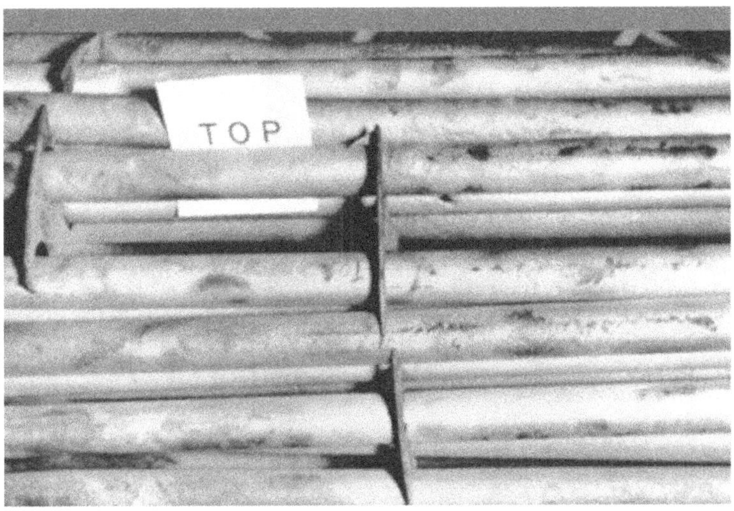

Adapted from API Division of Refining, *Guide for Inspection of Refinery Equipment, Chapter II–Conditions Causing Deterioration or Failures,* 2nd ed. (API, 1973).

FIGURE 16-22 IMPINGEMENT ATTACK OF ALLOY 625 TUBESHEET AT LIGAMENTS AND TUBE ENDS OF CRUDE/OVERHEAD HEAT EXCHANGER

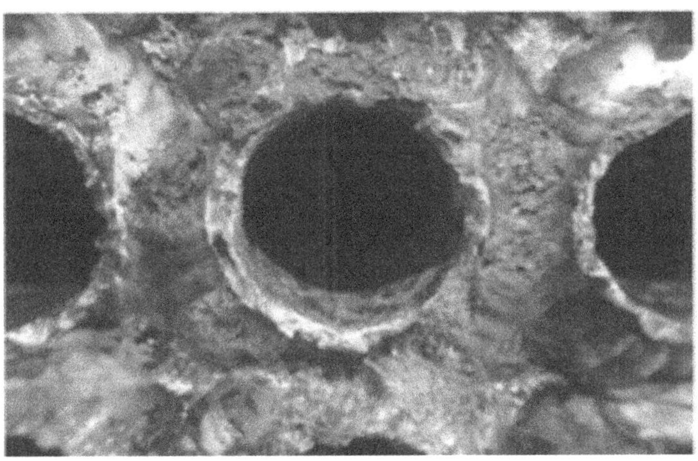

FIGURE 16-23 IMPINGEMENT ATTACK INSIDE CARBON STEEL CONDENSER TUBE AT INLET TUBESHEET

FIGURE 16-24 IMPINGEMENT ATTACK OF CARBON STEEL TUBE PLUG IN FIN-FAN OVERHEAD CONDENSER

16.10.6. General Corrosion. General acid corrosion in overhead systems is caused by dilute HCl acid and occurs primarily where overhead vapor temperatures have dropped to below the dew point after much of the steam has condensed. Dilute hydrochloric acid has a pH between 2 and 4, depending on chloride content. In contrast to most other types of overhead corrosion, general corrosion is fairly predictable. Once a suitable location has been found (for example, a line segment or spool piece), periodic ultrasonic thickness (UT) measurements can be used to predict the life of a component reasonably well. Alternatively, retractable weight-loss coupons or Corrosometer™ probes can be mounted through suitable, valved, access nozzles to monitor corrosion periodically or continuously. General corrosion requires that either a continuous water phase or dispersed water droplets is present throughout the vapor and naphtha phases. General corrosion is always found at some point in single-stage overhead systems. Despite anecdotal evidence to the contrary, it is found only in "wet" stages of two-stage overhead systems where water is continuously withdrawn at a receiver drum. Dilute acid attack of the inside of tubes in a fin-fan cooler for cooling overhead vapors from 110 °C (230 °F) to 102 °C (215 °F) resulted in tubercle-type deposits, as shown in Figure 16-25. These corrosion deposits were formed in situ due to poor pH control and were not carried into the tube from upstream corrosion.

FIGURE 16-25 CORRODED 25MM (1IN.) OD X 12BWG CARBON STEEL TUBE FROM FIN-FAN OVERHEAD CONDENSER

Other locations for acid corrosion include carbon steel lines, tube bundles, and shell areas of overhead heat exchangers. Often carbon steel baffles and tie-rods of titanium bundles are corroded by dilute acid shown in Figure 16-26.

FIGURE 16-26 CORRODED CARBON STEEL BAFFLES AND TIE-RODS OF TITANIUM CONDENSER BUNDLE (BUNDLE REINFORCED WITH COPPER STRAPS)

Acid attack of a carbon steel condenser shell usually occurs at the vapor/liquid interface in the form of grooves formed by condensate running down the shell, as shown in Figure 16-27. The grooves tend to be covered with corrosion products. To extend the life of this shell, two 6mm (1/4in.) thick carbon steel plates measuring 150mm (6in.) wide x 2m (78in.) long were welded on the shell exterior to reinforce the two areas where acid corrosion had occurred.

FIGURE 16-27 CORROSION AT VAPOR/LIQUID INTERFACE DUE TO ACID CONDENSATE

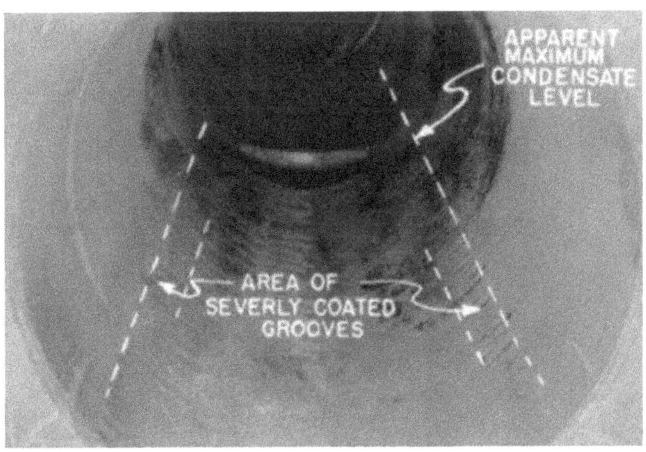

Not surprisingly, severe acid attack is found in overhead reflux pumps even though Cr-Ni cast iron impellers were used, as shown in Figure 16-28. The usual remedy is to use Type 316 stainless steel pumps per API A-8 in this service.

FIGURE 16-28 ACID CORROSION OF CR-NI CAST IRON IMPELLER FROM OVERHEAD REFLUX PUMP OPERATING AT 60 °C (140 °F)

16.10.7 Tube Vibration Damage. All thin-walled heat exchanger tubes can be subject to fretting wear at baffles holes and mechanical impact damage at locations halfway between baffles. Fretting wear occurs at loose-fitting baffle/tube support holes when the tubes rub continuously against the baffle/support while exposing new metal to corrosion. Midspan collision damage is found halfway between baffle plates and is caused by adjacent tubes colliding with each other. Both damage modes are caused by high vapor velocities resulting in excessive tube vibration and considerable acoustic noise and, therefore, are hard to ignore. Both problems occur when a tube bundle is retubed with tubes of reduced wall thickness. Unless vapor velocities can be reduced, the usual approach toward eliminating the vibration problem is to redesign the tube bundle with shorter baffle spacing, install various types of dampening devices between tubes (for example, Teflon™ strips), or both.

As a general rule, the likelihood and severity of tube vibration increases with decreasing tube-wall thickness, increasing baffle spacing, and increasing flow velocity. Baffle spacing of water-cooled heat exchangers is specifically designed for Admiralty brass tubes, which used to be the standard choice for cooling water service. Vibration issues are seen (and heard) primarily when tubes are upgraded with thin-walled 18BWG or 20BWG titanium tubes to keep costs down. Vibration issues rarely occur with carbon steel heat exchangers with 10 or 12BWG tubes for longer tube life. Vibration issues also rarely occur with various alloy tube bundles that have 14BWG tubes. Fretting wear can cut through the wall of titanium tubes in a matter of hours, although most failures occur over a period of several months and sometimes years. Figure 16-29 shows a typical example of through-wall fretting wear. Metallographic examination of the tube metal often reveals the presence of mechanically deformed metal grains in addition to corrosion damage. The presence of deformed metal grains confirms that vibration instead of vapor bypassing through the baffle holes is the root cause of the damage seen.

FIGURE 16-29 FRETTING WEAR OF TITANIUM HEAT EXCHANGER TUBE AT A STAINLESS STEEL BAFFLE HOLE

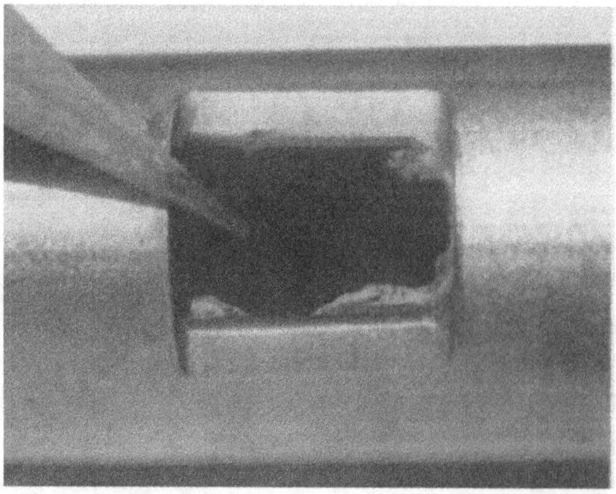

Midspan collision damage, like fretting wear at baffle holes, can damage titanium tubes in a relatively short time (hours or days). The tube exterior surfaces often show a number of 50-100mm (2-4in.) long flats that appear to have been hammered into the metal surface by adjacent tubes, as shown in Figure 16-30 (upper view). This constant hammering eventually leads to metal fatigue and stress cracking in aggressive environments, as shown in Figure 16-30 (lower view).

FIGURE 16-30 MIDSPAN TUBE COLLISION DAMAGE CAUSED BY ADJACENT VIBRATING TUBES

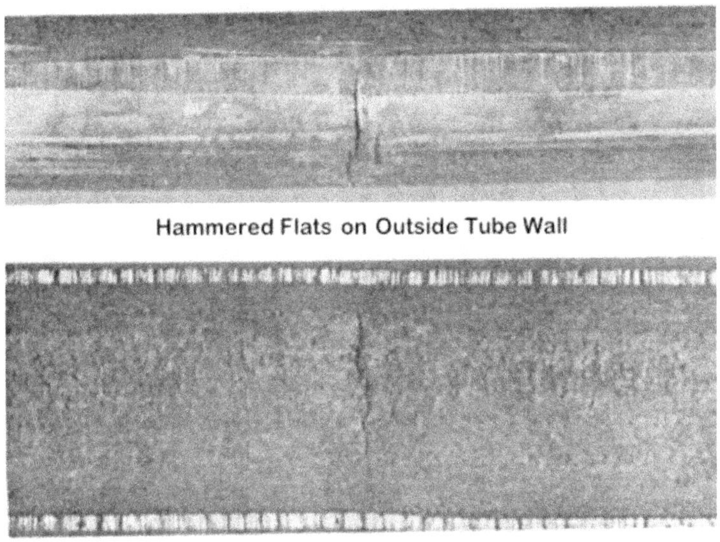

Hammered Flats on Outside Tube Wall

Through-wall Cracking on Inside Tube Wall

16.11 Controlling Overhead Corrosion

Various methods for controlling overhead corrosion are used and discussed in detail in Chapters 17 through 19 of the *Guide,* covering neutralizer injection, corrosion inhibitor injection, and water injection, respectively. While necessary, they tend to address the symptoms and not the causes of overhead corrosion. In general, corrosion in vacuum overhead systems is far less severe than corrosion in atmospheric overhead systems, because a major portion of corrosive or potentially corrosive compounds has evolved in the upstream preheat exchangers and fired heaters and traveled up the preflash and atmospheric columns, respectively. In addition, certain other options should be considered to minimize overhead corrosion, as follows:

16.11.1 Alloy Upgrading. The standard metallurgy may have to be upgraded with more corrosion resistant alloys. In the past, the usual remedies were cladding or weld overlaying corroded areas with Alloy 400. That alloy was found to be unreliable at temperatures above 120 °C (250 °F and selective alloying with Alloy 625 or Alloy C-276 may be required in some overhead systems.

16.11.2 Repairing or Applying Exterior Insulation. Uninsulated pipe and vessel walls can be expected to be significantly colder than adjacent insulated surfaces and may result in highly localized shock condensation. However, applying exterior insulation probably only makes sense for vapor lines at relatively low temperatures in the 105-120 °C (220-250 °F) range, unless needed for personnel safety.

16.11.3 Raising Column Top Temperatures. If possible, the upper column temperatures should be increased at least 10 °C (18 °F) degrees above the dew point to prevent acid condensation. Process flows or other operating conditions may have to be modified to move the dew point and minimize localized acid condensation. Vapor velocities may have to be lowered by enlarging overhead lines.

16.11.4 Keeping Process Additives Out of Dry Stages of Overhead Systems. Ammonia and liquid organic neutralizers should not be injected into the dry stage of a two-stage overhead system to avoid salt deposit formation and avoid sending excess neutralizer and neutralizer hydrochloride salts into the column top with reflux. The same applies to some filming amine corrosion inhibitors.

16.11.5 Reconfiguring Heat Exchanger Banks. Heat exchanger banks may have to be reconfigured to accommodate a different cooling medium to reduce tube side corrosion and resulting leaks. Additional heat exchangers may have to be provided to adjust temperatures or flow velocities. Piping may have to be provided to allow bypassing heat exchangers for regular cleaning or repairs during operation.

16.11.6 Modifying Corrosion Control Measures. Injection points may have to be relocated to avoid impingement problems. A different neutralizer product may have to be chosen to move the salt point of neutralizer hydrochlorides at least 14 °C (25 °F) degrees) below the dew point to minimize salt deposit formation. Corrosion control measures should take into account the different types of corrosion seen in dry and wet stages of two-stage overhead systems, as listed in Table 16-5.

TABLE 16-5 TYPICAL METHODS FOR CONTROLLING CORROSION IN DRY AND WET STAGES OF TWO-STAGE OVERHEAD SYSTEMS

	Dry	Wet
Cause of corrosion		
HCl	no*	yes
H_2S	no	yes
Chloride deposits	yes	no
Sulfide deposits	no	no
Corrosion product deposits	no	yes
Materials of construction		
Vapor line	CS	CS, Alloy 400
Condenser tubes	CS**	Alloy 400, Ti,
Alloy		
Condensate lines	CS	CS
Drums	CS	CS, Alloy 400
External Insulation and steam tracing	yes	no
Injection of		
Neutralizer	no	yes
Corrosion inhibitor	no	yes
Wash water	no***	yes

* Shock condensation on condenser tubes possible
** With shock condensation, use Monel, TI, Hast.C
*** Intermittent water wash permitted, provided that water draw (boot) is added to reflux drum

16.11.7 Controlling Air Leaks. Air leaks into the vacuum section can cause severe corrosion where none was seen before. UT and helium testing can be used to find leaks. The leakage rate can be established by measuring the nitrogen content of off-gas. Flange bolts should be hot-tightened after startup. Flange leaks can be detected by wrapping plastic sheathing around flanges, taping, and pressuring with helium since the process is under vacuum. Valve-stem seals should be greased with silicon grease to ensure proper sealing.

16.11.8 Reducing Water Side Problems in Condensers. Protective organic coatings can be applied in channel boxes and on tubesheets to reduce waterside corrosion. In addition, sacrificial anodes are often installed to prevent pitting at coating holidays. Waterside corrosion and fouling of condenser tube bundles is sometimes caused by water starvation due to elevated position, competition from other water users, or wrong dimensioning of restriction orifices.

16.12 Steam Ejectors

Steam ejectors are made of carbon steel or stainless steel castings. The number of ejector stages depends on the level of vacuum required in accordance with Table 16-6. Depending on the crudes charged and desired end products, vacuum section of crude units and upgraders have 2-4 ejector stages. As a general rule, no wash water is injected into vacuum overhead systems, but neutralizer injection may become necessary. A neutralizer such as ammonia or one of the liquid organic neutralizers can be used to raise the pH of overhead water condensate from a precondenser used upstream of the ejectors. The recommended pH range is 6-7 for most systems. On some units, neutralizer injection is accompanied by filming amine inhibitor injection. However, inhibitor injection may lead to fouling problems, if insufficient hydrocarbon liquids are condensed for removal of excess inhibitor. Neutralizer injection points should be located to avoid impingement problems at downstream elbows and ensure that neutralizer hydrochloride salt deposits are washed away by steam condensate.

16.12.1 Corrosion Problems. Corrosion problems in ejector systems include erosion corrosion of the steam ejectors, corrosion of the precondenser and ejector condensers, and corrosion of the seal drum (hot well or vacuum sump). Related problems are loose pipe threads that provide a leak path for ejector motive steam that, over time, can destroy the threaded joint and perforate the pipe at the connection. Hydrogen blistering due to acid corrosion in the presence of hydrogen sulfide (H_2S) is usually confined to the shell of the seal drum. To prevent blistering, the shell may have to be clad or weld overlaid with Alloy 400.

16.12.2 Fouling Problems. Fouling found in precondensers, ejector intercondensers and ejector aftercondensers is likely to occur on both the shell side (vapor side) and tube side (cooling-water side). Fouling increases the vapor load to the downstream ejectors and can cause those ejectors to break operation. With vacuum overhead vapors normally on the shell side, fouling occurs on the outside surfaces of condenser tubes and is caused by salt deposits that form when acids are neutralized by neutralizer injection. In the presence of steam, these salts cause underdeposit corrosion. In the absence of an oil layer, rusting of wet ejector and condenser components occurs when exposed to air during extended shutdown periods. The rust layer eventually flakes off and fouls downstream components. Sulfur deposits have been found in vent gas compressors. Polysulfides have been blamed for pitting corrosion throughout, but tend to form protective iron polysulfide films in some services. Since cooling water is normally on the tube side, waterside fouling is found on the inside surfaces of condenser tubes and in channel boxes. Waterside fouling is caused primarily at water velocities below 1m/sec (3FPS) and at elevated cooling water temperatures in the summer months when calcium carbonate scale precipitation can be a problem at temperatures above 60 °C (140 °F).

16.12.3 Erosion Problems. Steam ejectors are sized by their motive nozzle and diffuser throat diameters, as shown in Figure 16-31. Manufacturers certify the motive nozzle and diffuser throat diameters and guaranty ejector performance. When routine inspections of these sections indicate a 7% increase in cross sectional areas due to erosion, ejector performance has deteriorated and replacement may become necessary. Ejector erosion corrosion problems are usually caused by wet steam and excessive steam velocities. As a general rule, the onset and severity of erosion by water droplets are proportional to the 4th power of velocity. Therefore, a small change in steam velocity can have a significant effect on the erosion damage seen in ejectors. Other ejector problems can be attributed to poor deaeration and neutralization of motive steam that can lead to gouging corrosion of the nozzle's diverging diffuser section. Steam deaeration and neutralization with morpholine, diethylaminoethanol (DEAE), or cyclohexylamine also help reduce steam condensate corrosion downstream of the ejectors.

TABLE 16-6 NUMBER OF EJECTOR STAGES FOR DIFFERENT VACUUM REQUIREMENTS

Number of Stages	Vacuum Quality	mm Hg	Pa
0	Atmospheric pressure	760	1.0×10^5
2	Low vacuum	760 to 25	1×10^5 to 3×10^3
3	Medium vacuum	25 to 1×10^{-3}	3×10^3 to 1×10^{-1}
4	High vacuum	1×10^{-3} to 1×10^{-9}	1×10^{-1} to 1×10^{-7}

FIGURE 16-31 PRINCIPAL COMPONENTS OF STEAM EJECTOR

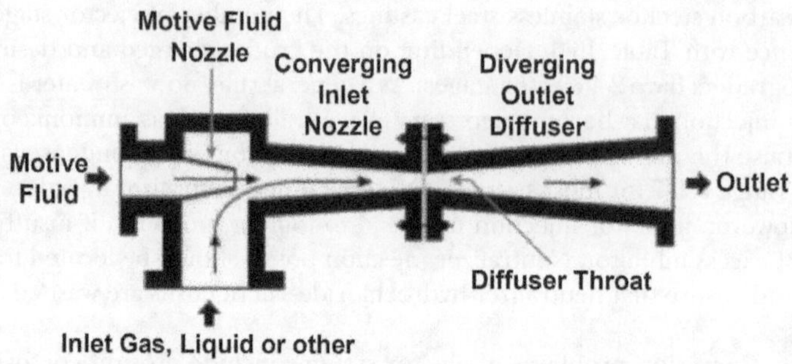

Adapted from J. R. Lines, "Understanding Ejector Systems Necessary to Troubleshoot Vacuum Distillation," *Hydroc. Proc.* 73, 10 (1994).

Ejector erosion can be seen when ejector nozzles and diffusers are visually inspected during shutdown. Etched, striated patterns and lake-type erosion corrosion of the diverging section occurs until the nozzle mouth is actually worn out, as seen in Figure 16-32. The inlet diffuser section of ejectors shows erosion corrosion due to droplet impingement and in severe cases can also show substantial metal loss and corresponding built-up of corrosion products. Since droplet impingement is caused by wet steam, ejector exhaust temperatures can be used to determine if the steam is wet. Steam exhaust temperatures should normally range from 120 °C (250 °F) to 150 °C (300 °F) and substantially lower exhaust temperatures indicate that the steam is too wet. To solve wet-steam problems, all lines leading to the ejector should be well insulated. A steam trap should be installed upstream of the steam connection to each ejector. In some instances, a steam super heater may be required. Depending on the severity of erosion corrosion in the ejectors, standard cast steel ejectors may have to be replaced with cast stainless steel ejectors.

FIGURE 16-32 EROSION CORROSION DAMAGE ON INSIDE OF SMALL STEAM EJECTOR

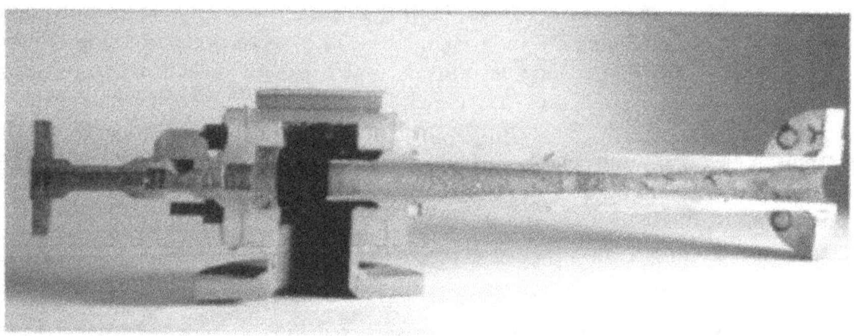

Adapted from J. R. Lines, "Understanding Ejector Systems Necessary to Troubleshoot Vacuum Distillation," *Hydrocarbon Processing* 73, 10 (1994).

16.13 Overhead Inspection Methods

Aside from visual inspection during turnarounds, onstream or offstream inspections methods include ultrasonic thickness (UT) testing and radiography.

16.14 Overhead Repair Methods

Corroded carbon steel components can be repaired or replaced in kind, depending on the severity of corrosion. Corroded components may have to be selectively upgraded or protected by overlaying with Alloy 625 or Alloy C-276. In many applications, replacement or overlaying with Alloy 400 does not provide sufficient long-term improvement. Depending on the severity of corrosion, carbon steel or Admiralty brass tubes in water service may have to be replaced with titanium tubes.

16.15 Corrosion Monitoring

Corrosion monitoring is discussed in some detail in Chapter 5 of the *Guide*. Regarding overhead systems, the following methods and locations are recommended. The preferred monitoring location in the atmospheric section is the vertical rundown piping between the last set of water-cooled coolers and condensers and the receiver drum where water condensate is withdrawn. The preferred monitoring location in the vacuum section upstream of the ejectors is the vertical rundown piping between the precondenser and the receiver drum (seal drum, hot well, or vacuum sump). On units without a precondenser, corrosion is monitored downstream of the ejectors in the vertical rundown piping between the interstage or aftercondensers and the receiver drum. In all cases, either retractable Corrosometer™ probes or weight-loss coupon probes can be used. For long probe life, Corrosometer™ probes with wire elements are preferred over probes with tubular elements. Probes with water-cooled tubular sensing elements and special water-cooled weight-loss specimens are also available to force water shock condensation in overhead vapor lines. For best sensitivity, sensing elements or weight-loss coupons made of carbon steel are used regardless of the overhead metallurgy. A desirable goal is to control carbon steel corrosion to less than 0.1mm/y (4mpy). Alloy elements or coupons are only used to evaluate alternative metallurgies.

16.16 Chapter Summary

The root cause of overhead corrosion in both atmospheric and vacuum sections is hydrogen chloride (HCl) formed by hydrolysis of residual magnesium chloride ($MgCl_2$) and calcium chloride ($CaCl_2$) in desalted crude. Besides hydrochloric acid (HCl), other chemicals can contribute to overhead corrosion such as ammonia (NH_3) and hydrogen sulfide (H_2S). When neutralizers are injected into an overhead system for pH control, corrosive salt and corrosion deposits form and may have to be removed by water injection. Overhead corrosion occurs at vapor temperatures between approximately 93 °C (200 °F) and 175 °C (350 °F), depending on the unit. Affected components include piping and other components of the overhead systems for preflash, atmospheric, and vacuum columns.

In the atmospheric section, corrosion is controlled by limiting the amount of residual salt in desalted crude and reducing the corrosivity of overhead water condensate by neutralizer and corrosion inhibitor injection. In the vacuum section, corrosion is controlled by neutralizer injection into both process and motive steam. Vacuum sections have 2-4 ejector stages and corrosion problems include erosion of the steam ejectors, corrosion of the precondenser and ejector condensers, and corrosion of the seal drum (hot well or vacuum sump). Ejector erosion problems are usually caused by wet steam, excessive steam velocities, poor steam deaeration, and ineffective neutralization and can lead to gouging of the ejector nozzle's diverging diffuser section.

Minimum corrosion of carbon steel occurs in the 5.5-6.5 pH range. Alloy 400 is used for selective alloy cladding or weld overlaying at temperatures below 120 °C (250 °F). Alloy C-276 is used for selective alloy cladding or weld overlaying regardless of temperature. Two types of two-stage (two-drum) condensing systems are used, depending on vapor temperatures and condensation profiles. In dry stages or components, corrosion is caused by various deposits. In wet stages or components, corrosion is caused by dilute hydrochloric acid (HCl).

Most overhead systems use two or three types of heat exchangers, including crude/overhead heat exchangers, air-cooled fin-fan coolers and condensers, and water-cooled coolers and condensers. Velocity limits are recommended for carbon steel vapor lines to control impingement attack. Methods for controlling overhead corrosion are discussed in more detail in Chapters 17, 18, and 19 of the *Guide*.

16.17 Read More

1. S. H. Bernhardt et al., "Condensation of Immiscible Mixtures, Heat Transfer-Tulsa," *AIChE Symposium Series* 68, 118 (1970?).
2. J. A. Biehl and E. A. Schnake, "Corrosion in Crude-Oil Processing – Low pH vs. High pH," *Proc. API* 37, III (1957).
3. J. A. Biehl and E.A. Schnake, "Processing Crude Oil at Low pH," *Proc. API* 39, III (1959).
4. A. S. Couper, "Corrosion Control in Crude Oil Distillation Overhead Condensers," *Proc. API* 44, III (1964).
5. S. G. Ewing, "Electrochemical Studies of the Hydrogen Sulfide Corrosion Mechanism," *Corrosion* 11, 1 (1955).
6. R. T. Foley, "The Role of the Chloride Ion in Iron Corrosion," *Corrosion* 26, 2 (1970).
7. J. Gutzeit, "Corrosion of Steel by Sulfides and Cyanides in Refinery Condensate Water," *MP* 7, December (1968).
8. J. Gutzeit, "Corrosion and Fouling of Admiralty, Aluminum, and Steel Tubes in Open Recirculating Cooling Water Systems," *MP* 4, July (1965).
9. R. J. Hafsten and K. R. Walston, "Use of Neutralizers and Inhibitors to Combat Corrosion in Hydrocarbon Streams," *Proc. API* 35, III (1955).
10. C. B. Hutchison and W. H. Hughes, "Oxidation Reduction Potential as a Control Criterion in Inhibition of Refinery Sulfide Corrosion," *Corrosion* 17, 11 (1961).
11. R. D. Merrick and T. Auerbach, "Crude Unit Overhead Corrosion Control," *MP* 22, 9 (1983).
12. E. Mohsen, "Crude Unit Operation / Corrosion are Linked," *Oil & Gas J.*, April 17 (1978).
13. NACE Standard RP0170 (latest revision), "Protection of Austenitic Stainless Steel from Polythionic Acid Stress Corrosion Cracking During Shutdown of Refinery Equipment," (Houston, TX: NACE).
14. K. Reiser, "The Control of Corrosion in Refinery Distillation Units," *J. Institute of Petroleum* 53, 527 (1967).
15. J. R. Richmond and G.S. Browne, "Automatic pH Control of Crude Column Overhead Streams," *MP* 21, 11 (1982).
16. J. R. Rue, "Cold Tower Aqueous Corrosion: Causes and Prevention," paper no. 211, CORROSION/90 (Houston, TX: NACE, 1990).
17. C. J. Scherrer et al., "Crude Units: Focusing on Corrosion of Initial Condensation Equipment," paper no. 66, CORROSION/81 (Houston, TX: NACE, 1981).
18. D. E. Simon II, *State-of-the-Art Survey of Nonchromate Cooling Water Corrosion Inhibitors*, manual no. 4 (Columbus, OH: Materials Technology Institute, 1980).
19. J. Gutzeit, "Effect of H_2S on Crude Tower Overhead Corrosion," paper presented at State-of-the-Art Session on Effect of Crude Slate on Crude Unit Corrosion (NACE FCW, 1991).
20. J. Gutzeit, "External Heat Loss Causes Internal Vessel Corrosion," *MP* 33, March (1994).
21. J. Gutzeit, "Controlling Crude Unit Overhead Corrosion – Rules of Thumb for Better Crude Desalting," paper no. 07567, CORROSION/2007 (Houston, TX: NACE, 2007).
22. J. Gutzeit, "Controlling Crude Unit Overhead Corrosion by Improved Desalting," *Hydroc. Proc.* 87, 2 (2008).
23. N. P. Hilton et al., "Refinery Reduces Crude Overhead Corrosion through Analysis and Online Control," paper no. 4347, CORROSION/2014 (Houston, TX: NACE, 2014).
24. J. R. Lines, "Understanding Ejector Systems Necessary To Troubleshoot Vacuum Distillation," *Hydroc. Proc.* 73, 10 (1994).
25. A. Veld, "Rapid Failure of a Copper/Nickel Overhead Condenser Bundle," *MP* 52, 4 (2006).
26. NACE Publication 34109, 2009 ed., *Crude Unit Distillation Column Overhead Corrosion* (Houston, TX: NACE, 2009).

17 Overhead Neutralizer Injection

In This Chapter

17.1 First Glance

With high levels of hydrogen chloride (HCl) in vapor lines, heat exchanger tube bundles, and separator drums, severe acid corrosion, fouling, and underdeposit corrosion can be expected in most overhead systems. Unless the HCl level is sufficiently reduced by desalting and caustic addition to the desalted crude, a neutralizer has to be added to elevate the pH of any water condensate. Depending on the extent of corrosion issues and the type of neutralizer used, neutralizer is added to overhead systems by injecting into:

- The vapors coming off the top of a column.
- A slipstream of naphtha taken off the reflux line to the column. This slipstream, in turn, connects to the vapor line coming off the top of a column.
- Wash water that is injected into the vapor line to control fouling by ammonium chloride ($NH_3{\cdot}HCl$), organic neutralizer hydrochlorides ($R\text{-}NH_2{\cdot}HCl$), and various corrosion deposits.

- A dedicated water dilution line that, in turn, connects to the vapor line coming off the top of a column.
- A dedicated steam line that connects to the vapor line coming off the top of a column or elsewhere as needed.

As was discussed in detail in Chapter 16 of the *Guide*, the ultimate goal is to reduce corrosion by delivering neutralizer to all areas where HCl acid corrosion is likely to occur. At the same time, overhead water pH is controlled at 5.5-6.5 to minimize corrosion and fouling throughout the overhead system.

17.2 Chemistry and Properties of Neutralizers

Several types of neutralizers are used to control acid corrosion and maintain pH control: 1. liquid organic neutralizers; 2. ammonia available as concentrated aqueous solutions and compressed gas; and 3. combinations of liquid organic neutralizers, ammonia, or certain bulk alkaline organic chemicals (primarily MEA).

17.2.1 Liquid Organic Neutralizers. Liquid organic neutralizers were initially applied as a substitute for ammonia to avoid fouling and underdeposit corrosion by ammonium chloride ($NH_3{\cdot}HCl$) salt. While liquid organic neutralizers were used for many years against carbon dioxide (CO_2) corrosion by water condensate, their application in overhead systems was limited primarily by cost. Before the implementation of so-called "low-pH operation" in the late 1950s and early 1960s, the pH of overhead water condensate was routinely raised to 7.5-8, and often higher. This meant, of course, that the amount and cost of the required amount of liquid organic neutralizer was considerably higher than it is now, and certainly much higher than ammonia. Upon dilution and injection into overhead vapors, liquid organic neutralizers form droplets that react with hydrogen chloride (HCl) vapor to form a mist of neutralizer hydrochlorides ($R\text{-}NH_2{\cdot}HCl$). Any residual neutralizer droplets disperse in overhead naphtha or dissolve in water condensate to react with entrained or dissolved HCl, as shown in Figure 17-1. For example, the neutralization reaction between a liquid organic neutralizer such as ethylamine ($C_2H_5NH_2$) and HCl can be expressed in most simplistic terms, as follows:

$$C_2H_5NH_2 + H_2O \Rightarrow C_2H_5NH_2{\cdot}HOH \quad \text{(Eq.17-1)}$$
$$C_2H_5NH_2{\cdot}HOH + HCl \Rightarrow C_2H_5NH_2{\cdot}HCl\downarrow + H_2O \quad \text{(Eq.17-2)}$$

Liquid organic neutralizer hydrochlorides can deposit on metal surfaces upstream of the dew point and cause underdeposit corrosion if left in place. All neutralizer hydrochloride salts are soluble in water, but insoluble in naphtha. One notable exception is MOPA-hydrochloride,[1] which is sufficiently hygroscopic to liquefy and dissolve in naphtha, but can still cause underdeposit corrosion when present in corrosion deposits. Most organic neutralizer hydrochloride salt deposits can be removed by water washing. Excess neutralizer droplets tend to wet metal surfaces and neutralize any acid droplets that condense on these surfaces. Excess neutralizer droplets also dissolve in water condensate downstream of the dew point.

Some commercial liquid organic neutralizers are proprietary formulations while others are patented blends and their composition is known in general terms. Relevant properties of typical compounds (used either alone or added to commercial products) and their hydrochloride salts if available are summarized in Table 17-1. As a rule, cost and a ready and reliable supply are often the most important criteria used by suppliers for selecting the components of commercial formulations. Not counting the solvent carrier, most liquid organic neutralizers are:

- Short-chain (C_2 through C_6) amines and polyamines.
- Liquids with boiling point (BP) temperatures in the 115 °C (240 °F) to 175 °C (350 °F) range.
- Soluble in water and some polar solvents.
- Miscible, but not soluble in naphtha and many hydrocarbons.
- Immiscible and insoluble in heavy hydrocarbons, unless specifically formulated for that purpose (so-called hydrocarbon-soluble products).

Most liquid organic neutralizers form hydrochloride salts with melting point (MP) temperatures in the 110 °C (230 °F) to 200 °C (400 °F) range. In contrast, some common liquid organic neutralizers, such as methoxypropylamine (MOPA), form hydrochloride salts with relatively low MP temperatures while others, such as ethylenediamine (EDA), form salts with relatively high MP temperatures.

1. Methoxypropylamine hydrochloride.

FIGURE 17-1 DISTRIBUTION OF LIQUID ORGANIC NEUTRALIZER FROM VAPOR TO WATER PHASE

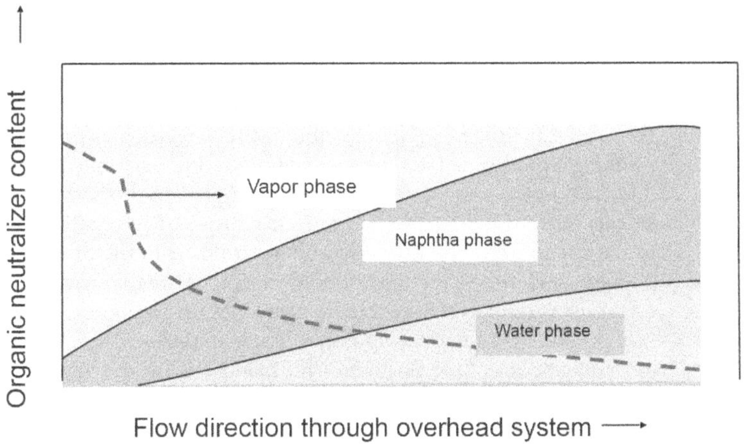

TABLE 17-1 COMPOUNDS IN LIQUID ORGANIC NEUTRALIZER FORMULATIONS

		Neutralizer			Neutralizer·HCl Salt	
Compound	**MW**	**BP, OC (OF)**	**Basicity**	**V/L**	**MP, OC (OF)**	**BP, OC (OF)**
Ammonia (for comparison)	17	35 (-31)	$1.8 \cdot 10^{-5}$	346	340 (645)	520 (968)
Ethylenediamine (EDA)	60	115 (240)	$5.1 \cdot 10^{-4}$	1.0	300 (570) +	-
Methoxypropylamine (MOPA)	89	118 (245)	$1.3 \cdot 10^{-4}$	30.0	100 (212)	300 (570)
Morpholine	87	130 (265)	$2 \cdot 10^{-6}$	15.1	174 (346)	-
Diethylamine	73	56 (133)	-	-	230 (446)	320 (608) +
Tripropylamine	143	155 (312)	-	-	-	-
Monoethanolamine (MEA)	61	170 (339)	$2.8 \cdot 10^{-5}$	1.0	85 (185)	-
Dimethylethanolamine (DMEA)	89	134 (273)	$2.4 \cdot 10^{-5}$	5.4	57 (135)	-
Diethylaminoethanol (DEAE)	117	163 (325)	-	-	131 (266)	-
Dimethylisopropanolamine	101	125 (255)	$5.6 \cdot 10^{-5}$	22.5	110 (230)	-
N-methylmorpholine	101	115 (240)	-	-	204 (400)	270 (518)
Cyclohexylamine	99	135 (274)	-	-	204 (400)	-
Ethoxylateamine blend	-	{135-177 (275-350)}	-	-	{100-103 (212-216)}	210 (410)
Trimethylamine	59	3 (37)	$6.3 \cdot 10^{-5}$	173	283 (541)	-

17.2.2 Ammonia. Ammonia (NH_3) is the oldest and still the most commonly used neutralizer for overhead systems worldwide. Even when not injected intentionally, enough NH_3 can evolve from organic nitrogen compounds in some crudes to make further addition unnecessary. Enough ammonia can also transfer from desalter water to desalted crude when stripped sour water or overhead water condensate is used as desalter water. In some cases, the desalter pH may have to be lowered by acid addition to decrease the amount of NH_3 going overhead. Ammonia is a gas with a BP of -35 °C (-31 °F). Upon addition to an overhead system, ammonia reacts with hydrogen chloride (HCl) vapor to form a mist of ammonium hydrochloride ($NH_3 \cdot HCl$) immediately downstream of the addition point. The neutralization reaction between ammonia and hydrogen chloride (HCl) can be expressed in most simplistic terms, as follows:

$$NH_3 + H_2O \Rightarrow NH_3 \cdot HOH \quad \text{(Eq.17-3)}$$

$$NH_3 \cdot HOH + HCl \Rightarrow NH_3 \cdot HCl\downarrow + H_2O \quad \text{(Eq.17-4)}$$

Depending on the overhead chloride content and temperatures, ammonium chloride mist forms sticky, salt deposits in upper columns and overhead systems upstream of the dew point. The salt sublimes upon heating and has a MP of 340 °C (645 °F) and BP of 520 °C (970 °F). Concentrated ammonium hydrochloride solutions have initially a pH of 5.3, but become increasingly acidic as ammonia evolves. The residual salt deposits have pH values near 1.0, as shown in Figure 17-2; they are completely soluble in water, but insoluble in naphtha. Ammonia when used as a neutralizer can easily elevate the pH of water condensate into the preferred 5.5-6.5 range. If too much ammonia is injected in the presence of hydrogen sulfide (H_2S) in overhead vapors, hydrogen sulfide distributes to the water phase and causes sour-water corrosion, as shown in Figure 17-3. The resulting iron sulfide (FeS) deposits can foul downstream components.

17.2.3 Bulk Alkaline Organic Chemicals. A third category of neutralizers consists of certain bulk organic chemicals, such as monoethanolamine (MEA). They are not considered liquid organic neutralizers in the usual sense and are unsuitable for many applications. However, like ammonia, they are sometimes used in combination with liquid organic neutralizers to lower the overall cost of the neutralizer product. MEA is readily available on the open market and may already be used elsewhere in the plant (for example, in sour-gas treating units). For comparison, MEA too is listed in Table 17-1 along with the properties of its hydrochloride salt. One problem with using MEA is the formation of MEA-hydrochloride deposits that have caused severe fouling and underdeposit corrosion in atmospheric columns and elsewhere.

FIGURE 17-2 DECREASE IN PH OF $NH_3 \cdot HCl$ SOLUTIONS DUE TO NH_3 EVOLUTION

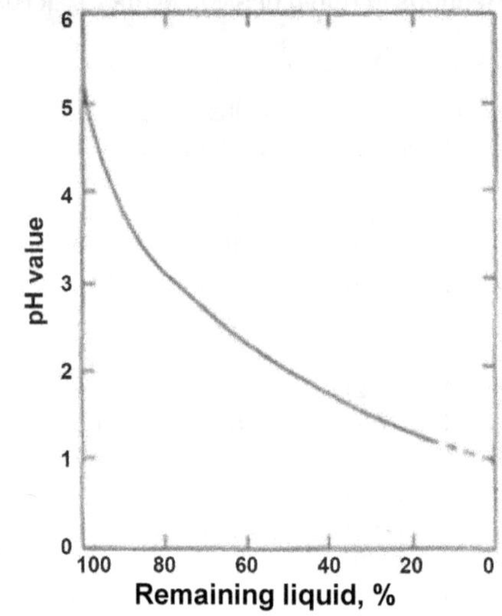

FIGURE 17-3 RELATIONSHIP BETWEEN NH_3 AND H_2S CONTENT OF OVERHEAD WATER CONDENSATE

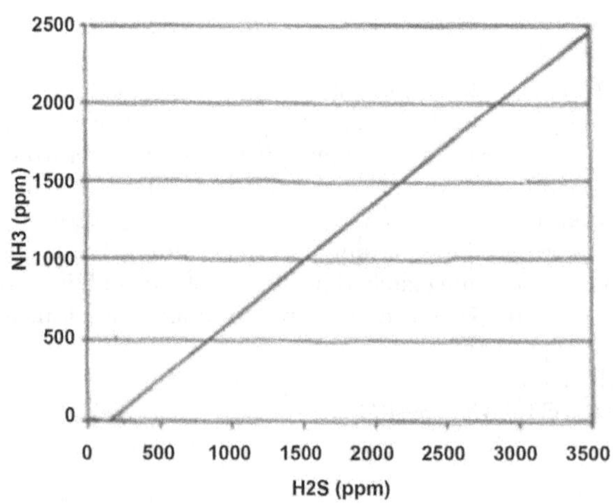

17.3 Organic Neutralizers and Corrosion Inhibitors

Liquid organic neutralizers (neutralizing amines) are often confused with filming-amine corrosion inhibitors (filmers) since, in a manner of speaking, both reduce or inhibit corrosion. However, as far as their structure and function are concerned, there are very pronounced differences:

- Most organic neutralizers are short-chain (C_3 to C_6) amine molecules, liquid at ambient conditions and used to neutralize HCl and other acids in overhead systems. They are properly called liquid organic neutralizers to differentiate them from solid organic neutralizing compounds.
- Most filming-amine corrosion inhibitors are long-chain (C_{12} to C_{24}) amine molecules, solid at ambient conditions and have to be dissolved in a solvent carrier to make a product suitable for injection. Inhibitors form a protective film on metal surfaces that repels corrosive condensate as it forms.

There are other differences in regard to their respective properties:

- Most liquid organic neutralizers are water-soluble. They can be formulated, however, to make a naphtha-base or naphtha-dispersible product for injection into predominantly naphtha streams.
- Most filming-amine corrosion inhibitors are water-insoluble. They can be formulated, however, to make a water-base or water-dispersible product for injection into predominantly water condensate or wash water streams.
- Some filming-amine inhibitors are prepared by reacting long-chain amine molecules with acid molecules to make a water-soluble inhibitor. However, these amine salts tend to be less effective as corrosion inhibitors than the original amine molecules.

17.4 Common Neutralizer Products

Contrary to information provided by process additives suppliers, almost any neutralizer including ammonia can be used to reduce overhead corrosion, provided the hydrogen chloride (HCl) content of overhead vapors is controlled reasonably well by desalting and/or caustic addition to the desalted crude and the neutralizer is correctly diluted and injected. Ammonia (NH_3) is the oldest and still the most commonly used neutralizer for overhead systems worldwide and used at many plants that are unwilling to pay for liquid organic neutralizers.

17.4.1 Liquid Organic Neutralizers. Most liquid organic neutralizers are purchased from the same process additives suppliers that also supply desalting chemicals, filming-amine corrosion inhibitors, and similar process additives. Major suppliers also provide technical advice and support services for their products at no additional cost. Liquid organic neutralizers for specific plant applications can vary significantly in terms of type of amine components, solvent carrier (base), and concentration of active ingredients, as follows:

Amine Components

- Some proprietary neutralizers consist of mixtures of several primary amines.
- Other proprietary neutralizers are reaction products or by-products from other processes and contain amine compounds with a range of molecular weights.
- Depending on availability of amine constituents, some products do not necessarily have the same composition throughout the life of the product.
- Still other neutralizers are intentionally formulated or reformulated to satisfy a broad spectrum of application requirements or to meet specific cost limitations.

Solvent Carrier (Base)

- Most products are water-soluble liquids and supplied without further dilution.
- Some products are both water-soluble and naphtha-miscible liquids and supplied without further dilution.
- Other products are available diluted with water or a suitable hydrocarbon carrier (often aromatic naphtha or kerosene) to meet specific application requirements.

Concentration of Active Ingredients

- Most products are supplied full-strength, but their basicity can differ from product to product depending on the amine constituents.
- Other products are supplied diluted to meet specific strength requirements.

When liquid organic neutralizers are selected for plant applications, certain properties need to be considered that can determine the success or failure of an application. Examples include the BP temperature of the neutralizer, the strength and basicity of the neutralizer, the solubility of the neutralizer in naphtha and light hydrocarbons (propane and butane), the MP temperature of the neutralizer hydrochloride salt, and the solubility of the neutralizer hydrochloride salt in naphtha and water condensate, respectively.

17.4.2 Ammonia. Ammonia is readily available in various forms on the open market. As a general rule, a 26% NH_3 aqueous ammonia solution in bulk containers or compressed, anhydrous ammonia gas in steel cylinders is chosen for plant applications. For ease of handling and reliable control during injection, aqueous ammonia solution in bulk containers is usually preferred. No copper alloys in any form or shape should be used with ammonia since copper alloys are rapidly attacked by ammonia vapors, solutions, and residues. When ammonia is injected for neutralization, Admiralty brass tubes should not be used in tube bundles of overhead condensers. Excess ammonia readily attacks Admiralty brass bundles in the form of localized corrosion and pitting. When Admiralty brass bundles are opened to the atmosphere for cleaning, any residual ammonia in deposits can cause stress-corrosion cracking of the tubes. As a precautionary measure, brass bundles should be immediately washed down with dilute sulfuric acid solution to neutralize residual ammonia and prevent cracking.

17.4.3 Advantages of Liquid Organic Neutralizers over Ammonia. One important advantage is the fact that process additives suppliers can supply and deliver liquid organic neutralizers in portable bulk containers. Most can also provide turnkey injection facilities on skids or pallets along with technical advice and support services for their products. In contrast, it is usually left up to the plant to purchase ammonia on the open market and design and assemble a suitable ammonia injection facility. Compared to ammonia, most liquid organic neutralizers:

- Are easier and safer to ship, store, and handle.
- Are less concentrated and injection rates are easier to adjust and maintain.
- Have BP temperatures in the 115 °C (240 °F) to 175 °C (350 °F) range and usually remain liquid after injection (except diethylamine which vaporizes).
- Readily absorb hydrogen chloride (HCl) vapor upon contact.
- Readily neutralize hydrochloric acid droplets as the acid condenses.
- Provide additional corrosion protection by wetting corroding metal surfaces with a film of alkaline solution that resists further corrosion.
- Contrary to some published information, do not attack Admiralty brass tubes in overhead heat exchangers.

17.4.4 Advantages of Ammonia over Liquid Organic Neutralizers. No process additives supplier likes to use ammonia for overhead pH control, unless pressed to do so by the plant. One issue is that supplier sales personnel are reluctant to recommend a product that they don't sell. Another issue is that in many cases supplier personnel is unfamiliar with the ins and outs of ammonia injection. Still, ammonia is used at many locations worldwide that, for one reason or another, lack convenient access to liquid organic neutralizers. Other locations, including those of some major petroleum companies, have developed special know-how over the years and, without good and convincing reasons are unwilling to change from ammonia to the much more expensive alternative of liquid organic neutralizers. Specifically, NH_3 offers a number of distinct advantages over liquid organic neutralizers:

- NH_3 gas distributes readily and evenly with steam and hydrocarbon vapors throughout various components and piping branches of the typical overhead system.
- Excess NH_3 is neither returned to the column top with naphtha reflux, nor sent into downstream units with naphtha product. Instead, NH_3 stays with overhead vapors and flare gas unless dissolved and removed with water condensate.
- NH_3 is much cheaper than any liquid organic neutralizer on the market. It is also much cheaper than any bulk alkaline organic chemical such as MEA that is sometimes added to liquid organic neutralizers to reduce the overall cost of neutralizer injection. Based on actual comparison tests, neutralization with NH_3 at the time of the tests cost approximately 10% of the cost of using MEA, and 1 to 3% of the cost of using various commercial liquid organic neutralizers.[2]
- NH_3 can be just as effective as liquid organic neutralizers, if sufficiently diluted and correctly injected to compensate for its higher basicity and strength in commercial products. Again, based on actual comparison tests, NH_3 and liquid organic neutralizers have similar characteristics with respect to pH buffering[3] at different hydrogen sulfide (H_2S) concentrations. Both can provide nearly the same amount of pH buffering in the critical pH 5-7 region when adjusted for equal strengths at pH 6.

2. Unpublished data based on 1980 cost figures.
3. PH buffering is defined as Δ pH/Δ neutralizer volume. From a pH control viewpoint, this value should be as small as possible, but from a cost viewpoint this value should be as large as possible.

17.5 Neutralizer Dilution Prior to Injection

For proper injection and dispersion throughout overhead vapor lines, all neutralizers should be diluted with a suitable solvent. As a general rule, liquid organic neutralizers are diluted at least 1:10 to 1:30 with either naphtha or water depending on the product, and injected through a dedicated line and injection nozzle regardless of whether injected into water, naphtha reflux, steam, or directly into an overhead vapor line after batch dilution. Instead of piping NH_3 directly to an overhead vapor line, it is first diluted with steam by metering (pumping) aqueous ammonia solutions from a tank; by boiling off anhydrous ammonia from a heated tank; or by discharging gaseous ammonia from manifolded cylinders. Alkaline bulk chemicals such as MEA are diluted with water. Even though sometimes ignored for lack of suitable equipment or access nozzles, dilution of neutralizers prior to injection improves control over the amount added and helps prevent gunk formation in overhead lines. Dilution also helps reduce corrosion just downstream of the injection point and at elbows, tees, and vessel inlets by undiluted neutralizer injected into hot vapors at high flow velocities. Depending on the neutralizer, plant practice, and available injection equipment and piping, one of the following methods of dilution is normally used:

Liquid Organic Neutralizers
- Batch mixing with naphtha or water in a day tank.
- Metering into wash water or a dedicated dilution water line (except so-called hydrocarbon soluble products).
- Metering into a naphtha reflux slipstream.

Aqueous Ammonia
- Batch mixing with water in a supply tank and metering into a dedicated steam line.

Anhydrous Ammonia
- Boiling off into a dedicated steam line.

Ammonia Gas
- Discharging banks of manifolded, compressed gas cylinders into a dedicated steam line.

Bulk Alkaline Chemicals (MEA)
- Batch mixing with water in a day tank.
- Metering into wash water or a dedicated dilution water line.

If aqueous ammonia is to be distributed by injection into a wash water or dilution water line instead of steam, the supply tank and delivery lines will have to be heated, heat traced, and insulated at plants located in the northern part of the United States or Canada. While concentrated, 26% NH_3 aqueous ammonia solution has a freezing point of -61 °C (-78 °F) and does not require heated storage and handling facilities, dilution to less than 20% NH_3 increases the freezing point to above -36 °C (-34 °F) as listed in Table 17-2. If MEA is to be diluted by injection into wash water or dilution water, the delivery lines may also have to be heat traced and insulated at plants located in the northern part of the United States or Canada. However, hot MEA solutions are corrosive to carbon steel and can cause alkaline stress corrosion cracking (ASCC) at the elevated temperatures of heat-traced components.

TABLE 17-2 FREEZING POINT OF AQUEOUS AMMONIA SOLUTIONS

Composition, wt. %		Freezing Point	
NH_3	NH_4OH	°C	°F
5.0	10.3	-6	21
10.0	20.6	-13	8
15.0	30.9	-23	-10
20.0	41.2	-36	-34
25.0	51.5	-56	-70
26.0	53.5	-61	-79
28.0	57.6	-72	-100

17.6 Organic Neutralizer Injection into Overhead Lines

Liquid organic neutralizers should never be metered into the column feed, top of the column, naphtha reflux line to the column, or top pump-around circuit to avoid plugging column trays with neutralizer hydrochloride deposits. Instead, the neutralizer solutions are injected undiluted into wash water (where used) or diluted into an overhead vapor line close to the column outlet elbow, as shown in Figure 17-4. The injection point should be located far enough downstream from the elbow to prevent neutralizer from dropping back into the column when pressure or flow reversals occur. Liquid organic neutralizers should not be mixed with corrosion inhibitors or injected together through the same injection line or nozzle to prevent gunk formation. Individual control of the two types of chemicals is necessary because the neutralizer addition rate should be varied in response to pH changes, while the inhibitor rate should remain constant at a set minimum rate.

No additives should be injected into a "dry" first stage of a two-stage (2-drum) Type 2 overhead system where no water condenses and collects in the first-stage reflux drum. Without water condensate, neutralizer injection cannot be controlled with pH measurements. Also, unless some means for periodically or continuously removing neutralizer hydrochloride salts are provided, the salts become trapped in a loop formed by the first stage and the column top. This may require adding a water boot at the drum and converting the dry first stage to wet operation by periodic or ongoing water washing. Otherwise, the neutralizer injection point should be moved downstream to the vapor inlet line of the second stage. Any resulting corrosion due to acid shock condensation in the bypassed first stage is likely to be less severe than underdeposit corrosion due to neutralizer hydrochloride deposits. Should excessive corrosion do occur in a dry first stage, the following options can be considered:

- Adjusting the desalter operation and/or caustic injection procedures to reduce the amount of hydrogen chloride (HCl) vapor going overhead.
- Implementing process changes to raise process temperatures and minimize acid shock condensation.
- Upgrading the metallurgy.

FIGURE 17-4 BASIC LAYOUT OF ORGANIC NEUTRALIZER INJECTION SYSTEM FOR SINGLE-STAGE OVERHEAD SYSTEMS

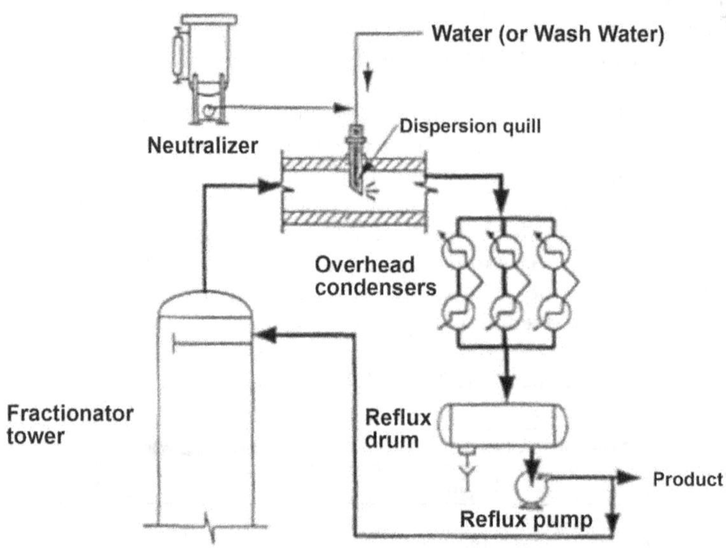

Adapted from W.L. Witzig et al., "The Design and Operation of Process Chemical Treatment Facilities," paper no. 64, CORROSION/81 (Houston, TX: NACE, 1981).

17.6.1 Location of Injection Points. A number of factors should be considered before neutralizer injection is implemented or changed. For best dispersion and maximum effectiveness, injection points should be located as far upstream of problem areas as practical. If possible, neutralizer should be injected upstream of the corrosion inhibitor injection point to ensure optimum inhibitor performance at near-neutral pH values. A major portion of the neutralizer should be injected into the wash water to minimize wash water corrosion. Neutralizers are only effective downstream of an injection point and despite some claims neutralizer does not somehow work itself upstream. Multiple injection points may have to be provided to ensure even neutralizer distribution into various piping branches and protect each of several parallel banks of heat exchangers, coolers, and condensers.

17.6.2 Corrosion Downstream of Injection Point. Undiluted liquid organic neutralizer can cause localized pipe wall corrosion for up to 1.5-3m (5-10ft) downstream of an injection point. Corrosion can occur at different distances from an injection point in horizontal or vertical legs of straight piping. Once identified, susceptible areas may have to be upgraded with alloy material. Droplet impingement corrosion can occur when liquid organic neutralizers are injected at vapor velocities above 9m/s (30FPS). To avoid impingement problems, injection points should be located at least 10 pipe diameters upstream of the nearest elbows, tees, or vessel inlets. Again, once identified, susceptible areas may have to be upgraded with alloy material.

17.6.3 Injection System for Liquid Organic Neutralizers. A typical neutralizer injection system consists of a bulk container for the neutralizer and enough piping to connect the container to an injection nozzle at a dilution water, steam, or reflux slipstream line and then to an injection nozzle at the overhead vapor line. If these lines are not already in place, they will have to be provided. The detailed piping diagram in Figure 17-5 shows many, if not all, of the required components starting at the bulk tank. They include primarily positive block valves, pressure indicators, and strainers at pump inlet. Also needed is a positive-displacement, proportioning, or metering pump with spare, pump bypass, outlet piping, and block valves. Flow meters, sampling points, and positive block valves at inlet nozzles to dilution water, steam, or reflux slipstream lines are a must. Positive block valves should not be overlooked to prevent neutralizer leaking into the system during outages.

FIGURE 17-5 FACILITIES FOR INJECTING LIQUID ORGANIC NEUTRALIZER INTO A REFLUX DILUTION LINE AND OVERHEAD SYSTEM

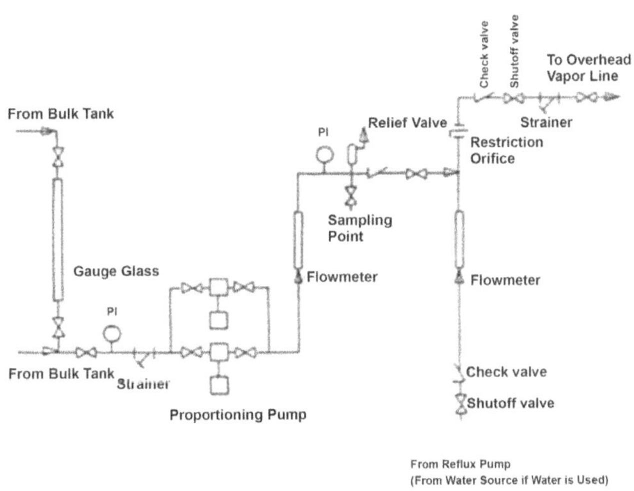

Adapted from J. Gutzeit, "Issues with Injection Facilities for Process Additives Or Wash Water," paper no. 591, CORROSION/96 (Houston, TX: NACE, 1996).

17.6.4 Injection Nozzles. Simple tee connections are inadequate for injecting liquid organic neutralizer and should not be used. In most cases, liquid organic neutralizer is injected through a quill-type or hydraulic spray-type nozzle. Each type has its advantages and disadvantages, which should be kept in mind. Injection quills can provide excellent dispersion and are least likely to cause issues over the long run, if properly designed and oriented in the line. Hydraulic spray nozzles can provide excellent dispersion when first installed, but will eventually clog with particles and assorted debris.

A strainer is usually necessary to delay nozzle clogging, but will have to be periodically cleaned or replaced. Instead of a permanently mounted nozzle, a retractable Alloy C-276 quill or spray nozzle is recommended for insertion to the center of the vapor line. The nozzles are inserted via a retractable holder with packing gland and fully opening gate valve, as shown in Figure 17-6. The quill is secured in place with special brackets and two threaded rods to prevent the quill from being accidentally removed, or pushed through the packing gland by process pressure. "Witness marks" on the pipe indicate the direction and depth of the quill in a vapor or reflux slipstream line. The quill should be piped to a solid inlet line rather than a flexible (corrugated) stainless steel connector to prevent chloride stress corrosion cracking (CSCC). No copper alloy tubing or fittings should ever come in contact with concentrated liquid organic neutralizer, or neutralizer to which ammonia has been added.

FIGURE 17-6 RETRACTABLE INJECTION NOZZLE FOR ONSTREAM REPLACEMENT OF QUILL OR SPRAY NOZZLE

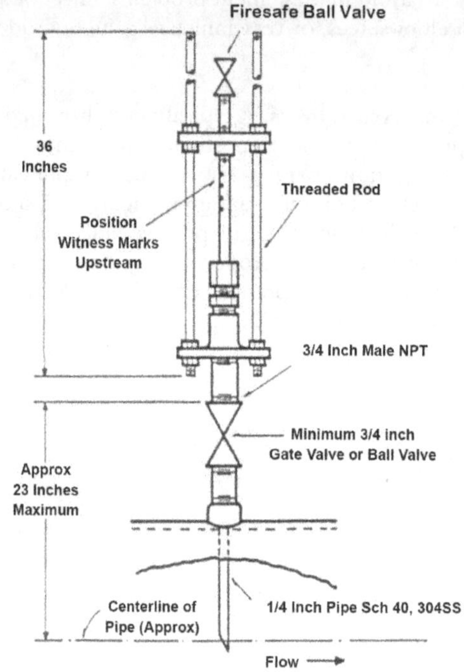

Adapted from W.L. Witzig et al., "The Design and Operation of Process Chemical Treatment Facilities," paper no. 64, CORROSION/81 (Houston, TX: NACE, 1981).

17.7 Ammonia Injection into Overhead Lines

Like liquid organic neutralizers, ammonia should never be added to the column feed, top of the column, naphtha reflux line to the column, or top pump-around circuit to avoid plugging column trays with ammonia hydrochloride (NH_3-HCl) deposits. Instead, ammonia is injected into a dedicated small-bore steam line that connects to the overhead line after the column outlet elbow via a simple tee connection. NH_3 should not be added to corrosion inhibitor solutions or injected through the same injection line or nozzle to prevent gunk formation. Moreover, individual control of both chemicals is necessary because the ammonia addition rate should be varied in response to pH changes while the inhibitor rate should remain constant at a set minimum rate.

As in the case of liquid organic neutralizers, ammonia should also not be added into a "dry" first stage of a two-stage (2-drum) Type 2 overhead system where insufficient water condenses and collects in the reflux drum. Without water condensate, the ammonia addition rate cannot be controlled with pH measurements and the same concerns and remedies apply.

17.7.1 Location of Injection Points. A number of factors need to be considered before ammonia addition is implemented. For best dispersion and maximum effectiveness, the steam connection points should be located as far upstream of problem areas as practical, and upstream of any inhibitor injection point to ensure optimum inhibitor performance. Multiple connections may have to be provided to ensure even distribution of ammonia into various piping branches of the typical overhead system. Multiple connections may also have to be provided to ensure enough ammonia is available to protect each of several parallel banks of heat exchangers or condensers.

17.7.2 Corrosion Downstream of Injection Point. Injected ammonia solutions can cause localized pipe wall corrosion for up to 1.5-3m (5-10ft) downstream of an injection point. Corrosion can occur at different distances from an injection point in horizontal or vertical legs of straight piping. Once identified, susceptible areas should be upgraded with Alloy C-276 material. Droplet impingement corrosion can occur when aqueous ammonia is injected at vapor velocities above 9m/s (30FPS) too close to a downstream elbow.[4] To avoid impingement problems, injection points should be located at least 10 pipe diameters upstream of the nearest elbows, tees, or vessel inlets. Once identified, susceptible areas may also have to be alloy upgraded.

17.7.3 Injection System for Ammonia. Depending on location and amount required, ammonia is pumped out of a bulk container of aqueous ammonia solution using a positive displacement or proportioning pump, boiled off a bulk container of anhydrous ammonia, or obtained as metered gas from a bank of manifolded compressed ammonia cylinders, as shown in Figure 17-7. A typical system also includes enough piping to connect the containers to a dilution steam line and then to the overhead vapor line. If these lines are not in place, they will have to be provided. The diagram shows most required components starting at the ammonia bulk tanks and manifolded gas cylinders. They include primarily control valves, pressure indicators, and steam supply facilities. Also a positive-displacement or proportioning pump with spare and pump bypass and outlet piping and block valves. Positive block valves should not be overlooked to prevent ammonia from leaking into the system during outages. No copper alloy tubing or fittings should ever come in contact with ammonia.

Most plants prefer using aqueous ammonia solution because it is easier and safer to handle than other forms of ammonia. Steam is used instead of water to dilute and disperse ammonia in an overhead vapor line. It is important to keep all ammonia supply tanks and cylinder banks with their delivery systems warm enough with heat tracing and insulation to prevent auto-refrigeration and condensation of saturated ammonia gas exiting the tank. When ammonia condensation takes place, slugs of liquid ammonia accumulate in the delivery line and make steady pH control impossible. Steam injection not only dilutes and disperses the ammonia, but also decreases the ammonia partial pressure sufficiently to prevent condensation. Manifolded ammonia cylinders also need to be kept sufficiently warm; they are often placed in a heated enclosure to prevent auto refrigeration and blockage of the valves.

FIGURE 17-7 BASIC FACILITIES FOR INJECTING AMMONIA INTO A STEAM DILUTION LINE AND OVERHEAD SYSTEM

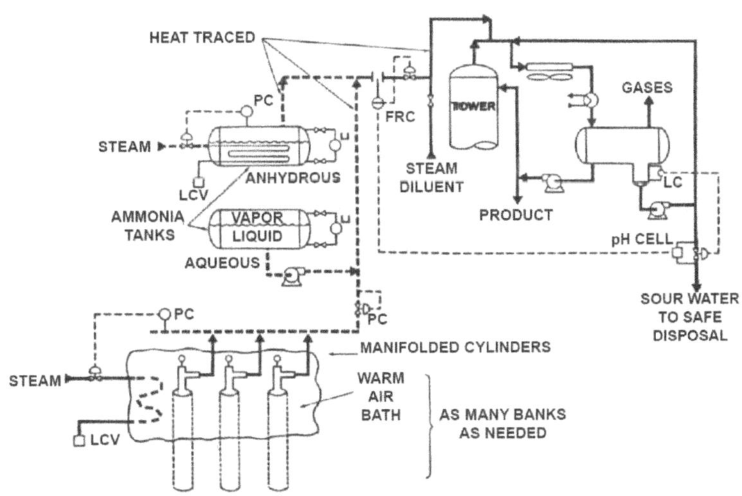

4. In 1988, a leak in a 200mm OD (8IPS) overhead vapor line of a domestic refinery released approximately 9,000Kg (20,000lbs) of C_3 hydrocarbons. A vapor cloud formed and ignited, causing major damage and multiple casualties. Subsequent investigation determined that the leak was caused by ammonia droplets impinging on the outer radius of an elbow located only 1-2 pipe diameters downstream of an ammonia injection point.

17.7.4 Injection Nozzles. Simple tees are adequate for connecting the ammonia lines and the steam feeder to the overhead vapor line. The connection lines should be self-draining. No copper alloy tubing or fittings should ever come in contact with ammonia.

17.8 Neutralizer Injection into Wash Water

When wash water is injected into an overhead system to remove salt or corrosion deposits, hydrogen chloride (HCl) vapor is dissolved in the water and the resulting dilute hydrochloric acid can cause accelerated corrosion downstream of the injection point. Adding a neutralizer to the wash water not only dilutes the neutralizer as necessary, but also prevents this type of corrosion from occurring. Published data have shown that wash water evaporation in overhead lines is delayed because:

- Water droplets are surrounded by a film of naphtha, which interferes with mass transfer.
- High vapor velocities provide insufficient time for wash water to become hot enough to evaporate before it enters the first set of heat exchangers.
- As a result, some process additives suppliers have concluded that adding liquid organic neutralizers to wash water can interfere with neutralizer vaporization and neutralizer vapors are unavailable to neutralize HCl in vapor areas.

However, considering the degree of flow turbulence in overhead vapor lines, most vapor areas are likely to be wetted by a spray of wash water downstream of the injection point. Since most liquid organic neutralizers have high BP temperatures, they are unlikely to become vaporized, even when not added to wash water. In fact, lack of vaporization is one of the reasons for choosing a liquid organic neutralizer over ammonia in the first place. Since wash water injection often leads to accelerated corrosion downstream of the injection point, adding neutralizers to wash water helps prevent this type of corrosion.

17.9 Optimum Treatment Levels

Enough neutralizer should be added to overhead systems to maintain the pH of water condensate from the receiver drum between 5.5 and 6.5. The correct amount is usually determined by trial-and-error by obtaining consecutive pH readings at 30-minute intervals while the unit is operating. The neutralizer rate is adjusted after each reading until the correct pH range is obtained. Alternatively, several water samples are obtained while the unit is briefly operating without any neutralizer addition. These samples are then titrated with diluted neutralizer solution to pH 6 to obtain the correct amount of neutralizer. In either case, enough readings should be obtained to establish a scatter band of pH readings, which allows setting a reasonably accurate injection rate. However, constantly adjusting the neutralizer injection rate can become counter-productive, as the pH is likely to drift through a wide range between readings.

For optimum pH control, continuous pH monitoring and automated neutralizer addition should be considered, as are used, for example, with recirculating cooling water systems. It should be kept in mind that at pH values below 5.0 acid corrosion increases rapidly with increasing hydrogen ion (H^+) concentration. At the same time, iron sulfide deposits are loosened and likely end up in downstream components. Many filming-amine corrosion inhibitors are no longer able to protect against the increased corrosion severity at low H values while others react with HCl to form filming-amine chloride salts that are less protective than the original inhibitor. At pH values above 7.0, neutralizer consumption and cost increase rapidly, leading to deposit lay-down and underdeposit corrosion in various components. Sour-water corrosion of carbon steel and Admiralty brass tubes also increases significantly due to the higher bisulfide (HS^-) content of water condensate. Ammonia stress corrosion cracking of Admiralty brass tubes is likely to occur when ammonia-containing deposits are exposed to air during turnarounds.

17.10 Neutralizer Injection Problems

Many neutralizer-related issues are the result of incorrect applications leading to fouling, corrosion, or contamination problems. Incorrect applications are usually caused by selecting the wrong product in the first place, failing to sufficiently dilute the product, and then injecting the product at the wrong location. Fouling problems are caused by gunk formation when undiluted neutralizer and corrosion inhibitor are injected together into an overhead vapor line or injected through the same line or nozzle for dilution with a reflux slipstream. Impingement corrosion occurs when neutralizer is injected too close to downstream elbows and tees. Corrosion also occurs when the injection rate is slow to response to chloride swings due to desalter upsets or organic chloride contamination. Contamination problems occur when excess neutralizer is used because of poor pH control. Other potential problems are dissolved or entrained neutralizer and neutralizer hydrochloride salts in overhead product streams, dissolved nitrogen compounds in feed streams to downstream process units, various issues in downstream wastewater treatment plants, and residual liquid organic neutralizers in wastewater that cannot be stripped out in the sour water stripper due to foaming.

17.11 Properties of Neutralizer Hydrochloride Salts

All neutralizers except MOPA form crystalline, sticky and, in some cases, hygroscopic salt deposits. Typical salt characteristics include excellent water solubility; slight or no solubility in many hydrocarbon liquids including naphtha; melting (or sublimation points) ranging from 110 °C (230 °F) to 200 °C (400 °F) except for EDA-hydrochloride that melts between 300 °C (570 °F) and 350 °C (660 °F); and boiling (or decomposition) points ranging from 110 °C (230 °F) to 110 °C (700 °F) depending on the product.

17.11.1 Issues with Neutralizer Hydrochloride Salts. All neutralizer hydrochloride salts, including organic neutralizer hydrochlorides, can cause severe and widespread underdeposit corrosion when the salts deposit in dry areas of column trays, overhead vapor lines, overhead heat exchangers, coolers, and condensers, and in separator drums. Neutralizer hydrochloride salts tend to dissolve in water droplets entrained in naphtha, return to the column as reflux, and foul the upper trays or work their way down the column. They cause fouling in downstream operating units (for example, naphtha desulfurizers) due to carryover of neutralizer hydrochlorides with various hydrocarbon streams and end up contaminating product streams.

17.11.2 Effect of Salt Point on Corrosion. Neutralizer hydrochlorides form salt deposits at temperatures below their salt point and above the dew point. Hydrochloride salt deposits are sticky and/or hygroscopic and usually become highly corrosive in the presence of moisture and steam. With heat exchanger tubes, the resultant underdeposit corrosion is often misdiagnosed as acid corrosion due to shock-condensation. Salt point and dew point can be used to determine qualitatively, if salt deposit formation and underdeposit corrosion are likely to occur in a given component (or stage) of an overhead system. As a general rule, underdeposit corrosion is likely to occur in a component at temperature 14 °C (25 °F) degrees below the salt point and 14 °C (25 °F) degrees above the dew point. While this general rule of thumb may help resolve some issues caused by salt deposition, many questions remain regarding the accuracy of calculated salt point and dew point data and the effect of ongoing fluctuations of process conditions on these data. Similar limitations apply as those discussed in the Chapter 16 of the *Guide*.

17.11.3 Salt Point of Ammonium Hydrochloride. The salt point of ammonium hydrochloride (NH_3-HCl) depends on the partial pressures of ammonia and hydrogen chloride in an overhead system. The partial pressures are readily calculated from overhead flow rates or from the concentrations in overhead water condensate. Once the partial pressures have been determined, the dissociation constant K_p is calculated and the salt point obtained from the data in Table 17-4. The extent of deposition at various temperatures can then be determined from Figure 17-8.

TABLE 17-4 NH_3-HCI SALT POINT AS A FUNCTION OF DISSOCIATION CONSTANT

Dissociation Constant K_p, PSIA^2	Salt Point Temperature, °C (°F)	Dissociation Constant K_p, PSIA^2	Salt Point Temperature, °C (°F)
10^{-1}	250 (480)	10^{-7}	102 (215)
10^{-2}	215 (420)	10^{-8}	83 (180)
10^{-3}	193 (380)	10^{-9}	65 (150)
10^{-4}	162 (325)	10^{-10}	51 (125)
10^{-5}	146 (295)	10^{-11}	40 (105)
10^{-6}	121 (250)	10^{-12}	24 (75)

FIGURE 17-8 NH_3-HCl SALT DEPOSITION AS A FUNCTION OF DISSOCIATION CONSTANT AND TEMPERATURE

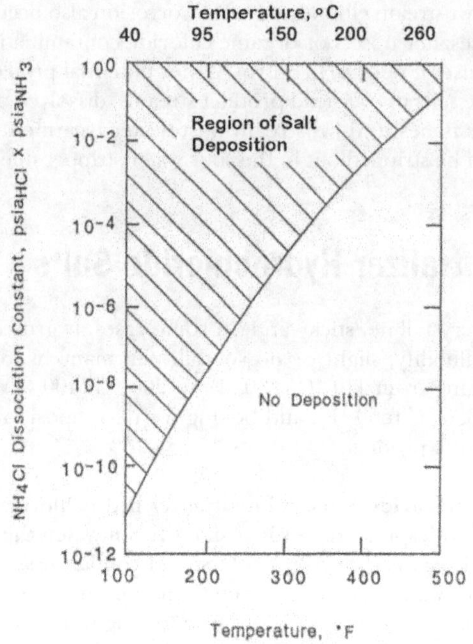

17.11.4 Salt Point of Organic Neutralizer Hydrochlorides. The salt point of liquid organic neutralizer hydrochlorides depends on the partial pressures of the neutralizer and hydrogen chloride (HCl) in the overhead system. As in the case of ammonium hydrochloride, the partial pressures are readily calculated from overhead flow rates or from concentrations in overhead water condensate. Once the partial pressures have been determined, the dissociation constant K_p is calculated. In most applications, however, the salt point can only be determined after the process additives supplier has provided the appropriate K_p versus salt point data for his product(s). Several process additives suppliers use software programs to predict both dew point and the salt point of ammonium hydrochloride and various liquid organic neutralizer hydrochlorides, as shown in Figure 17-9. These programs are used primarily as a sales tool to help convince plant technical staff to switch to another neutralizing chemical. Since they are considered proprietary, it is not possible to evaluate or rate the accuracy or relevancy of the programs.

FIGURE 17-9 SCREEN DATA OF SOFTWARE PROGRAM FOR PREDICTING DEW POINT AND NEUTRALIZER SALT POINT

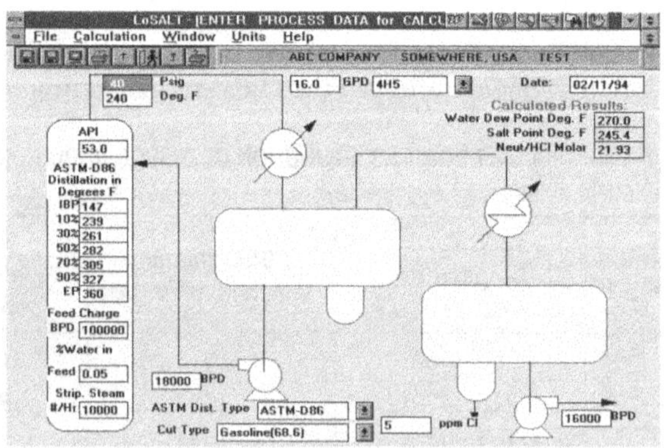

Courtesy of Betz Process Chemicals, The Woodlands, TX.

17.12 Measuring the pH of Water Condensate

In view of the importance of accurate pH control, the pH of overhead condensate should be monitored continuously with the readout available on the control board. Turnkey installations are available, but the following caveats need to be considered when selecting a system for installation:

- Some sort of "salt bridge" is usually necessary to isolate the electrode chamber from overhead condensate water.
- An automatic device for cleaning pH electrodes on a regular schedule may be necessary to prevent the electrodes from being coated when a filming amine corrosion inhibitor is present in naphtha droplets entrained in the water.
- An activated carbon filter may be necessary to remove traces of inhibitor-containing naphtha before the water enters the electrode chamber.
- Saturated calomel reference electrodes are prone to foul, unless the vent opening is isolated from the atmosphere by nitrogen blanketing to keep hydrogen sulfide (H_2S) out.
- Suppliers of industrial pH measuring equipment should be consulted regarding the availability of turnkey equipment suitable for petroleum plant environments.

On many crude units, pH is still measured manually once a shift, using an electrometric pH meter, color discs, or even pH paper. It should go without saying that only an electronic pH meter can be relied upon to provide reasonably accurate and precise readings. This assumes, of course, that the meter is properly maintained and calibrated daily at two pH values with suitable buffer solutions. In the absence of continuous pH monitoring equipment, the extent of pH swings throughout a typical operating shift or day can be established only by periodically drawing water samples every 30 minutes and immediately measuring the pH. Ideally and with good pH control the pH should stay within +/− 0.5 pH units. What often happens, however, is that pH continuously drifts upwards or downwards in a sinusoidal band of +/− 2.0pH units or more, as successive shift operators adjust the neutralizer rate to stay within the recommended pH range. Additional information on measuring pH is provided in Chapter 5 of the *Guide.*

17.13 Services and Reporting

Neutralizer injection, like crude desalting, caustic addition, or inhibitor injection, plays a critical part in overhead corrosion control. Process additives suppliers are usually responsible for seeing that the neutralizer injection system is operating properly. The plant supplies and maintains the storage and mixing tanks for neutralizer solutions; provides and maintains the piping, pumps, and other components to facilitate neutralizer injection; and purchases sufficient bulk neutralizer solution. Some suppliers provide portable, turnkey injection systems mounted on a skid or pallet and obviously are responsible for those. Many suppliers carry spare injection pumps in the trunk of their car to allow speedy replacement in case of failure, but only plant personnel (pipe fitters, electricians) are usually allowed to implement changes. Suppliers are also responsible for trouble-shooting the neutralizer injection system any time upsets occur. More specifically, suppliers should coordinate neutralizer injection with desalter operation, caustic addition to desalted crude, and inhibitor injection. This includes measuring the corrosion of carbon steel in mixed overhead condensate going to the receiver drum, using Corrosometer™ or coupon probes. Most plants also require a written a daily report to the operating engineer or unit superintendent along with an annual report summarizing past overall performance and recommending potential improvement projects.

17.14 Chapter Summary

Neutralizers are injected into overhead systems to control hydrochloric acid corrosion by elevating the pH of water condensate. Commercial liquid organic neutralizers or ammonia are used for that purpose. Each has its advantages and disadvantages. Some commercial liquid organic neutralizers are proprietary formulations, while others are patented and their composition is known in some detail. The active ingredients usually consist of a short-chain nitrogen-base organic compound that is combined with similar compounds to produce a product of desirable characteristics and acceptable cost. The final products often are liquids that need to be further diluted prior to injection into an overhead system. Neutralizers for plant applications vary primarily in terms of solvent base and concentration. Most products are marketed in a water base, but some are available in a hydrocarbon base to make them soluble or at least dispersible in naphtha and similar hydrocarbons.

Commercial neutralizer solutions should never be injected neat as received and should be diluted prior to injection to ensure proper dispersion. In most cases, a neutralizer is injected continuously through an injection quill or hydraulic spray nozzle into water, steam, or a slipstream of naphtha reflux for dilution prior to injection into an overhead vapor line. Ammonia gas is injected into steam for dilution. Severe corrosion due to droplet impingement can occur when neutralizer solution is injected at vapor velocities above approximately 15m/sec (50FPS) which is often the case. To avoid this problem, injection points should be located at least 10 pipe diameters upstream of the nearest piping elbows, tees, or vessel inlets. Normally, neutralizer solutions are injected continuously to elevate the pH of water condensate to 5.5 to 6.5 at the first receiver drum following the injection point. Vapor line corrosion by neutralizer droplets can also occur just downstream of the injection point. Other potential neutralizer-related issues include fouling, gunk formation due to reactions between neutralizers and corrosion inhibitors, and contamination of overhead product streams by some neutralizers.

17.15 Read More

1. R. J. Hafsten and K. R. Walston, "Use of Neutralizers and Inhibitors to Combat Corrosion in Hydrocarbon Streams," *Proc. API* 35, III (1955).
2. R. D. Merrick and T. Auerbach, "Crude Unit Overhead Corrosion Control," *MP* 22, 9 (1983).
3. K. Reiser, "The Control of Corrosion in Refinery Distillation Units," *J. Institute of Petroleum* 53, 527 (1967),
4. J. R. Richmond and G. S. Browne, "Automatic pH Control of Crude Column Overhead Streams," *MP* 21, 11 (1982).
5. J. R. Rue, "Cold Tower Aqueous Corrosion: Causes and Prevention," paper no. 211, CORROSION/90 (Houston, TX: NACE, 1990).
6. NACE International Publication 34101, "Refinery Injection and Process Mixing Points, Task Group 174 on Refinery Additives Injection Facilities," (Houston, TX: NACE, 2001).
7. NACE International Publication SP0114-2014, "Standard Practice: Refinery Injection and Process Mix Points, Item No. 21175," (Houston, TX: NACE).
8. J. Gutzeit, "Refinery Process Additives A User's Viewpoint," *MP* 29, February (1990).
9. J. Gutzeit, "Problems with Injection Facilities for Process Additives and Wash Water," paper no. 591, CORROSION/96 (Houston, TX: NACE, 1996).
10. API Division of Refining, Inspection, Repair, Alteration, and Rerating of In-service Piping Systems, publication API 570 (Washington, DC: API, 1993).
11. V. K. Braden et al., "Crude Unit Overhead Corrosion Control," paper no. 585, CORROSION/98 (Houston, TX: NACE, 1998).
12. W. C. Fort, III et al., "Process Design of Injection Systems," paper no. 587, CORROSION/96 (Houston, TX: NACE, 1996).
13. A. C. Gysbers and W.J. Neill, Jr., "Results of Problematic Refinery Plant Injection Point Survey," paper no. 585, CORROSION/96 (Houston, TX: NACE, 1996).
14. S. E. Lehrer and J.G. Edmondson, "Advances in Neutralizing Amine Technology," paper no. 514, CORROSION/94 (Houston, TX: NACE, 1994).
15. P. R. Petersen, "Predicting the Effect of Neutralizing Amines on the pH Profile in Crude Unit Overhead Condensers," paper no. 166, CORROSION/88 (Houston, TX: NACE, 1988).
16. W. L. Witzig et al., "The Design and Operation of Process Chemical Treatment Facilities," paper no. 64, CORROSION/81 (Houston, TX: NACE, 1981).
17. P. R. Petersen et al., "Choosing a Neutralizing Amine Corrosion Inhibitor," ePTQ, Summer 2004.

18 Overhead Inhibitor Injection

In This Chapter

18.1 First Glance

The injection of filming-amine corrosion inhibitors (filmers[1]) into overhead systems is but one of several measures developed by the petroleum industry over many years to control overhead corrosion. Corrosion inhibitors control corrosion by maintaining a continuous barrier "film" between metal surfaces and corrosive condensate. With good desalting, caustic addition to desalted crude, chloride levels below 15PPM in overhead condensate, and pH control, corrosion inhibitor injection may well be unnecessary. In fact, some overhead systems in plants outside the United States seem to operate quite satisfactorily without inhibitor injection, but in general these are the exception to the norm. In practice, corrosion inhibitor injection should be considered insurance against corrosion problems, because despite all good intentions unforeseen desalter upsets can and do occur and neutralizer injection pumps can and do fail.[2]

1. The term "filmer" is commonly used in the industry, but by-and-large avoided in this Guide.
2. As one maintenance engineer famously put it: Half of the time the neutralizer pumps don't work, half of the time the inhibitor pumps don't work, and a quarter of the time both don't work.

18.2 Corrosion Inhibitor Chemistry and Properties

Like neutralizers, most commercial corrosion inhibitors are proprietary formulations while others are patented and their composition is known in some detail. Sometimes referred to by plant operators as "mouse milk," the active ingredient usually consists of an organic nitrogen-base compound that is reacted with a long-chain fatty acid. Typical nitrogen-base compounds used in corrosion inhibitors include imidazolines, primary amines, diamines, quaternized amines, alkyl pyridine, pyrimidine, and various dimer amido amines. Typical long-chain fatty acids (C_{18+}) in inhibitors include dimer/trimer acids, naphthenic acids, petroleum oxidates, and various other organic acids and esters.

The resultant corrosion inhibitors are long-chain (C_{12} to C_{24}) amines and polyamines that are solids at ambient temperatures. A suitable hydrocarbon carrier such as heavy aromatic naphtha, kerosene, or cycle oil is required to make a product that can be placed in drums, shipped, stored, and dispensed by pumping. The final products are often viscous liquids with boiling points (BP) in the 115 °C (240 °F) to 175 °C (350 F) range and decomposition temperatures in the 260 °C (500 °F) to 370 °C (700 °F) range. Some proprietary corrosion inhibitors consist of mixtures of more than one compound. All inhibitors need to be further diluted prior to injection. Upon dilution and injection into an overhead system, corrosion inhibitors form a mist of inhibitor droplets that, upon condensation, develop a more or less continuous corrosion barrier between the metal surface and corrosive condensate. The nitrogen end of each inhibitor molecule attaches itself to the metal surface through a chemisorption mechanism, while the hydrocarbon tail displaces the corrosive electrolyte sublayer normally present on corroding surfaces, as shown schematically in Figure 18-1. The resultant inhibitor barrier film is only 1 to 2 molecules thick.

FIGURE 18-1 DISPLACEMENT OF CORROSIVE ELECTROLYTE SUBLAYER BY CORROSION INHIBITOR FILM

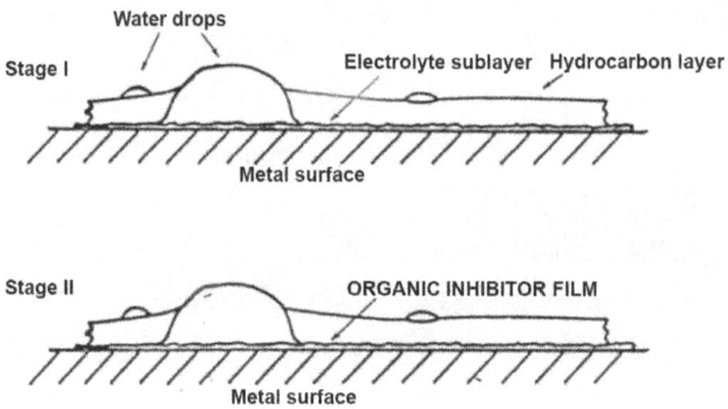

The following general characteristics of filming-amine corrosion inhibitors should be kept in mind:

- Tightly adhering corrosion deposits become part of and actually improve the barrier film.
- Loose corrosion deposits are removed from the metal surface due to the inhibitor's detergency properties.
- The barrier film is easily removed by flow turbulence and must be continuously renewed.
- Filming amine corrosion inhibitors are not volatile. When injected into a dry first-stage of a two-stage (2-drum) Type 2 overhead system, excess inhibitor can only leave dissolved in reflux and consequently ends up in the top of the column.
- The length and structure of the inhibitor molecule determines the effectiveness of the barrier film.
- Certain corrosive species can interfere with barrier film formation.

18.3 Comparison between Corrosion Inhibitors and Organic Neutralizers

Corrosion inhibitors are often confused with liquid organic neutralizers (neutralizing amines) and vice versa. In a manner of speaking, both types of chemicals "inhibit" corrosion. The differences are discussed in greater detail in Chapter 17 of the *Guide* and will not be repeated here.

18.4 Commercial Inhibitors

Most corrosion inhibitors are purchased from the same process additives suppliers that also supply desalting chemicals, liquid organic neutralizers, and various other process additives. Major suppliers also provide technical advice and support services for their products. As a rule, commercial products are formulated for specific metal, process, and temperature combinations and should not be used interchangeably without prior testing. In practice, many inhibitors for plant applications vary primarily in terms of solvent base (or carrier) and concentration of active ingredients.

18.4.1 Inhibitor Solvent Base. Most corrosion inhibitors for petroleum applications are dissolved in a hydrocarbon base. Some products are dissolved in a hydrocarbon base, but contain surfactants to make them water dispersible. Other products are available in special solvent combinations to lower the application cost. Products based on amine salts are dissolved in a water base for use in primarily aqueous streams. In theory, it should not matter what type of solvent base is used when a product is injected into an overhead vapor line. To ensure, however, that excess inhibitor leaves with one of the overhead condensate streams, the choice between a hydrocarbon base and water base (or water dispersible) product depends on the relative amounts of naphtha and water condensate available downstream of the injection point. As a general rule of thumb, hydrocarbon-base products should be used when overhead condensate consists of more than 75% naphtha as is typical in atmospheric overhead systems. Water-base or water-dispersible products are used when overhead condensate consists of more than 75% water as is typical in vacuum overhead systems.

18.4.2 Concentration of Active Ingredients. Most "standard products" contain approximately 18-22% active ingredients in a hydrocarbon or water base, respectively. The following exceptions will be found to apply:

- Concentrated (1.5x or 2x) versions containing up to 40% active ingredients are usually available and are offered primarily to reduce shipping, storage, and dilution costs.
- Products containing more than 40% active ingredients are usually too viscous to ship and difficult to dilute and consequently are not available.
- In the highly competitive corrosion inhibitor market, comparative product cost per kilogram or pound is often a good indicator of the amount of active ingredients in competitive products. To avoid comparing "apples and oranges," concentrations and injection rates of competitive products should be quoted in weight or volume percentages, but not both.

18.4.3 Other Considerations. Some corrosion inhibitor suppliers have recommended replacing hydrocarbon-base products with water-base products for injection with wash water. This concept was based on the premise that since corrosion is caused by aqueous condensate, only wash water can get the inhibitor to active corrosion sites. It ignored, however, the facts that: 1. Overhead corrosion can proceed without aqueous condensate; 2. Water condensation and resultant acid corrosion are likely to shift from one area to the next; and 3. Insufficient wash water is usually injected to distribute a water-base inhibitor throughout an entire overhead system.

Periodically, combination corrosion inhibitor/organic neutralizer products are offered for applications where only a single dilution line and injection nozzle is available. These products prove to be useless for the simple reason that corrosion inhibitors require a minimum or "maintenance" injection rate to remain effective, while organic neutralizers need to be injected at varying rates to maintain pH control.

18.5 Inhibitor Dilution Prior to Injection

Commercial corrosion inhibitor solutions should never be injected neat[3] into an overhead system. Neat corrosion inhibitor solutions should be diluted prior to injection to ensure proper dispersion and avoid fouling problems. Depending on the solvent base of the corrosion inhibitor, the following apply when corrosion inhibitors are to be diluted prior to injection:

3. The term "neat" refers to single liquids, undiluted or mixed with anything else.

Hydrocarbon-Base Products

- Are soluble in most hydrocarbons, but may require the addition of aromatic components to dissolve completely.
- May or may not be soluble in light hydrocarbons (propane, butane).
- Are insoluble in water.

Water-Dispersible Hydrocarbon Base Products

- Are dispersible, but not soluble in water.
- Are soluble in most hydrocarbons, but may require addition of aromatic components to dissolve completely.

Water-Base Products

- Are soluble in water.
- Are insoluble in most hydrocarbons.

Standard commercial products should be diluted at least 1:10 and preferably 1:30 with a suitable solvent. Dilution improves dispersion and avoids the corrosion seen when neat inhibitor droplets impinge on elbows and tees of overhead lines at temperatures above approximately 120 °C (250 °F). Neat inhibitor can also form corrosive deposits in hot vapor lines and in crude versus overhead heat exchangers. For example, neat corrosion inhibitor deposits have corroded through overhead lines to the point of leakage in a matter of 1-2 years, sometimes repeatedly until the problem was recognized. As shown in Figure 18-2, deposits of neat corrosion inhibitor underneath the shell inlet-nozzle area of crude versus overhead heat exchangers not only interfered with proper vapor flow, but also penetrated some tubes to the point where crude ended up in the reflux drum.

FIGURE 18-2 CORROSION INHIBITOR FOULING OF CRUDE/OVERHEAD HEAT EXCHANGER BUNDLE UNDER SHELL INLET NOZZLE

Corrosion inhibitors are usually diluted using one of the following methods, depending on the type of product involved:

Standard Versions of Hydrocarbon-Base Products

- Batch mixing in day tank with a suitable hydrocarbon (heavy aromatic naphtha, kerosene, cycle oil) prior to injection into a vapor line.
- Continuous injection into a slipstream of naphtha reflux prior to injection into a vapor line.

Dilution of Water-Base or Water-Dispersible Products

- Batch mixing in day tank with water prior to injection into a vapor line.
- Continuous injection into dilution water or wash water prior to injection into a vapor line.

Dilution of Concentrated Versions of Standard Products

- Concentrated versions of standard commercial products are highly viscous liquids and require proportionally greater dilution ratios than standard products.

As a general rule, corrosion inhibitors should not be premixed or injected together with liquid organic neutralizers through the same line or nozzle to avoid possible gunk formation. Moreover, individual control of both chemicals is necessary because the inhibitor rate should remain constant at a set minimum rate while the neutralizer rate should be varied in response to pH. Corrosion inhibitors should never be diluted or injected with wash water, unless a water base or water dispersible product is used.

18.6 Inhibitor Injection into Overhead Lines

Corrosion inhibitors should never be injected into the column feed, top of the column, naphtha reflux line to the column, or top pump-around circuit to avoid plugging column trays with inhibitor deposits. Corrosion inhibitors are first diluted by injection into a reflux slipstream that, in turn, is injected into an overhead vapor line close to the column outlet elbow, as shown in Figure 18-3. The injection point should be located far enough downstream from the elbow to prevent inhibitor droplets from falling into the column when pressure or flow reversals occur. Inhibitor should never be injected into a "dry" first stage of a two-stage (2-drum) Type 2 overhead system because excess inhibitor becomes trapped in a loop formed by the first stage and the column top. Since inhibitors are not volatile, the only way inhibitor can leave is with reflux returning to the column top or with product to downstream units. This means the inhibitor injection point should be moved downstream to the vapor inlet line to the second stage. Corrosion due to acid shock condensation in the bypassed first stage is less of a problem than corrosion under concentrated inhibitor deposits. Should excessive corrosion do occur in a dry first stage, the following options can be considered:

- Adjusting the desalter operation and/or caustic injection procedures to reduce the amount of hydrogen chloride (HCl) vapor going overhead.
- Implementing process changes to raise process temperatures and minimize shock condensation.
- Alloy upgrading the metallurgy.

FIGURE 18-3 BASIC LAYOUT OF CORROSION INHIBITOR INJECTION SYSTEM FOR SINGLE-STAGE OVERHEAD SYSTEMS

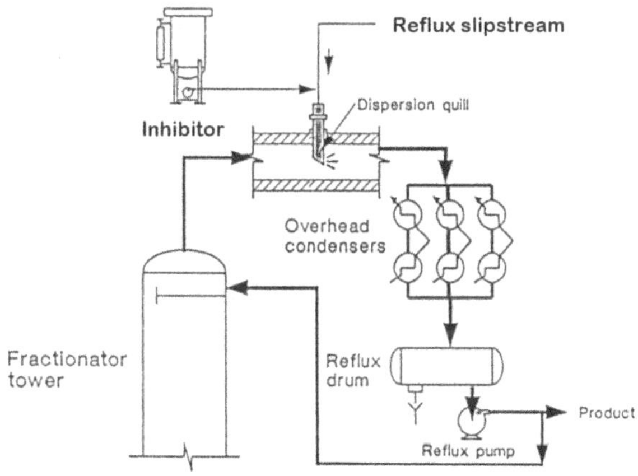

Adapted from W.L. Witzig et al., "The Design and Operation of Process Chemical Treatment Facilities," paper no. 64, CORROSION/81 (Houston, TX: NACE, 1981).

18.6.1 Location of Injection Points. A number of factors need to be considered before inhibitor injection is implemented. For best dispersion and maximum effectiveness, injection points should be located as far upstream of problem areas as practical. If possible, a neutralizer should be injected upstream of the inhibitor injection point to ensure optimum inhibitor performance at near-neutral pH values. An inhibitor is only effective downstream of an injection point and despite some claims cannot somehow protect upstream components. Multiple injection points may have to be provided to ensure even distribution into various piping branches and protect each of several parallel banks of heat exchangers and condensers.

18.6.2 Corrosion Downstream of Injection Point. Injection of corrosion inhibitors into straight sections of piping at temperatures above 120 °C (250 °F) can cause localized pipe wall corrosion for up to 1.5-3m (5-10ft) downstream of the injection point. The exact locations where attack occurs differ for horizontal and vertical legs of straight piping. Once identified by ultrasonic thickness (UT) measurements, affected areas may have to be overlaid with a corrosion resistant alloy such as Alloy C-276. Droplet impingement corrosion can occur when diluted inhibitors are injected into an overhead line at vapor velocities above 9m/s (30FPS). To avoid impingement problems, injection points should be located at least 10 pipe diameters upstream of the nearest elbows, tees, or vessel inlets. Once identified, susceptible areas may have to be upgraded with alloy material. Neat inhibitor can also form corrosive deposits in hot vapor lines and in crude versus overhead heat exchangers.

18.6.3 Injection System for Corrosion Inhibitors. A typical inhibitor injection system consists of a bulk container for the inhibitor and enough piping to connect the container to an injection nozzle at the reflux slipstream line and then to an injection nozzle at the overhead vapor line. If these lines are not already in place, they will have to be provided. The detailed piping diagram in Figure 18-4 shows many, if not all, of the required components starting at the bulk tank. They include primarily positive block valves, pressure indicators, and strainers at pump inlet. Also needed is a positive-displacement, proportioning, or metering pump with spare, pump bypass, outlet piping, and block valves. Flow meters, sampling points, and positive block valves at inlet nozzles to dilution water, steam, or reflux slipstream lines are a must. Positive block valves should not be overlooked to prevent inhibitor from leaking into the system during outages.

18.6.4 Injection Nozzles. Simple tee connections are inadequate for corrosion inhibitor injection and should never be used. In most cases, diluted corrosion inhibitor is injected continuously through a hydraulic spray nozzle or injection quill. Each type has its advantages and disadvantages that should be kept in mind:

- Hydraulic spray nozzles can provide excellent inhibitor dispersion when first installed. Despite claims to the contrary, however, spray nozzles eventually clog with corrosion inhibitor particles, gunk, and other debris. A strainer is absolutely necessary to delay the onset of clogging as much as possible and needs to be bypassed and cleaned periodically to ensure proper flow.
- Injection quills similar to those used for neutralizer injection also can provide good inhibitor dispersion and are likely to cause the fewest problems over the long run, if properly designed and oriented in the line.

FIGURE 18-4 FACILITIES FOR INJECTING CORROSION INHIBITOR INTO A REFLUX DILUTION LINE AND OVERHEAD SYSTEM

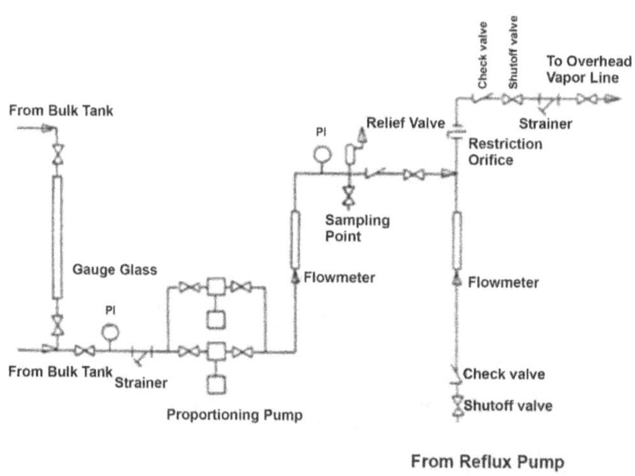

Adapted from J. Gutzeit, "Issues with Injection Facilities for Process Additives or Wash Water," paper no. 591, CORROSION/96 (Houston, TX: NACE, 1996).

Instead of a permanently mounted nozzle, a retractable Alloy C-276 spray nozzle or quill is recommended for insertion to the center of the vapor line. The nozzle is inserted via a retractable holder with packing gland and fully opening gate valve, as shown previously in Figure 16-6 for neutralizer injection. Removable nozzles should be secured in place with special brackets to prevent them from being accidentally removed, or pushed through the packing gland by process pressure. Witness marks indicate the direction and depth in the overhead line. Nozzles should be piped up to a solid inlet line rather than flexible (corrugated) stainless steel connection to avoid chloride stress corrosion cracking (CSCC) at the corrugations. Unless a neutralizer is injected into wash water, corrosion inhibitor should be injected downstream from a neutralizer injection point to ensure maximum effectiveness. As a general rule:

- Most corrosion inhibitors perform best at pH above 4-5.
- Multiple injection points may have to be provided to ensure corrosion inhibitor distribution into various piping branches of the typical overhead system.
- Multiple injection points may also have to be provided to ensure enough corrosion inhibitor is available to protect each of several parallel banks of heat exchangers.
- Most, if not all corrosion inhibitors, are not volatile and unable to leave with vapors.
- Most, if not all corrosion inhibitors are unable to provide inhibition of vapor areas except when these areas are continuously wetted by spray.

18.7 Optimum Treatment Level

Based on the results of an API survey shown in Figure 18-5, standard corrosion inhibitor products containing 18–22% active ingredients in a hydrocarbon base should be injected continuously at a rate of 6-8mg/L by volume. This amount is based on the volume of naphtha condensed in the first receiver drum following the injection point.[4] As a general rule, lower injection rates tend to be ineffective while higher rates waste money. Standard water base products containing 18–22% active ingredients in a water or hydrocarbon base should also be continuously injected at a rate of 6-8mg/L by volume, but based on the volume of water condensed in the first receiver drum following the injection point. In all cases, concentrated versions should be injected at proportionally lower rates than those used for the corresponding standard products.

18.8 Inhibitor Pretreatment

Optimum corrosion inhibitor treatment levels are film maintenance levels only. Film maintenance levels are insufficient for forming a protective film on new equipment and piping or on equipment and piping that was cleaned and opened for inspection. Film maintenance levels are also insufficient for re-forming a protective film whenever corrosion inhibitor injection was interrupted for more than 8hr during normal operations. To form or re-form a protective inhibitor film, a pretreatment level 2-3 times higher than the normal treatment level is usually required over a period of 1-2 weeks. Even though pretreatment is so important, the fact remains that it is often forgotten or ignored during the burst of startup activities following a shutdown. This is unfortunate because in at least some instances insufficient corrosion inhibitor addition during the pretreatment period accounted for the fact that protection was simply not achieved.

FIGURE 18-5 API SURVEY ON CORROSION INHIBITOR CONCENTRATION IN CRUDE COLUMN OVERHEAD SYSTEMS

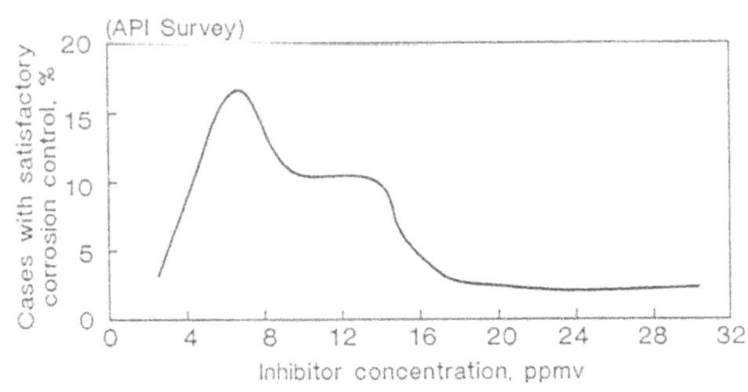

Adapted from D.L. Burns, R. L. Hildebrand, and P.D. Thomas, "Corrosion Inhibitors in Refinery Process Streams," *Proc. API* 40, III (1960).

4. Note that 8-10mg/L by volume is roughly equivalent to 1.3-1.7L/120 m^3 (1.5qt/1,000 barrel) liquid or 0.7-0.8L/28M m^3 (0.7-0.8 qt/ 1MM SCFT vapor.

18.9 Inhibitor Limitations

By their very nature and the mechanism of protection, most corrosion inhibitors are ineffective:

- At vapor velocities above approximately 15m/s (50fps) because the film is eroded away.
- At temperatures above 150 °C (300 °F) because the rate of desorption exceeds the rate of adsorption.
- Against underdeposit corrosion by ammonium chloride (NH_3•HCl), organic neutralizer hydrochloride (R - NH_2•HCl), or corrosion deposits because the inhibitor cannot penetrate to the metal surface.
- At pH values below 3 because the film thickness becomes insufficient to resist the increased corrosion rates.
- Against hydrochloric (HCl) acid because the inhibitor reacts with free acid to form a filming-amine chloride salt that is less protective than the original inhibitor.

18.10 Inhibitor Injection Problems

Vapor line corrosion by neat corrosion inhibitors can occur at temperatures above 120 °C (250 °F) because corrosion inhibitor droplets can partially dissociate and revert to the acid form. Neat corrosion inhibitors typically has the consistency of molasses and can easily form gunk that can corrode an overhead vapor line to the point of penetration within a matter of 1-2 years, as noted earlier. In one such case, the inhibitor injection quill first corroded and then broke off. This allowed inhibitor droplets to deposit and produce a fanlike corrosion pattern, as shown in Figure 18-6 and Figure 18-7. Corrosion penetrated the vapor line just downstream of the injection point after only 6 years. The average corrosion rate was 1.5mm/y (60mpy), but was probably higher once the nozzle had broken off. Chalk line and figures show the extent of pipe wall thinning and resultant wall thicknesses (in inches) along a half segment of the line downstream of the missing injection nozzle. Unpublished laboratory data confirmed that corrosion of carbon steel in several neat commercial corrosion inhibitor solutions increases with increasing temperature, as shown in Table 18-1.

FIGURE 18-6 CORROSION OF INJECTION QUILL AND RUPTURE OF VAPOR LINE DUE TO NEAT CORROSION INHIBITOR AT 150 °C (300 °F)

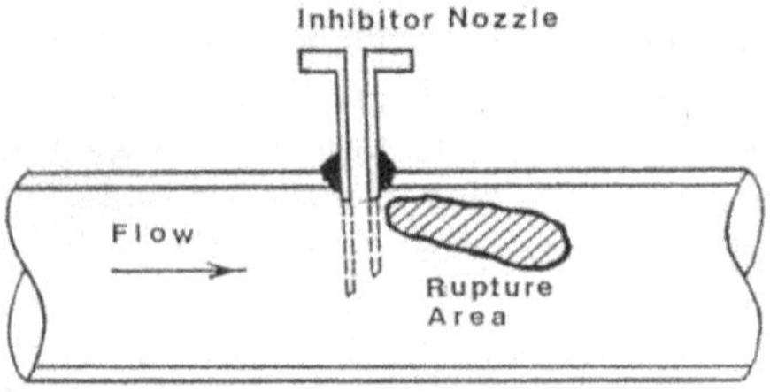

Adapted from J. Gutzeit, "Corrosion Inhibitor Can Cause Corrosion," *MP* 32, 7 (1993).

FIGURE 18-8 WASHED-OUT SECTION OF OVERHEAD VAPOR LINE JUST DOWNSTREAM OF BROKEN OFF INJECTION QUILL

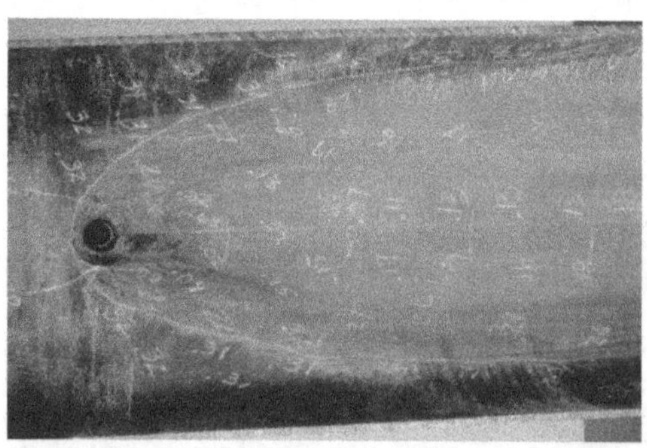

Adapted from J. Gutzeit, "Corrosion Inhibitor Can Cause Corrosion," *MP* 32, 7 (1993).

TABLE 18-1 CORROSION OF CARBON STEEL IN NEAT CORROSION INHIBITOR SOLUTIONS

Temperature		Corrosion Rate	
°C	°F	mm/y	mpy
93	200	0	0
120	250	0.1 - 0.25	4 - 10
177	350	0.5 - 1.0	20 - 40
260	500	1.0 - 5.0	40 - 200

Other inhibitor related problems can be summarized, as follows:

Incorrect Product or Injection

- Incorrect product for application.
- Insufficient dilution.
- Incorrect injection point location.

Corrosion Inhibitor Fouling Problems

- Gunk formation due to interaction between neat inhibitor and undiluted neutralizer when both are mixed together, or injected through the same quill.
- Gunk formation when neat inhibitor and undiluted neutralizer are injected too close together into the same reflux slipstream.
- Fouling when the detergent action of corrosion inhibitors loosens and removes corrosion deposits that end up in downstream components.
- Polymerization of neat corrosion inhibitor left in heat-traced injection lines during a turnaround.

Corrosion Inhibitor Corrosion Problems

- Pitting corrosion due to insufficient inhibitor concentration.
- Corrosion of carbon steel injection quills.
- Droplet impingement corrosion when corrosion inhibitor is injected too close to downstream elbows and tees at velocities greater than 7.5-9 m/s (25-30FPS).

Corrosion Inhibitor Contamination Problems

- Contamination of overhead product streams with corrosion inhibitor.
- Deterioration of WSIM[5] rating of jet fuel may require cutting back on corrosion inhibitor when jet fuel is run.
- Contamination of water condensate with naphtha containing traces of corrosion inhibitor prevents accurate pH readings because the corrosion inhibitor films the pH electrode.
- Interference with naphtha/water separation in receiver drums.
- Nitrogen contamination of feed streams to downstream process units.

18.11 Services and Reporting

Corrosion inhibitor injection, like crude desalting, caustic addition, or neutralizer injection, is a critical part of overhead corrosion control. Process additives suppliers are usually responsible for seeing that the corrosion inhibitor injection system is operating properly. Plants usually supply and maintain storage and mixing tanks for inhibitor solutions; provide and maintain piping, pumps, and other components to facilitate inhibitor injection; and purchase sufficient bulk inhibitor solution. Some suppliers can provide portable, turnkey, injection systems and obviously are responsible for those. Many suppliers carry spare injection pumps in the trunk of their car to allow speedy replacement in case of failure, but only plant personnel (pipe fitters, electricians) are usually permitted to do the actual installation. Suppliers are also responsible for trouble-shooting the inhibitor injection system any time upsets occur. More specifically, sup-

5. Water Separometer Index Modified per ASTM D2550-69.

pliers should coordinate inhibitor injection with desalter operation, caustic addition to desalted crude, and overhead neutralizer injection. This includes measuring the corrosion of carbon steel in mixed overhead condensate going to the overhead separator drum, using Corrosometer™ or coupon probes. Most plants also require a written a daily status report to the operating engineer or unit superintendent along with an annual report summarizing past overall performance and recommending potential improvement projects.

18.12 Chapter Summary

Filming-amine corrosion inhibitors are injected into overhead systems to help control corrosion by maintaining a continuous corrosion barrier between metal surfaces and corrosive condensate. Most commercial corrosion inhibitors are proprietary formulations. The active ingredient usually consists of an organic nitrogen-base compound that is reacted with a long-chain fatty acid. The final products often are viscous liquids that need to be further diluted prior to injection.

Corrosion inhibitors for plant applications vary primarily in terms of solvent base and concentration of active ingredients. Most products are dissolved in a hydrocarbon base, but some are available in a water base or contain surfactants to make them water dispersible. Standard products usually contain 18-22% active ingredients. Commercial inhibitor solutions should never be injected neat into an overhead system and should be diluted prior to injection to ensure proper dispersion and avoid fouling problems. In most cases, the inhibitor is injected continuously through a hydraulic spray nozzle or injection quill into a slipstream of naphtha reflux for dilution prior to injection into a vapor line.

Severe corrosion due to droplet impingement can occur when a corrosion inhibitor is injected at vapor velocities above approximately 15m/s (50FPS), which is often the case. To avoid impingement problems, injection points should be located at least 10 pipe-diameters upstream of the nearest elbows, tees, or vessel inlets. Standard inhibitor solutions in a hydrocarbon base should be injected continuously at a rate of 6-8 ml/L by volume based on the volume of naphtha condensed in the first receiver drum following the injection point. Lower injection rates tend to be ineffective, while higher rates waste money.

Normal corrosion inhibitor treatment levels are film maintenance levels. Film maintenance levels are often insufficient for forming a protective film on new equipment and piping, or on equipment and piping that was cleaned and opened for inspection. Most inhibitors are ineffective at vapor velocities above 15m/s (50FPS), at temperatures above 150 °C (300 °F), and against corrosion by salt or corrosion deposits. Corrosion by neat corrosion inhibitor droplets can occur along the bottom of overhead vapor lines, and under the shell inlet nozzle of heat exchangers at temperatures above 120 °C (250 °F). Other inhibitor-related problems include fouling, gunk formation due to interaction between the inhibitor and a neutralizer, contamination of overhead product streams by inhibitor, and deterioration of WSIM ratings of jet fuel.

18.13 Read More

1. J. Gutzeit and J. M. Johnson, "Corrosion Inhibitors for Petroleum Refining and Petrochemical Operations," paper no. 452, CORROSION/89 (Houston, TX: NACE, 1989).
2. J. Gutzeit and J. M. Johnson, "Corrosion Inhibitors for Petroleum Refining and Petrochemical Operations" in *Reviews on Corrosion Inhibitor Science and Technology,* A. Raman and P. Labine, eds. (Houston, TX: NACE, 1993).
3. J. Gutzeit, "Problems with Injection Facilities for Process Additives and Wash Water," paper no. 591, CORROSION/96 (Houston, TX: NACE, 1996).
4. C. Baumann and C. Scherrer, "Evaluation of Inhibitors for Crude Topping Units," *MP* 18, 11 (1979).
5. D.B. Brooks, "Military Research on Jet Fuel Contamination," *Proc. API* 43, III (1963).
6. D. L. Burns, R. L. Hildebrand and P. D. Thomas, "Corrosion Inhibitors in Refinery Process Streams," *Proc. API* 40, III (1960).
7. N. Hackerman, "Review of Corrosion Inhibition Science and Technology – Historical Perspectives," *MP* 29, 2 (1990).
8. J. Gutzeit, C. Samans, and F. B. Bulls, "Evaluating Refinery Process Side Corrosion Inhibitors," *MP* 12, 7 (1973).
9. J. Gutzeit, "Corrosion Inhibitor Can Cause Corrosion," *MP* 32, 7 (1993).

10. J. Gutzeit, "Refinery Process Additives: A User's Viewpoint," *MP* 29, 2 (1990).
11. G. L. Scattergood, "Corrosion Inhibitors for Crude Oil Refineries," *VOLUME 13 - CORROSION, ASM Metals Handbook*, 9th ed. (Metals Park, OH: ASM, 1987).
12. P. J. Stone, Corrosion Inhibitors for Oil and Gas Production, *VOLUME 13 - CORROSION, ASM Metals Handbook*, 9th ed., (Metals Park, OH: ASM, 1987).
13. NACE International Publication 34101, "Refinery Injection and Process Mixing Points," Task Group 174 on Refinery Additives Injection Facilities (Houston, TX: NACE, 2001).
14. NACE International Publication SP0114-2014, "Standard Practice: Refinery Injection and Process Mix Points," Item No. 21175, (Houston, TX: NACE, 2014).
15. W. C. Fort, III et al., "Process Design of Injection Systems," paper no. 587, CORROSION/96 (Houston, TX: NACE, 1996).
16. C. Gysbers and W.J. Neill, Jr., "Results of Problematic Refinery Plant Injection Point Survey," paper no. 585, CORROSION/96 (Houston, TX: NACE, 1996).

19

Overhead Water Washing

In this Chapter

19.1 First Glance

Many overhead systems operate quite satisfactorily without water injection of any kind, while others rely on intermittent water injection depending on the plant operators' perceived need to improve heat transfer in certain problem heat exchangers. In general, wash water injection into overhead systems should not be necessary, if the crude charge is properly desalted and contains less than 3.5mg/L or roughly 1PTB salt. Barring unforeseen desalter upsets, overhead water condensate can then be expected to contain less than 20mg/L chloride.[1] No ammonium chloride ($NH_3{\cdot}HCl$) deposits are likely to form as long as the dew point exceeds approximately 110 °C (210 °F) in accordance with equilibrium data shown in Chapter 16 of the *Guide*. A safety margin may have to be added to the dew point to ensure that no deposits remain. Provided relevant equilibrium data are made available by process additives suppliers, similar conclusions can be reached when liquid organic neutralizers are injected instead of, or in addition to, ammonia (NH_3). When heat transfer unexpectedly deteriorates due to fouling, it usually means that the crude is no longer properly desalted or the frequency of desalter upsets has increased. Since fouling of overhead heat exchangers is usually caused by salt

1. Refer to Chapter 9: *Desalter Operation for Improved Corrosion Control.*

and corrosion deposits, continuous water injection becomes necessary to remove these deposits and at the same time minimize underdeposit corrosion.

19.2 Purpose of Water Washing

The principal objective of water injection is to restore heat transfer and decrease the pressure drop in overhead heat exchangers, coolers, and condensers by removing any salt and corrosion deposits. This means dissolving ammonium hydrochloride ($NH_3 \cdot HCl$) and neutralizer hydrochloride ($R\text{-}NH_2 \cdot HCl$) deposits that have built up in dry areas of an overhead system and at the same time preventing tube failures due to underdeposit corrosion. As a general rule, water injection is insufficient for extracting significant amounts of hydrogen chloride (HCl) from overhead vapors. While water injection does dilute hydrochloric acid (HCl) as it forms, dilution depends primarily on the water injection rate and the extent to which the water does not evaporate after injection. Even tenfold dilution at best elevates the pH by one unit and most likely is offset by the increased volume of corrosive condensate throughout the system. For example, even the theoretical decrease in corrosion from 11mm/y (450mpy) at pH 2.5 to 4mm/y (150mpy) at pH 3.5, while significant, is still unacceptable. Therefore, no matter how well planned and implemented, water injection by itself is no substitute for other overhead corrosion control measures. As a general rule, enough water must be used for water washing to be effective. Insufficient water only increases the amount of corrosive condensate in an overhead system without removing the deposits. Therefore, the final decision regarding water washing often depends not only on the quality, but also on the quantity of available water at the refinery or upgrader. After years of experimentation with various injection schemes, most plants prefer to cut back on makeup water and use recirculated overhead water condensate for water washing. Depending on local conditions, however, other types of water are also used for water injection. Listed in order of decreasing desirability and likely increasing cost, they include:

- Vacuum column overhead water condensate.
- Stripped sour water.
- Treated boiler feed water.
- Treated steam condensate.
- Potable water (city water).

The following should not be used under any circumstances no matter how readily available or cheap:

- River, lake, or pond waters, because of potential problems with oxygen, silt, and biomatter contents.
- Cooling tower blowdown, because of potential problems with deposit formation with high solids, calcium, or silica contents.
- Water condensate from the vapor recovery (light ends) sections of FCCU or hydrotreating units, because of potential problems with sour water corrosion due to the high ammonia (NH_3) and hydrogen sulfide (H_2S) contents.
- Brackish or sea water, because of potential problems with salt water corrosion, deposit formation with high chloride and solids contents, and marine biomatter.

19.3 Water Injection Procedures

If wash water is to be injected intermittently to remove fouling in a problem heat exchanger, a purpose-designed distributor pipe is usually installed in the exchanger. Otherwise, water is injected continuously through a simple pipe connection or injection quill. Each method has advantages and disadvantages:

- Simple pipe tee connections may not be best choice for vapor streams, but are satisfactory at the high vapor velocities seen in many overhead vapor lines.
- Slotted injection quills can provide good water dispersion and are likely to cause the fewest problems over the long run. The quill should be made from Alloy C-276 and the outlet slots should be directed upstream for best dispersion. They should also be oriented to avoid droplet impingement on the opposite pipe wall.
- In some cases, existing sample connections on overhead vapor lines can be used, provided they are designed to prevent droplet impingement on the opposite pipe wall.

Several other factors should be kept in mind when implementing water injection. For optimum dispersion and maximum effectiveness, injection points should be located as far upstream of likely problem areas as practical. Except for unusual flow reversals due to pressure surges, water does not somehow work itself upstream; corrosion problems upstream of the injection point should be addressed by other means. As a general rule:

- Water injection into straight sections of pipe can cause localized pipe wall corrosion for up to 1.5-3m (5-10ft) downstream of an injection point. The exact locations where attack occurs differ for horizontal and vertical legs of straight piping and once identified by ultrasonic thickness (UT) measurements may have to be overlaid with a corrosion resistant alloy.
- Water injection into straight sections of vertical lines requires at least 15m/s (50FPS) vapor velocity to keep water from falling down the line. In contrast, water injection into straight sections of horizontal lines at vapor velocities greater than 15m/s (50FPS) can result in severe corrosion due to droplet impingement. To avoid impingement problems, injection points should be located at least 10 pipe diameters upstream of the nearest piping elbows, tees, or vessel inlets.
- Multiple injection points may have to be provided to ensure even water distribution into various piping branches of the typical overhead system. Multiple water injection points may also have to be provided to ensure enough water is available for water washing each of several parallel banks of heat exchangers.
- The first overhead receiver drum downstream of the water injection point must have sufficient boot capacity to handle the increased water volume to avoid water carryover with naphtha reflux and/or product ("wet naphtha"). If necessary, a remote boot or separate water coalescing drum may have to be added downstream of the receiver drum.

19.4 Typical Wash Water Installations

A basic design of a recirculating wash water system consist of a suitable pump and enough piping to connect the water boot of the overhead receiver to the pump inlet and the pump outlet to the injection point at the overhead vapor as shown in Figure 19-1. No pump is needed if the water is obtained from outside sources and is not to be recirculated. A complete design includes many, if not all, of the following components, listed in sequential order starting at the water source:

- Water inlet piping (carbon steel throughout).
- Positive block and ball check valves at pump inlet or water supply line.
- Centrifugal water pump and spare, if recirculated overhead water condensate is used (stainless steel pump internals to minimize corrosion in case of desalter upsets).
- Pump bypass piping and valves (carbon steel throughout).
- Pump outlet piping (carbon steel throughout).
- Positive flow indicator, followed by restriction orifice.
- Ball check and positive block valves at inlet to overhead vapor line to keep vapors from backing up into water line in case of pump outage.
- Retractable Hastelloy C-276 quill for insertion to the center of the vapor line, if injection quill is used.
- Some sort of indicator to positively identify direction of quill outlet with respect to vapor flow direction.
- Fully opening gate valve and packing gland to allow onstream removal of quill for cleaning or replacement.
- To prevent possible chloride stress corrosion cracking, no flexible (corrugated) stainless steel tubing should be used for connecting a removable quill to water piping.

As a general rule, at least a portion of the neutralizer used for pH control is injected into the wash water and to minimize corrosion downstream of the injection point.

FIGURE 19-1 SCHEMATIC OF RECIRCULATING WASH WATER SYSTEM USING OVERHEAD CONDENSATE

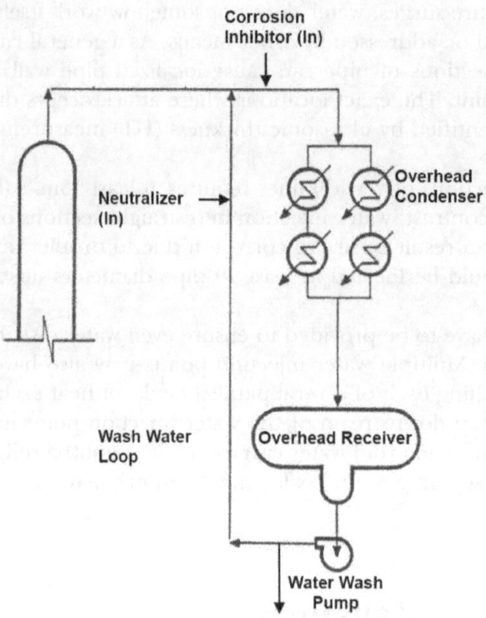

19.5 Effect of Wash Water Injection on Dew Point

Depending on the design and operating mode of the atmospheric section of the column, column top temperatures can range from 95 °C (200 °F) to 175 °C (350 °F) and absolute pressures from 2.7bar (40PSI) to 3.4bar (50PSI). Overhead vapor line velocities can range from 15m/s (50FPS) to 90m/s (300FPS), and sometimes even higher. While significant condensation is unlikely to occur upstream of the first set of heat exchangers, some water shock condensation may occur and cause localized corrosion in various segments of overhead vapor lines. Theoretically, water injection lowers overhead vapor temperatures, because sensible heat is removed as the water evaporates. In practice, however, at high vapor velocities and water rates, there is insufficient time for the wash water to evaporate before it enters the first set of heat exchangers. Moreover, water droplets become surrounded by a naphtha film which further delays evaporation. This means that the vapor temperature at the injection point often becomes the de facto dew point where initial condensate corrosion is likely to take place. Hence, a portion of the neutralizer used for pH control is injected into the wash water to prevent acid corrosion and impingement attack downstream of the injection point.

19.6 Optimum Water Injection Rate

As a general rule, overhead water condensate is recycled continuously from the receiver drum and/or water coalescing drum for injection upstream of the first heat exchanger with known fouling and corrosion problems. The following should be kept in mind:

- A typical water injection rate is approximately 5wt. % of the total overhead vapor rate. This amount is somewhat more than the amount of dilution and stripping steam (usually 3wt. %) that condenses in the overhead system.
- High water injection rates up to 15wt. % of the total overhead vapor rate have been used successfully in some overhead systems to overcome major salt deposit problems due to neutralizer injection into dry stages or components. As result, these stages are converted to wet stages.
- While effective for that purpose, large volumes of water can cause impingement corrosion problems and result in water entrainment problems in naphtha reflux and overhead product streams.
- High water injection rates can be expected to upset crude feed and overhead heat balances.

- Insufficient water rates are counterproductive, because dissolved salts or suspended corrosion deposits are only moved about or partly left behind as the water evaporates.

With insufficient receiver capacity, continuous water washing may not be practical because any excess water will return to the column entrained in reflux and corrode the top of the column. In that case, it is probably better to inject a relatively small amount (1-2% volume of the total overhead condensate) upstream of each heat exchanger over a period of 5-10 minutes on a regular schedule of 1-3 times/week. A purpose-designed distributor pipe can be installed in the shell to distribute wash water along the entire length of the tube bundle. The actual injection time and schedule are based on the results of sequential chloride analyses of the water discharge during the injection period. Water injection is discontinued when no further changes in chloride content are measured. Since intermittent wash water injection requires only a relatively small volume of water, it is sometimes more convenient to use fresh water, boiler feed water, or steam condensate instead of recycled overhead water condensate.

19.7 Problems with Water Injection

The biggest issue is the fact that a previously "dry" stage of a two-stage two-drum overhead system has no means for water removal. Consequently, wash water is sent with reflux to the column top unless a suitable water boot is provided at the reflux drum to allow water removal. Alternatively, a water coalescer may have to be added. Other potential problems include:

- Previously dry components of a single-stage overhead system are likely to experience acid corrosion, unless a neutralizer is added to the water.
- Dissolved air in some types of wash water can produce air pockets that become locations for accelerated pitting corrosion and sulfur deposit formation.
- Poor water quality can cause scale and salt deposit problems.
- Overhead vapor lines can experience localized wall thinning at elbows and tees due to droplet impingement and need to be monitored regularly with UT and radiography.
- Corroded segments need to be replaced at regular intervals, or protected by overlaying with Hastelloy C-276 or a similar corrosion-resistant alloy.
- Crude preheat is lost to the injected wash water. This effect is usually small, because the water volume is small compared to the crude volume passing through the overhead/crude heat exchangers.
- With high water injection rates of up to 10 wt. % of overhead vapors, the amount of crude preheat can be expected to be reduced by roughly 10 percent. To determine by how much wash water injection affects heat exchangers operating above design, a material and energy balance should be made.
- Unless recycled overhead water condensate is used, water disposal may become an issue.

19.8 Optimum Monitoring Frequency

The wash water rate should be monitored continuously on the operator's board. If possible, pH of overhead water condensate should also be monitored continuously to record pH excursions due to desalter upsets or organic chloride contamination. Chloride levels should be checked at least once/shift. Corrosion rate should be monitored continuously with a Corrosometer™ probe ahead of the overhead receiver drum. Water hardness should be checked once/week to detect cooling water leaks.

19.9 Chapter Summary

Wash water injection into crude unit overhead (overhead) systems should not be necessary if the crude charge is properly desalted. Many overhead systems operate satisfactorily without water injection, while others rely on intermittent water injection to clear critical bottleneck exchangers. Wash water injection prevents underdeposit corrosion and restores heat transfer of fouled heat exchangers. As a general rule, enough water must be used to be effective. Insufficient water is likely to increase corrosion by increasing the amount of corrosive water condensate. Therefore, water selection depends not only on quality, but also on quantity of available water.

Many plants prefer using recirculated atmospheric water condensate for water washing. Depending on local conditions, other types of water are also used, including vacuum overhead condensate, stripped sour water, or treated boiler feed water. Wash water is injected continuously through a simple pipe connection or an injection quill. Each method has advantages and disadvantages that should be kept in mind. For best dispersion and maximum effectiveness, injection points should be located as far upstream of likely problem areas as practical. Severe corrosion due to droplet impingement can occur when water is injected at excessive vapor velocities.

19.10 Read More

1. R. H. Carlton, "Continuous Injection of Overhead Receiver Water Controls Refinery Condenser and Exchanger Corrosion by Ammonium Chloride," *MP* 2, 1 (1963).
2. T. D. Whitehead and P.A. Ramachandran, "An Analysis of Water Wash Practices Used for Corrosion Control in Condensable Hydrocarbon Systems," paper no. 318, CORROSION/91 (Houston, TX: NACE International, 1991).
3. NACE International Publication 34101, "Refinery Injection and Process Mixing Points," Task Group 174 on Refinery Additives Injection Facilities (Houston, TX: NACE, 2001).
4. NACE International Publication SP0114-2014, "Standard Practice: Refinery Injection and Process Mix Points," Item No. 21175 (Houston, TX: NACE, 2014).
5. W. C. Fort, III et al., "Process Design of Injection Systems," paper no. 587, CORROSION/96 (Houston, TX: NACE, 1996).
6. C. Gysbers and W.J. Neill, Jr., "Results of Problematic Refinery Plant Injection Point Survey," paper no. 585, CORROSION/96 (Houston, TX: NACE, 1996).
7. J. Gutzeit, "Problems with Injection Facilities for Process Additives and Wash Water," paper no. 591, CORROSION/96 (Houston, TX: NACE, 1996).

Index

D

E

F

H

I

K

L

T

U

V

W

www.ingramcontent.com/pod-product-compliance
Lightning Source LLC
LaVergne TN
LVHW060032210926
848481LV00013B/3843
9781575903309